Cotton

Second edition

TROPICAL AGRICULTURE SERIES

The Tropical Agriculture Series, of which this volume
forms part, is published under the editorship of
Gordon Wrigley

ALREADY PUBLISHED

Tobacco *B C Akehurst*
Sugar-cane *F Blackburn*
Tropical Grassland Husbandry *L V Crowder and H R Chheda*
Sorghum *H Doggett*
Sheep Production in the Tropics and Sub-Tropics *Ruth M Gatenby*
Rice *D H Grist*
The Oil Palm *C W S Hartley*
Cotton *John M Munro*
Cattle Production in the Tropics Volume 1 *W J A Payne*
Spices Vols 1 & 2 *J W Purseglove* et al.
Tropical Fruits *J A Samson*
Bananas *R H Stover and N W Simmonds*
Agriculture in the Tropics *C C Webster and P N Wilson*
Tropical Oilseed Crops *E A Weiss*
An Introduction to Animal Husbandry in the Tropics
    G Williamson and W J A Payne
Cocoa *G A R Wood and R A Lass*

Cotton

Second edition

John M Munro

Longman Scientific & Technical

Copublished in the United States with
John Wiley & Sons, Inc., New York

Longman Scientific & Technical,
Longman Group UK Limited,
Longman House, Burnt Mill, Harlow,
Essex CM20 2JE, England
and Associated Companies throughout the world.

Copublished in the United States with
John Wiley & Sons, Inc., 605 Third Avenue, New York, NY 10158

First published 1987

British Library Cataloguing in Publication Data
Munro, John M. (John Meiklejohn)
  Cotton. – 2nd ed. –
  (Tropical agriculture series)
  1. Cotton growing – Tropics
  I. Title   II. Prentice, A. N.   III. Series
  633.5'1'0913   SB251.T73
  ISBN 0-582-46346-7

Library of Congress Cataloging-in-Publication Data
Munro, John M., 1912–
  Cotton.

  (Tropical agriculture series)
  Rev. ed. of: Cotton / A. N. Prentice. 1972.
  Bibliography: p.
  Includes index.
  1. Cotton.   2. Cotton growing.   I. Prentice, A. N.
Cotton.   II. Title.   III. Series.
SB249.M86   1987   633.5'1'0913   86–20140
ISBN 0-470-20782-5 (Wiley, USA only).

Set in Linotron 10/11 pt Times
Produced by Longman Singapore Publishers (Pte) Limited
Printed in Singapore

Contents

Preface

This series deals with tropical agriculture, and thus excludes the three major producers of cotton – the United States, Russia and China. In the first edition of this book, Alec Prentice restricted himself in the main to cotton in Africa; the publishers wanted to broaden the scope of the second edition to cover the whole of the tropics and subtropics – a formidable task for any one author. While I have had the privilege of visiting a number of cotton-growing countries outside Africa, these visits have mostly been of short duration and widely spaced; my knowledge of the organization and problems of the cotton industry in these countries is therefore limited, and some of the information given may be out of date. I apologize, therefore, for any errors and omissions which those with a more intimate knowledge may find in the text.

Most of my time on cotton has been spent on research stations with limited library facilities. This has had two effects: firstly the references quoted are biased towards the work of the Cotton Research Corporation, and secondly I have included some more general information which I found difficult to obtain from the limited sources available.

The conditions under which cotton is grown are so diverse that it is impossible to provide a ready-made recipe for success. The procedures in each case must be tailored to suit the particular circumstances, which can only be done by a study of the crop in the field. Regular observation of what is actually happening to the plants themselves is necessary to identify the factors limiting production, and the reasons why one farmer is more successful than his neighbours. I have tried to provide sufficient background to enable the reader to study his crop intelligently, with a knowledge of what has been done in other countries.

In planning this new edition, I have had in mind the agricultural officer who needs a more specialized knowledge of the cotton crop than can be provided in a general course on tropical agriculture. He may need this knowledge for his work in the field, and wants to know how other people are tackling similar problems; he may have to formulate an agricultural policy from research findings and the reports of visiting specialists, and must have sufficient knowledge of the crop to select the recommendations which will have the greatest impact on the local cotton industry; or he may be a consultant who finds that cotton is an important crop in a development project, and needs to know enough about the crop to cater for its special needs in formulating his proposals.

Among those who have helped me in the preparation of this book, I would like to thank especially Graham Matthews of Imperial College for his comments and suggestions regarding Chapters 11 and 12; Alan Cameron of the University of Nottingham Library for his help with articles and illustrations from the Cotton Growing Review; Peter Flatters of the Plant Protection Division of I.C.I. for his help with the colour plates; and above all the Cotton Research Corporation for providing me over thirty years with the opportunity to study such a fascinating and rewarding crop.

John M Munro
March 1986

Acknowledgements

We are grateful to the following for permission to reproduce copyright material:

Agricultural Research Corporation for fig 10.8 (H. G. Farbrother, 1969); American Association for the Advancement of Science for fig 4.7 (C. K. McClelland, 1916); the Editor, 'American Journal of Botany' for fig 14.3 (L. L. Phillips, 1966); Anmadu Bello University for Table 15.1 after (H. E. King, 1957); British Man-Made Fibres Federation for Table 19.2 (BM-MF, 1979); Cambridge University Press for figs 4.2 (A. B. Hearn, 1972b), 6.2 after (A. B. Hearn, 1972a), 8.1 after (E. Jones, 1976), 16.1 after (W. B. Mercer & A. D. Hall, 1911), Tables 8.1 after (E. Jones, 1972), 13.1 (M. H. Arnold, N. L. Innes & S. J. Brown, 1976); Centre d'Etude de l'Azote for Tables 8.3 part, 8.4 (J. G. Geus, 1973), Ciba-Geigy AG for Table 11.3 after (Ciba-Geigy, 1972); Commonwealth Agricultural Bureaux for Table 8.2 (A. B. Hearn, 1981); the Editor, 'Cotton Outlook' for fig 17.9, Table 17.5 (ICAC, 1983); Food & Agriculture Organisation of the United Nations for figs 10.1, 10.9 (J. Doorenbos & W. O. Pruitt, 1977); the Controller of Her Majesty's Stationery Office for fig 19.1 part, Table 11.1 (C. E. C. 1948); International Cotton Advisory Committee for figs 2.1, 2.5, 2.7, 2.8, 19.1 part, 19.5, Tables 2.1, 11.1 part, 17.8 part 19.3 (ICAC 1959–83); Liverpool Cotton Association for figs 2.1, 2.2, 19.1 part (J. A. Todd, 1921), Table 17.7 (Weekly Circular); Longman Group UK Ltd for figs 3.1, 4.6a (A. N. Prentice, 1972, Table 3.5 (J. W. Purseglove, 1968); Macmillan Publishing Co for Table 8.3 part (L. Tisdale & W. L. Nelson, 1975); Ministry of Agriculture (Lilongwe) for figs 12.1, 12.4, 12.6 (Malawi, 1976); National Cotton Council for fig 19.7 (NCC, 1965, 1983), Tables 7.3 (G. A. Buchan et. al. 1980 + T. Whitwell et. al. 1981), 9.1 (J. R. Mauney, 1966), 13.2 parts (L. S. Bird, 1966), 15.2 (J. C. Delouche, 1981), 17.8 (B. R. Hise & D. E. Ethridge, 1980); Oxford University Press for figs 4.4, 4.5, part Tables 3.2, 3.3 (Hutchinson et. al. 1947), fig 10.1 (Jameson, 1970); Pakistan Central Cotton Committee for Table 3.4 (M. A. Mirza &

A. L. Shaikh, 1980; Pergamon Press Ltd for figs 3.2 (J. B. Hutchinson, 1962), 4.10 (A. B. Hearn, 1980), Tables 7.1 (J. N. Hawtree, 1980), 7.2 adapted (M. B. Green et. al. 1979); the Author, N. Ramachandran for fig 15.1 (N. Ramachandran, 1983); Royal Meteorological Society for fig 10.6 (J. B. Hutchinson et. al. 1958a); Shirley Institute for fig 18.4 (G. G. Clegg, 1931); the Editor, 'Soil Science' for fig 10.7 part (M. B. Russell, 1939); Texas A & M Publications for part Tables 3.2, 3.3 (P. A. Fryxell, 1979); Textile Economics Bureau Inc for data in Table 9.1 (Textile Organon, 1983):

We are also indebted to the following copyright holders for permission to reproduce photographs:

British Cotton Growing Association for figs 17.2, 17.8; Cambridge University Press for figs 12.7, 13.2a, 14.5, 14.6 (S. J. Brown, 1976), 18.2 (W. L. Fielding 1948); the Author, Dr. G. W. Cathey & Pergamon Press Ltd for fig 17.1a + b (G. W. Cathey, 1980); Ciba-Geigy for fig 17.2a; Courtaulds plc for fig 19.2; IIC for figs 4.13b, c,d. 18.1; The International Institute for Cotton for figs 4.13, 18.1, 19.4; the Author, Dr. J. Logan & the Editor 'Annals Appl. Biol.' for fig 13.2b (J. Logan, 1958); Longman Group UK Ltd for fig 17.3 (A. N. Prentice, 1972); Shirley Trust from fig 4.14; The Textile Institute for fig 18.3.

Chapter 1

The origin of cotton

Textile fibres

The production of cloth has been one of the world's most important industries since prehistoric times. The processes in the modern factory remain basically the same as those used in the earliest civilizations: the raw fibres are spun by twisting them into thread, which is woven or knitted into cloth. The fibres used for spinning must cling together to give a satisfactory thread; woollen fibres have a natural curl, and man-made fibres may be twisted or crimped. In cotton the secret of success lies in the structure of the fibre, which ripens to form a twisted ribbon (see Ch. 4).

Wool, silk and flax were used for spinning and weaving long before cotton became important. Cotton fabrics have been found dating back to 3500 BC, but their use in the civilizations of Meso-potamia, Babylon and Egypt is not recorded until about 500 BC (Hutchinson *et al.* 1947); in these early references, Watt (1907) emphasizes the difficulties of translation, as the words used were often applied loosely to different kinds of fibre.

Evolution of cotton lint

The basic problem of the origin of the cultivated cottons is that of the origin of the convoluted lint hairs. Various theories have been advanced to explain the selective value of lint in the evolution of the species, but there is no convincing evidence that it is of any use to a cotton plant growing wild in its natural habitat.

Hutchinson *et al.* (1947) argued that it is only as a source of raw material for the textiles of civilized man that it has any selective advantage, and concluded that the origin as well as the distribution of the cultivated cottons is intimately connected with the utilization of their lint by man. Fryxell (1979) agrees that 'the cultivated species have been virtually transformed at the hand of man', but considers it probable that lint production and species differentiation took place before man took an interest, and that the four linted species 'were all domesticated independently in four different sets of circumstances'.

In the light of increased knowledge of the distribution and relationships of the primitive cottons, Hutchinson accepts that the Asiatic species and races probably differentiated before they were domesticated (Santhanam and Hutchinson 1974). It may be supposed that the cultivated *herbaceum* cottons were domesticated from the Arabian and Baluchistan race *acerifolium*, and that *arboreum* was first brought into cultivation in Gujerat or Sind.

Tetraploid cotton

The New World tetraploid cottons arose from a cross between two diploid species; cytological studies and interspecies crosses have shown that the two sets of chromosomes in the tetraploids have affinities, one with the Old World species and one with the American wild species (see Ch. 3). The sterile hybrid would become a fertile allopolyploid by chromosome doubling. All the New World diploids are lintless, although they carry seed hairs, so the lint-bearing genes found in the tetraploids must have come from the Old World diploid parent. There has been much speculation, still unresolved, on how, when and where the two diploid species could have met.

The tetraploids are indigenous in the New World, and those found in the Old World are known to have been imported from America since its discovery by Europeans; it is practically certain, therefore, that the cross took place in America, and the linted diploid from the Old World must have been transported across the ocean. To explain how this happened, various authorities have invoked Wagener's theory of continental drift, land bridges and routes via Antarctica and the Bering Straits (Purseglove 1968). Harland (1939) postulated a Pacific land bridge in late Cretaceous or early Tertiary times; Hutchinson *et al.* (1947) considered the objections to this theory to be insuperable, and put forward the idea that the Old World cottons were introduced to America much more recently by civilized man. Stebbins (1947) suggested that they reached America by way of China and Alaska. Purseglove (1960, 1963) suggested that *Gossypium herbaceum* could have reached South America in Tertiary times via the Antarctic, retreating northward as glaciation advanced. Fryxell (1965) showed that cotton seeds can survive floating in sea water for at least a year with undiminished viability, and can thus be distributed by ocean currents. Purseglove (1968) agreed that the most likely explanation was that cotton seeds floated across the Atlantic from Africa to South America.

Early cotton textiles

'*Gossypium herbaceum* var. *africanum* may be regarded as a wild ancestor of the domesticated plants included in *G. herbaceum* – or at least as a model of such an ancestor' (Fryxell 1979), but the development of cotton textiles appears to have taken place, not in Africa, but in the Indus valley in what is now Pakistan. Trade routes were open between Africa and India at that time, and linted cotton may well have been introduced to India as a curiosity, used first as a trimming or for embroidery on linen and woollen fabrics (Hutchinson *et al.* 1947). The earliest known cotton fabrics in the Old World belong to the Indus civilization, indicating that the development of cotton as a major new raw material took place in Sind; they were found during excavations at Mohenjo-Daro in Sind (Pakistan) at levels which are dated at approximately 3000 BC (Gulati and Turner 1928).

The fragments discovered at Mohenjo-Daro were evidently made by competent craftsmen, and not by people experimenting clumsily with a new art, or with an unfamiliar raw material. In all hair characters that could be measured, the Mohenjo-Daro cotton was within the range of Indian cottons of the present day, so it is certain that the major changes involved in the evolution of lint were complete at that time. (Hutchinson *et al.* 1947.)

In the New World, tetraploid cotton seeds and primitive cotton fabrics have been identified in excavations by Bird and Mahler in Northern Peru at Huaca Prieta, dated about 2500 BC (Hutchinson, J. B. 1959). The earliest record of cotton is in southern Mexico, where a specimen of a large-bolled cotton plant was found and dated about 3500 BC (Smith 1968). The differentiation into varieties within the species shows that the tetraploids must have originated long before these dates; their origin has been placed at various times from Cretaceous to the Recent geological systems.

Between 4000 and 3000 BC we thus have linted cotton species being used to make cotton fabrics – diploids in India and tetraploids in Mexico and Peru. Their distribution was a gradual process on a small scale up to the eighteenth century AD. As the use of cotton increased, selection took place for more desirable field characters, and the tall, unproductive, perennial cottons were replaced by the compact, heavy yielding, annual cottons of the present day. Seed dormancy and photoperiodicity were reduced or eliminated.

Spread of the tetraploid cottons

In the eighteenth century the imports of woven cotton goods from India created such a demand in Europe that much of the spinning and weaving industry in Lancashire, traditionally based on wool and flax, was diverted to the production of cotton goods (Trevelyan 1942). The invention of the spinning-jenny by James Hargreaves, the use of water power by Richard Arkwright, the combination of the two by Samuel Crompton in the spinning mule and the invention of the power loom by Edmund Cartwright, so increased the efficiency of the industry as to create a huge potential demand for raw cotton. The newly founded United States of America, in particular, were willing and able to extend the planting of cotton to meet this demand, and the invention of the saw gin by Eli Whitney in 1793 removed a major constraint on cotton production (see Ch. 2).

Long before this the tetraploid New World cottons had been distributed to many parts of the Old World, some as ornamental or garden plants like the perennial tree cottons and kidney cotton, some as crop plants like the *punctatum* cottons in the dry belt running across West Africa south of the Sahara, while Upland cottons were introduced into India, Cambodia and the Philippines with varying success. Even where the introductions proved a commercial failure, many of them still persist as wild species or weeds of cultivated land (Hutchinson *et al.* 1947).

Although cotton has been grown in Egypt since about the fourteenth century, it does not appear to have been an important crop and Egypt was importing cotton from Syria and Cyprus in 1592 (Watt 1907). The crop only assumed importance when cotton from the New World replaced the Old World Levant cotton about 1820. A perennial tree cotton, a tetraploid of the Peruvian type, had been grown in Egypt for more than two centuries as an ornamental garden shrub; it probably reached Egypt from South America by way of West Africa. Early in the nineteenth century a Franco-Swiss engineer, M. Jumel, employed by Mohammed Ali, the founder of the Khedivate, started to multiply seed that he found in a garden in Cairo; it was of much longer staple than the commercial crop, and quickly gained a reputation with the spinners. From 1820 onwards, Jumel cotton rapidly displaced the short-staple Asiatic cotton (*G. herbaceum*). Its success led to the importation and trial of other long-staple varieties, notably Sea Island, which became mixed with Jumel cotton and produced quick-maturing annual hybrids; one of these gave rise to the brown-linted Ashmouni stock, which had completely replaced Jumel cotton by 1887 (Balls 1912).

The American Civil War (1861–65) cut off supplies of cotton from America to Europe, and stimulated cotton growing in other parts of the world. Much of this was based on seed of the American tetraploid species, especially Upland varieties, which produce lint of markedly better length and fineness as well as better yields than the Old World diploid cottons. New varieties of American origin had already been developed in Egypt, while American Upland varieties were being grown in Turkey, Greece and in Africa south of the Sahara.

In India the American varieties were not successful, and there the expansion of cotton growing in the nineteenth and twentieth centuries was almost entirely in short-staple diploid cottons; only recently have Indo-American tetraploid varieties been developed to the stage that they are now rapidly displacing the indigenous diploids. Varieties from the USA have now spread to all parts of Central and South America, to Africa, Australia, Afghanistan, Thailand and Indonesia, and to many other countries.

Chapter 2

Production and consumption

The nineteenth century

Lancashire had already established a tradition of spinning and weaving wool and flax in the Middle Ages; the small arrivals of raw cotton from abroad – from the Levant, India and the New World – gravitated to where the skills in processing were available, and where the humid climate was particularly suitable for cotton spinning. There was opposition by the wool trade to the import of cotton goods, which led to an Act of Parliament in 1720, prohibiting the use and wear of printed or coloured calicos. This ban was apparently lifted about 1770, opening the way for the East India Company to bring in calico cloth from India, and the first Indian textile mill to produce such cloth was built near Calcutta in 1814. The imported cloth was cheaper than that produced in Britain, 'bringing cotton within the reach of the mass of the population and creating a popular demand so great that local manufacturers took to making cotton goods' (Trevelyan 1942). The early mills, using the machinery invented by Hargreaves, Arkwright and Crompton, spun the cotton into yarn, which was then sent out to handloom weavers in cottages or small workshops. As the industry developed, hand and horse power in the mills was replaced first with water and then with steam power, and the invention of power-operated looms completed the change-over from cottage industry to factory work (Miles 1980).

In the nineteenth century Britain became the leading purchaser and manufacturer of raw cotton, and the leading exporter of cotton goods. India was unable to meet the increased demand for cotton, and cultivation in Egypt only started systematically in 1821, but the USA was both willing and able to extend its production, and exported about 80 per cent of the crop to Britain. A major restraint had been removed by the invention of the saw gin by Eli Whitney in 1793, which allowed the separation of lint from seed to be speeded up and mechanized. By the time of the American Civil War

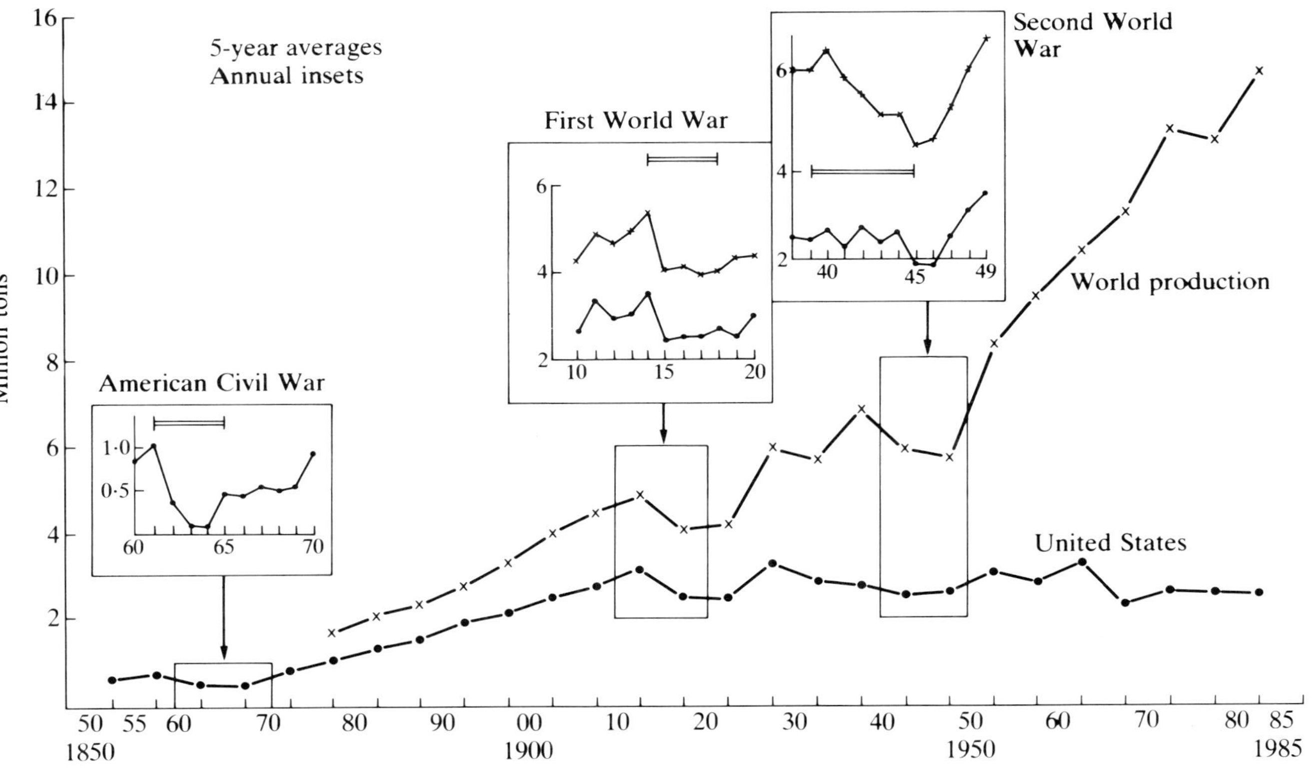

Fig. 2.1 World and USA cotton production, 1850–1984 (Todd 1921; CEC 1948; *Cott. Gr. Rev.*; ICAC).

(1861–65), cotton production was dominated by the United States, where more than three-quarters of the world crops were being grown (Fig. 2.1), and around 70 per cent of this was being exported to Britain. The Civil War reduced production in the USA from 4.5 million running bales in 1861 to 0.3 million in 1864, and effectively stopped the export of cotton to Britain (Todd 1921) where the industry struggled on with what it could buy from other sources. This had the double effect of stimulating cotton production in India and Egypt, and convincing the mill-owners of Lancashire that never again should they allow themselves to become so dependent on a single source of supply of raw cotton. They started to explore the possibilities of new areas for cotton growing in any likely region, but preferably in overseas British territories.

Among the various new areas to be explored, Queensland was one in which the production of lint reached 1000 tons in 1871, before fading away for another 50 years. An unsuccessful cotton growing venture in Natal was instrumental in bringing Cecil Rhodes to Africa about this time. Livingstone took an enthusiastic interest in the cotton he found growing in Zambesia in 1859–63, and businessmen reading the account of his travels 'were led to see Central Africa as a cotton-growing land that might make Lancashire independent of the United States. . . . Dr Livingstone magnified sporadic patches of cotton into rich fields beyond the Americas' best.' (Central African Archives 1952). West African merchants sent out machinery and started operations in many different parts, chiefly in the Gold Coast and around Lagos, and in Sierra Leone the archives show that a small amount of raw cotton was actually exported to the USA in 1864. American production, however, recovered quickly after the end of the war: cotton prices fell to their pre-war levels (Fig. 2.2) and many of the new projects found that, in the words of Mockler-Ferryman (quoted in Watt 1907) 'there was no market for West African cotton'.

The rise of the cotton industry in the nineteenth century had its darker side. The industrial revolution in Britain was achieved at the cost of great misery among the new class of factory workers, and the 'dark satanic mills' of the poet Blake were indeed largely the cotton mills. The plantations of the West Indies and the United States were manned by slaves transported from Africa; the slave trade completed a circuit for ships carrying cotton goods for export, slaves to the New World and raw cotton to Lancashire.

From 1900 to the Second World War

New sources of supply

The upward trend in cotton production continued without inter-

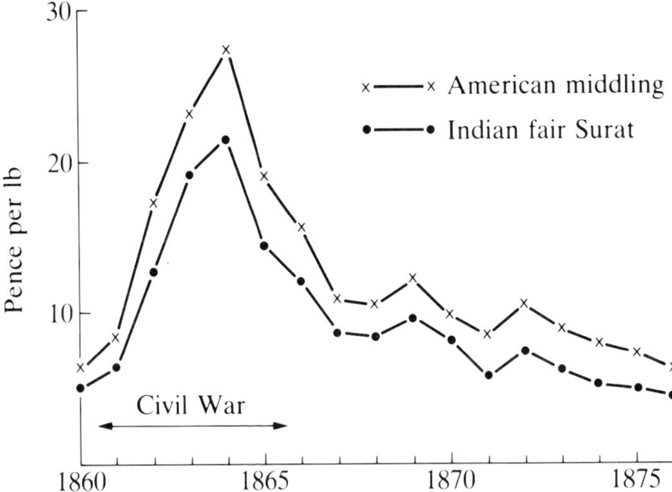

Fig. 2.2 Liverpool cotton prices during and after the American Civil War, 1861–65 (Todd 1921).

ruption from the American Civil War to the First World War; a new peak was reached at the start of the Second World War (Fig. 2.1), but the pattern of production and consumption was changing. The share of the USA in world production fell from about 66 per cent in 1900 to about 45 per cent in 1939. In fact, its production has remained fairly steady around an average of 2.75 million tons (range 2.4 to 3.6 m.) from the early 1900s; the increase came from other sources, notably Russia, India and Brazil (the early data for China are unreliable) but many of the smaller producers contributed to the increase (Fig. 2.3).

Agriculture in the smaller countries was becoming better organized. Departments of Agriculture were set up, which grew from small beginnings until they were able to give some attention to cotton growing and its problems; in many countries where cotton was a major export, boards were appointed to foster and regulate production and research stations were established to provide technical guidance and better varieties. These developments differed in speed and timing in different countries, Egypt, the Sudan and India leading the way, but the pattern in India may be taken as an example. Cotton Commissioners were appointed to develop the crop in certain provinces as early as 1863–64, the first being in Bombay, Behar and the Central Provinces; the commissioners had little or no specialist training. The Department of Agriculture was started early in the 1900s, and an Indian Cotton Committee was appointed in 1914; this was the forerunner of the Indian Central

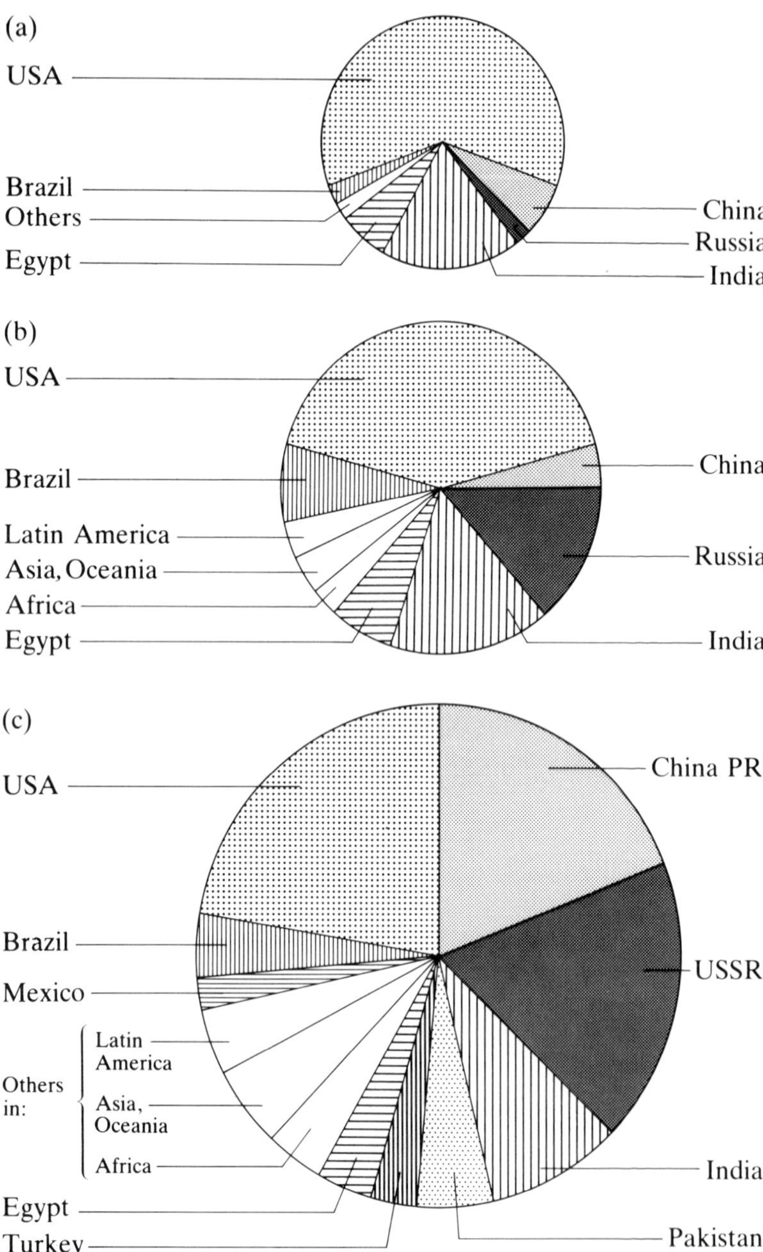

Fig. 2.3 Cotton production by countries in selected years: (a) 1902–03 4,000 million tons; (b) 1938–39 6,014 million tons; (c) 1981–82 15,360 million tons (Todd 1921, Windel 1946, ICAC 1983).

Cotton Committee (ICCC), which dealt with all aspects of cotton from field to factory from 1921 until the partition of India in 1947. In 1924 the ICCC set up the Cotton Technological Research Laboratory in Bombay, and the government established an Institute of Plant Industry at Indore with a section devoted to research on cotton (Chokey Singh 1980).

The USA equalled British consumption about 1900, and doubled the size of its industry between 1900 and 1945 to become by far the largest consumer of cotton (Fig. 2.4). Britain quickly recovered from the setback of the American Civil War and continued to expand until the outbreak of war in 1914, but it was facing more and more competition from newer mills in other countries. Except for a brief recovery to meet the demands of the Second World War, the industry in Britain has declined to a fraction of what it used to be. India, Russia, Japan and China all increased their consumption to about one-third of that of the USA; between the wars the Latin American countries, particularly Brazil, became important consumers; the continent of Europe took about the same amount as Britain and the USA in 1900, and its consumption remained at much the same level up to 1939.

Consumption figures are based on returns from the spinning

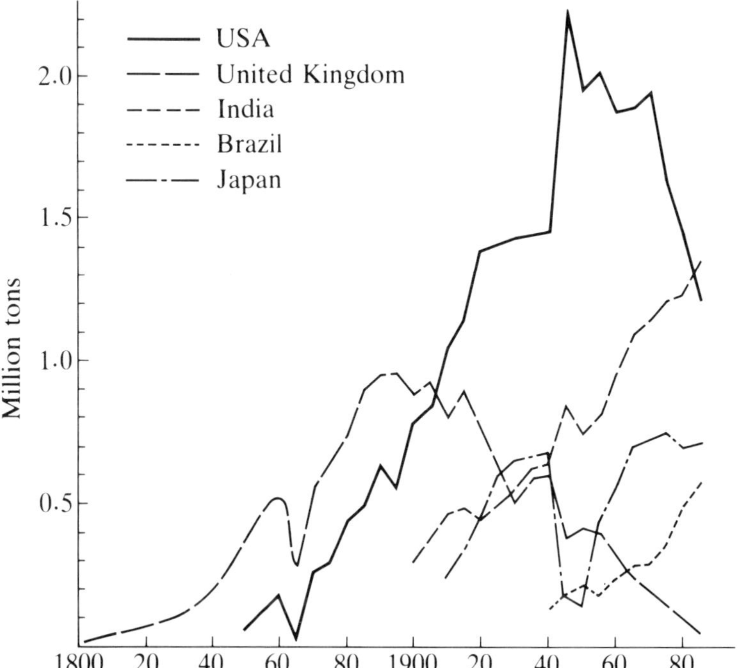

Fig. 2.4 Major consumers of cotton: 5-year averages, 1800–1984 (various sources).

mills, and these returns do not always cover all the spindles known to be installed in each country; it is assumed that the spindles not returned were consuming at the same rate as those which did make returns, but 'there is no evidence to show whether this assumption is correct or not' (Todd 1921). Statistics for China are missing or incomplete for quite long periods.

The Lancashire contribution to cotton

We have seen in Fig. 2.4 how the cotton industry in Britain rose and fell; during its heyday it made considerable contributions to cotton development in other parts of the world, both directly and indirectly, in production, manufacturing and processing.

Many individual efforts had been made in the nineteenth century to promote cotton production in new areas, but it was decided in 1902 that a joint effort would be more successful. A number of Lancashire cotton manufacturers joined together to form the British Cotton Growing Association (BCGA) for this purpose. In its early years it employed its own staff to supplement government services in the colonies; it arranged the import of cotton seed, conducted experiments and established cotton plantations to find out not only where cotton could be grown successfully, but where it could not. Its main effort has been in Africa, where the greatest successes have been in Nigeria (Cowley 1966), Uganda and Nyasaland (Malawi), but it has also assisted the cotton industries in the Punjab in India, the Sudan, Mesopotamia (Iraq) and the West Indies. One of its more lasting contributions has been the provision of ginneries, which had become its main interest by the time of the Second World War, and which have set standards of ginning and baling as good as any in the world.

Representations were made from Lancashire during the First World War for government assistance in the work of the BCGA, and in 1917 the Board of Trade appointed an Empire Cotton Committee to investigate the best means of developing cotton growing in the Empire. Among other things this committee emphasized the importance of research, and the Empire Cotton Growing Corporation (ECGC) was constituted under Royal Charter in 1921 as a permanent body to carry out the committee's recommendations. The ECGC was initially financed by the government and the cotton industry, the money being spent on supplementing the staffs of agricultural departments overseas, on establishing research stations in cotton growing areas, and on the education needed for these purposes. By 1945–46 the ECGC was publishing progress reports from experiment stations in Australia, South Africa, Rhodesia (Zimbabwe), Sudan, Uganda, Tanganyika (Tanzania), Nyasaland (Malawi), Nigeria and the West Indies, and had financed and staffed a Cotton Research Station in Trinidad from 1925 to 1944 for fundamental research on the genetics and physiology of cotton.

In Britain, nearly every firm and organization connected with the manufacture of cotton subscribed to the Cotton, Silk and Man-made Fibres Research Association, which set up the Shirley Institute in Manchester in 1919 to investigate the raw materials and finished products of the industry, and processes and treatments applied to them. Methods devised and tested at the Institute have since made significant contributions to all branches of the industry (see Ch. 18).

Individual contributions

During this period great advances were made in our knowledge of the cotton plant, in methods of breeding and in field experimentation. Among the many individuals who contributed to this knowledge, two are deserving of special mention.

Sir George Watt, who spent 30 years in India, first as Reporter on Economic Products to the government and later as Professor of Botany at Calcutta University, published *The Wild and Cultivated Cotton Plants of the World* (Watt 1907). There was a vast confusion of generic and specific names applied to cotton and its relatives; the specimens and species named by Linnaeus in 1753 and 1763 were all cultivated plants, and when wild species were later discovered 'they were named, on the standards of the five or six Linnaean types, the assumption being apparently accepted that there could be no other species' (Watt 1907). Watt redefined the cultivated species named by Linnaeus, and identified a range of wild species from both the Old and the New Worlds, providing a coherent classification on which to base the work of selection and improvement.

It comes as something of a surprise to find that Mendelian theory is only mentioned in the last paragraph of Watt's book, but the first English translation by W. Bateson of Mendel's *Experiments in Plant Hybridization* was only published in 1901. One of the first to study Mendelian segregation in cotton was W. Lawrence Balls in Egypt, while he was Botanist to the Khedivial Agricultural Society (1904–11) and later to the Department of Agriculture (1912–14). His detailed study of the development of the cotton plant and its reaction to the environment (Balls 1912 etc.) was an outstanding contribution to our knowledge which has stood the test of time. He was one of the first to appreciate the need for replication of treatments in field experiments and to calculate the probable error of his results; the Giza 'purity chequers' introduced in 1920 were elaborated from his target diagram tests (Brown 1953). Randomization and the analysis of variance used in modern methods of field experimentation had to wait for the publication in 1925 of *Statistical Methods for Research Workers* by R. A. Fisher.

Post-war developments

World cotton production has more than doubled from 6 million tons in 1938–39 to over 14 million in the last few years. The average production of individual countries over the five years 1977/78 to 1981/82, as recorded by the International Cotton Advisory Committee (ICAC, 1983), is given in Appendix B. Much of the increase has come from Russia and China, which now rival the United States as the major producers (Fig. 2.3c), but many of the smaller producers have shown spectacular increases. The United States is the only country to limit its production deliberately, with acreage restrictions, ceiling prices and marketing quotas applied off and on since 1950. Up to 1966, the United States held about half of the world's carry-over stocks from one season to the next (Fig. 2.5); it wanted to limit these stocks because of their expense and their

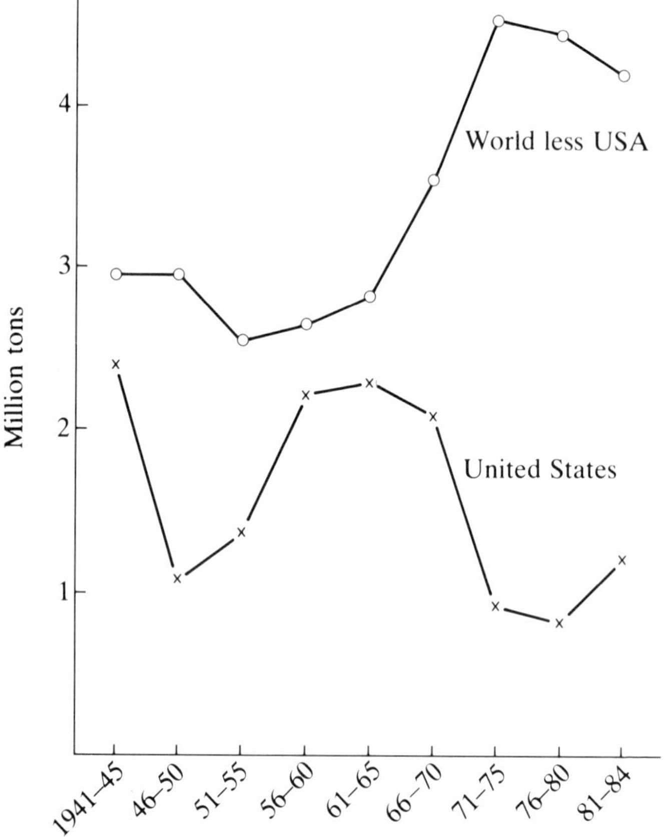

Fig. 2.5   World stocks of raw cotton; carry over at 1 August each year (ICAC).

effect on local and world prices, but it was remarkably successful in maintaining stable cotton prices during this period (see Fig. 17.9) by controlled release of the stocks.

In the last 40 years cotton production has been revolutionized by the new insecticides introduced since the war. The United States took the lead with the application of DDT to cotton in 1944, and a whole new range of synthetic organic products has appeared since then (see Ch. 12). Before the war, the success or failure of cotton depended largely on the level of insect attack, and cotton production was confined to areas where insect damage was not too severe. It is now possible to grow cotton successfully wherever the climate is suitable, using insecticides to control the insect pests, and production has spread to new areas where cotton was previously a failure. This is not to say that pest problems have all been solved, as some pests have proved intractable and pest control may become so expensive as to be uneconomic.

The expansion of cotton growing in Latin America is heavily dependent on the use of insecticides, and growers are now finding it necessary to limit the use of chemicals by methods of integrated pest control. Increases in production in East Africa and Nigeria were achieved without the use of insecticides, and their use even now is on a very limited scale. Red bollworm control in southern Africa, using spraying methods developed in Zimbabwe (see Ch. 12), has resulted in up to tenfold increases in production in the last 20 years, and similar methods applied in West Africa in the last 10 years have given similar increases.

Cotton production in West Africa owes much of its recent success to the work of the Institut de Recherches du Coton et des Textiles Exotiques (IRCT), in technical liaison with its commercial associate, the Compagnie Française pour le Développement des Textiles (CFDT). The IRCT was founded in 1946, with its central research station at Montpelier in France, where it is associated with other research organizations in the Groupement d'Etude et des Recherches pour le Développement de l'Agronomie Tropicale (GERDAT). It operates a string of research stations in the former French colonies in Africa, and has cooperated in cotton research with several other countries (Thailand, Paraguay, Nicaragua, Brazil, El Salvador and Iran). The CFDT acts as managing agent in West Africa, organizing the production, ginning and marketing of the cotton crop in the francophone countries, where production has increased and quality improved over the last 20 years (IRCT 1955, 1978).

In India and Pakistan insecticide spraying is a regular routine, but applications are still comparatively light. Regular spraying is practised in Australia. Development projects involving cotton now budget for insect control methods from the start.

A parallel revolution has taken place in the manufacture and consumption of cotton, as a result of the increasing competition from man-made fibres. This competition has slowed but not reversed the increase in cotton consumption: in the less-developed countries it is still rising, more than compensating for slightly reduced consumption in the United States and Europe. Treatments have been developed which incorporate easy-care qualities in cotton textiles, cotton is being blended with man-made fibres to produce new fabrics, and the natural advantages possessed by cotton have been emphasized (see Ch. 19). There is still a vast potential market for cotton if consumption world-wide is to reach the *per capita* levels of the developed countries.

Yields

Yields per acre or per hectare depend on two measurements – area and production – and errors in either of these will affect published yield figures. In Western Europe and the United States the area of each field is known from government or private surveys, and farmers are accustomed to making official returns of the area and yield of each crop. In the less-developed countries the situation is entirely different, and accurate figures can be very difficult to obtain. Cotton fields are often small and irregularly shaped, there are no official surveys sufficiently detailed to determine field size, and there are in any case often no permanent field boundaries such as fences, walls or hedges. There is also a problem in dealing with mixed cropping, although this is not so common with cotton as with food crops. Under these circumstances, total area is calculated from the estimated number of plots of cotton and the estimated area of each plot; both of these are likely to be inaccurate, especially the latter. In Uganda, for example, the annual Department of Agriculture estimates were compared with the results of a four-year census (1962–65), sampling 1 per cent of all land-holders, and it was found that the annual estimates of cotton acreage exceeded the census figures by a ratio of 1.5 to 1.0 (Jameson and Stephens 1970). As long as the discrepancy was fairly consistent and recognized, the 'Uganda acre', as it was called, could be used to measure changes in the pattern of agriculture from year to year, but it is misleading in any international comparison.

Figures for production by individual growers are generally unobtainable, although some data can be extracted from cooperative society records, but there is no guarantee that the whole crop was sold through one society. Ginnery returns give an accurate figure for district and national production, if the ginneries are well supervised by a marketing board or government department; but in Thailand, for example, the ginneries are privately owned and under no obligation to report their production figures to the government. Very few countries now export the whole of their cotton crop, so that official export figures must be supplemented by those for local consumption to give the total production. Reservations of this kind must be made before accepting the yield figures for various countries shown in Appendix B.

Israel has recorded higher yields than any other country for the last 20 years; its current five-year average (1978–82) is 1358 kg of lint per hectare, compared with its nearest rivals Australia with 1198 kg and Guatemala with 1187 kg. In general, yields are high in the Middle East (including Egypt) and in Latin America, while the lowest yields are found in Africa. This reflects to some extent the use of insecticides mentioned earlier. Separating out the three largest producers, average yields of lint (kg per ha) for 1978–82 were as follows (ICAC 1983):

United States	545
USSR	886
China PR	507
Remainder	307
World average	430

Geographical distribution

Cotton is generally regarded as a tropical crop, but at present about two-thirds of the world production comes from north of latitude 30 °N, where the three major producers, USA, USSR and China are all located; they therefore fall outside the scope of a book on tropical agriculture (Fig. 2.6). Small quantities of cotton are grown north of 40 °N in Bulgaria, Russia, China and Korea, but the summers are too short for anything but varieties which mature very quickly. About 10 per cent of the total crop comes from the southern hemisphere, ripening in May to July, while the remaining 25 per cent comes from the northern tropics up to 30 °N, mostly ripening in December to February. Outside the tropical belt, temperature rather than rainfall determines the cropping cycle, and north of 30° crops can only be grown in the summer months, ripening in September to November. Irrigated cotton in the tropics may be sown at any time of the year depending on the water supply; in the south of India, for example, sowing takes place from December to March and from July to October. This range of conditions helps to spread the supply of cotton throughout the year.

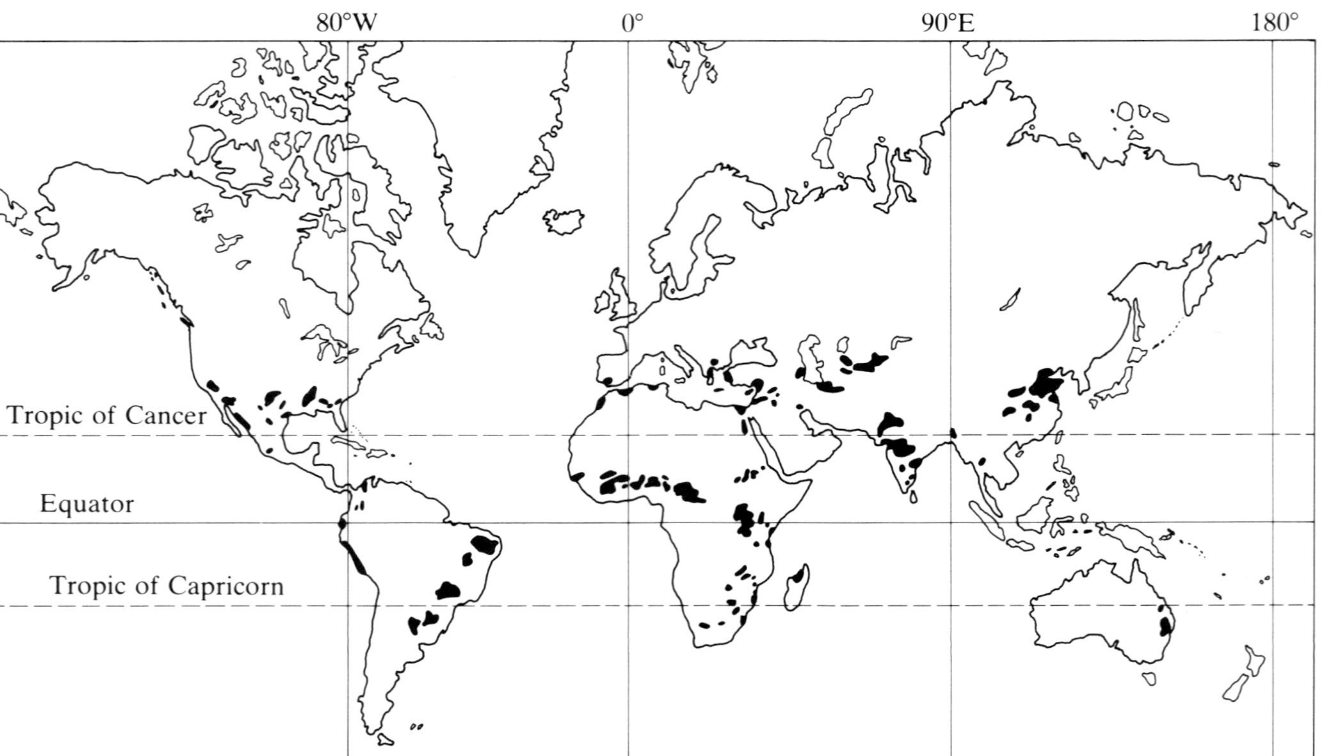

Fig. 2.6 Map showing areas in which cotton is grown.

Cotton-producing countries

More than half of the world's cotton crop is produced in three countries – the United States of America (USA), the Union of Soviet Socialist Republics (USSR) and the People's Republic of China (China PR). Production by countries, averaged over five seasons, is given in Appendix B, and the shares of the main producing countries are shown graphically in Fig. 2.3. The most striking development in the last 30 years has been the rapid expansion of production in the USSR and China PR, while production in the USA has been deliberately restricted by various government measures (Fig. 2.7).

India produces more than twice as much cotton as its nearest rivals, having shown a steady increase since partition in 1947. Its rivals in the Old World are Pakistan, Turkey and Egypt, and in the New World Brazil and Mexico; their development since the Second World War is shown in Figs. 2.7 and 2.8.

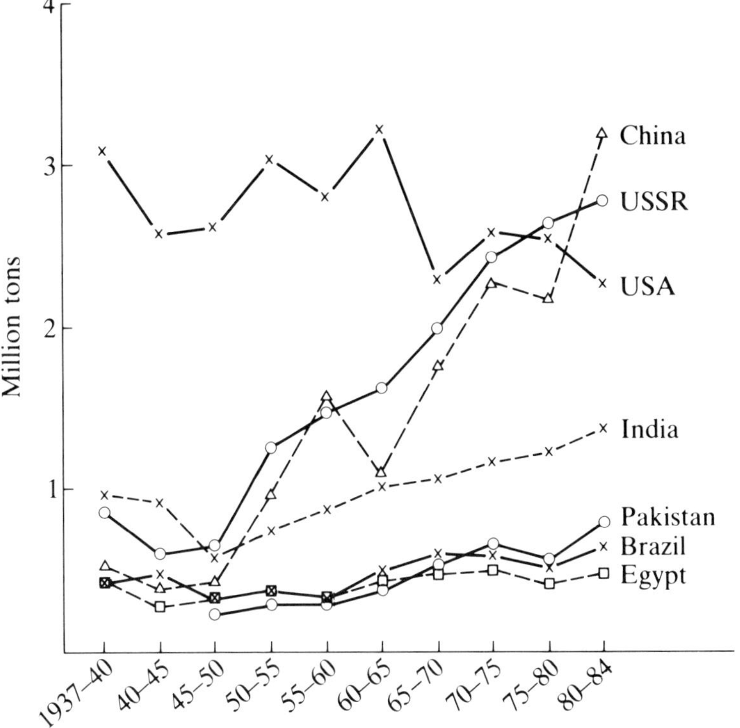

Fig. 2.7 Major producers of cottons; 5-year averages, 1937–84 (*Cott. Gr. Rev.*, ICAC).

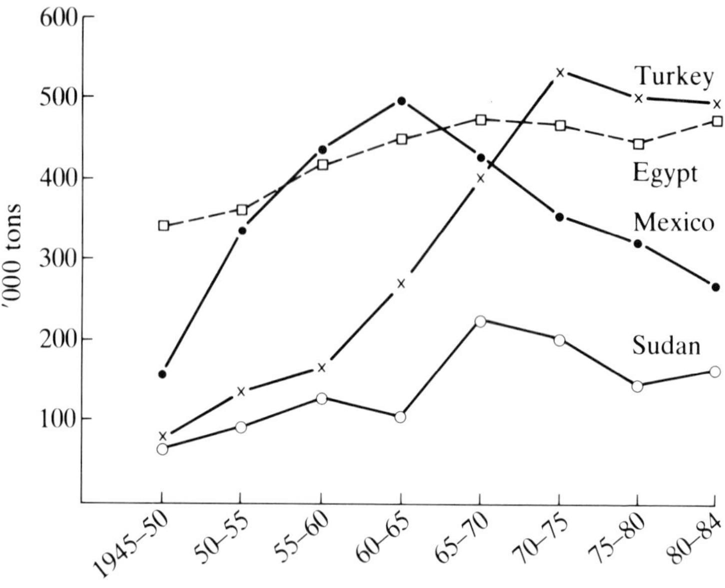

Fig. 2.8 Production trends in selected countries; 5-year averages, 1945–84 (*Cott. Gr. Rev.*, ICAC).

In Africa political troubles have played a major part in the pattern of production. The flourishing industries in Uganda, Tanzania, Zaïre, Mozambique and Nigeria have declined dramatically, while those in South Africa, Zimbabwe, Ivory Coast, Chad and Mali have taken their place as major producers south of the Sahara.

In the Middle East and Europe, the general pattern among the smaller producers has been an increase in production up to about 1970, with production remaining fairly steady since then; but there has been some decline in the Sudan, Iran and Spain while Israel and Greece have continued to expand. Among the smaller producers in Latin America, the fourfold increase in production in Paraguay in the last 10 years has been outstanding.

The statistics presented so far have dealt with total production, irrespective of staple length. The ICAC (1983) gives an estimate of production by staple length for each country, except for China PR, and Table 2.1 has been compiled from these statistics to show which countries produce either long-staple or short-staple cotton. About three-quarters of the world production is of medium staple, the proportion being similar in the two major producers, USA and USSR, and among the minor producers as a whole. This medium staple is typical of American Upland cotton, from which so many of the world's varieties have been derived.

Staple length has traditionally been measured in inches and

Table 2.1 *Cotton production by staple length (Five-year average 1977/78 to 1981/82)*

Staple	Short	Medium	Medium-long	Long	Extra-long	Total
	(1)	*(2)*	*(3)*	*(4)*	*(5)*	*(6)*
1000 metric tons						
United States	1	803	1,623	454	21	2,902
Brazil	—	43	499	27	—	569
Colombia	—	—	97	12	1	110
Ecuador	1	—	6	5	—	12
Mexico	1	15	310	11	—	337
Paraguay	—	—	78	10	—	88
Peru	—	—	1	68	22	91
Venezuela	—	—	7	7	—	14
Greece	—	—	90	38	—	128
Spain	—	—	29	21	—	50
India	341	692	171	39	93	1,336
Israel	—	2	61	11	4	78
Pakistan	33	404	200	11	—	648
Thailand	—	6	39	2	—	47
Turkey	—	—	408	95	—	503
Chad	—	7	25	5	—	37
Egypt	—	—	—	324	146	470
Madagascar	—	—	8	3	—	11
Morocco	—	—	—	—	6	6
Mozambique	—	—	15	3	—	18
Sudan	—	2	21	44	73	140
Tanzania	—	—	48	4	—	52
Uganda	—	—	4	5	—	9
Zimbabwe	—	—	46	7	9	62
Australia	—	1	78	4	—	83
USSR	—	—	2,258	383	95	2,736
Others	1	188	886	45	1	1,121
World (6)	378	2,163	7,008	1,638	471	11,658

(1) Under 13/16 in. (21 mm)
(2) 13/16 in. to 1 in. (21 to 25 mm)
(3) 1 1/32 in. to 1 3/32 in. (26 to 28 mm)
(4) 1 1/8 in. to 1 5/16 in. (28 to 34 mm)
(5) 1 3/8 in. (35 mm) and longer
(6) Excluding China PR
Source: ICAC 1983.

thirty-seconds of an inch; this measurement is still used in Britain, the United States and many other countries. In some countries, however, it is measured in metric units to the nearest mm, and occasionally in decimals of an inch. Equivalent staple lengths in inches and millimetres will be found in Table 17.4, while Table 2.1 gives the range of staple length in both units included in each of the broader classes of lint: short, medium, long, etc.

Long-staple cotton

Long-staple cotton (over 28 mm, 1⅛ in.) requires a regular supply of water to develop and mature the long, fine fibre, so that it is free from neps, irregularities in the spun yarn caused by tangles of immature fibres (see p. 363); much of it is therefore grown under irrigation, or in equatorial regions like the West Indies and Uganda where there is no prolonged dry season. Egypt was the first country to specialize in long-staple cotton: it started in 1820 with Jumel cotton (see Ch. 1); the first Ashmouni was selected about 1860 from a medley of types which almost certainly included early introductions of Sea Island, and this was followed in 1887 by Mit Afifi and in 1907 by Sakel. Extra-long staple is grown in the Nile delta, and long staple in Upper Egypt; the varieties have changed from the Sakel and Ashmouni of the 1920s, but the pattern remains the same.

The amount of long-staple cotton now grown in India is a considerable achievement, as early introductions from Egypt and the West Indies failed miserably under Indian conditions, and even American Upland varieties did not do well at first. The first success-ful adaptation of a Sea Island cotton was in 1955, but meanwhile a crossing programme between locally adapted *hirsutum* and *barbadense* varieties was under way and led to the release of the first interspecific hybrid Varalaxmi in 1972. There is now a range of long-staple varieties, including *hirsutum*, *barbadense* and various hybrid selections (Chokey Singh 1980).

Two types of long-staple cotton are grown in Peru. The bulk of the crop is Tanguis, a *barbadense* variety selected from a crop of the American Upland variety Suave by Sr Fermin Tanguis in 1908. The other type is Pima, an extra-long-staple variety bred in the United States (see below). The original Tanguis had a fairly coarse lint of extra-long staple and exceptional whiteness, and was resistant to *Verticillium* wilt. It constituted 91 per cent of the crop in 1933, but by then it had degenerated through contamination with the varieties it had replaced. The original qualities were restored from 1943 onwards by mass pedigree selection, described by Harland (1949), and several waves of improved seed have been sent out since then. About one-third of the Tanguis area is a ratoon crop, being grown as a perennial for three successive years, and cut back after each harvest (Bachini 1980). The cotton-growing areas are separated from each other by the coastal desert plain, each area being a delta irrigated from one of the rivers draining from the Andes, and temperatures are kept low by the cold Humboldt current from southern latitudes. Pima is grown in the two nothernmost valleys, those of the Piura and Chira rivers; some Del Cerro cotton is also grown.

Much of the production listed as long staple comes from Upland varieties grown under favourable conditions. A few of these varieties are known for their longer than average staple, such as Acala 1517 from California, BP52 and BPA from Uganda and MCU 5 and MCU 9 from India. Fifteen or twenty years ago many of the established breeders of Upland cotton in the United States put out high-quality counterparts of established varieties, of similar staple length but improved fibre strength, but the expected premium prices for lint of these varieties did not materialize.

Extra-long staple

Extra-long-staple cotton (over 35 mm, $1\frac{3}{8}$ in.) is a specialized product grown in only a few countries, from varieties of *G. barbadense* or crosses between *barbadense* and *hirsutum*. The Nile delta has already been mentioned, and the Gezira irrigation project in the Sudan was based on extra-long-staple cotton derived from the Egyptian Sakel variety. Several technical problems had to be solved: rotations and manuring; varieties resistant to bacterial blight and leaf curl; control of pink bollworm by seed treatment and field sanitation; control of jassid, whitefly and American bollworm by the use of insecticides; sticky lint caused by honeydew from aphis and whitefly, which still awaits a satisfactory solution (see Ch. 8, 12 and 13). Recent extensions to the area under irrigation at Khashm el Girba, Zeidab and Rahad, as well as the Tokar delta, have been sown with Upland varieties in the medium-long and long-staple categories; medium-staple raingrown crops are produced in the Nuba mountains, the Gedaref area and in Equatoria province.

Pima cotton is the extra-long-staple variety of the United States, Peru, Israel and Morocco. It was bred in the United States from the Egyptian (*barbadense*) variety Mit Afifi many years ago. Breeding work is now centred in Tempe, Arizona, and modern selections produce the 'quality cotton' of Arizona and New Mexico. Del Cerro is an offshoot of the Pima breeding programme, bred by a private grower in Arizona, which has proved to be quite adaptable in other countries, and is grown in Peru, Turkey and Zimbabwe (ICAC 1979). It differs from other long-staple varieties in that the plant habit is more like *hirsutum* than *barbadense*.

The longest and finest cotton in the world is grown without irrigation in the Windward and Leeward Islands of the West Indies. This is Sea Island cotton, which was developed in the states of Georgia, South Carolina and Florida in the nineteenth century; seed from South Carolina was imported into the British West Indies in 1903, where the sugar industry was meeting severe competition from European sugar beet and the development of sugar cane growing elsewhere. Cotton growing proved to be successful, either as a replacement for sugar or as a catch crop between the last ratoon and replanting. Meanwhile production of Sea Island cotton in the United States was almost wiped out by the boll weevil, falling as low as 10 bales in 1924, and leaving the West Indies as practically the only source of Sea Island cotton in the world. Demand for this extra-long staple has been very erratic since the First World War and there has been a constant danger of overproduction, even though the total output of half a dozen islands has rarely exceeded 1000 tons of lint (Lord 1945).

Short-staple cotton

Most of the world's short staple cotton (less than 21 mm, $\frac{13}{16}$ in.) is produced in India and Pakistan (see Table 2.1), typically from the Old World diploid species, *G. arboreum* and *G. herbaceum*, known collectively as 'desi' cottons. Since the last war the area sown with desi cotton has declined, while there has been a continuous increase in the area sown with *hirsutum* varieties in both countries:

Species	*India* [*] (*'000 ha*)		*Pakistan* [†] (*'000 ha*)	
	1947/48	*1975/76*	*1947/48*	*1975/76*
G. arboreum	2790	2080	202	155
G. herbaceum	1390	1480	—	—
G. hirsutum	140	3910	1035	1696
G. barbadense	—	90	—	—
Total	4320	7458	1237	1851

Sources: * Chokey Singh 1980, † Pakistan 1979

In Pakistan nearly all the cotton is grown under irrigation, and only about 5 per cent is now desi cotton (*G. arboreum*). Most of this is grown on the northern fringe of the cotton area, from Lahore to Rawalpindi in the Punjab, and Peshawar to Bannu in the North West Frontier Province. Smaller quantities are grown in Bahawalpur, while Sind desi comes from Nawabshah and adjoining areas of Khairpur.

There has been an eightfold increase in the area of irrigated cotton in India since 1947, but even now it represents only 25 per cent of the total cotton area. Under raingrown conditions the diploid cottons compete with Upland cotton, and are cultivated in all the cotton-growing states. There is no control over the type of seed the farmer chooses to plant, except in some project areas, but varieties are recommended by the All India Cotton Improvement Project for each area. The coarsest counts, *arboreum* cottons with a staple of less than 20 mm ($^{13}/_{16}$), are grown in the Punjab and

Haryana Provinces in the north, and in adjoining districts of Uttar Pradesh; *arboreum* and *hirsutum* varieties with staples between 20.5 and 24.5 mm are also grown in these provinces, while similar *herbaceum* varieties are grown in the Deccan area in the southern cotton zone. Great improvements in the quality of the Old World cottons have been made in India, with the *arboreum* varieties K8 and Saraswathi producing lint of 27–28 mm staple in the south, and *herbaceum* varieties up to 25.5 mm in Gujerat and south Rajasthan (Chokey Singh 1980).

Both species of diploid cotton are grown in China: *G. herbaceum* race *kuljianum* in the north-west of Sinkiang and western Kansu, and *G. arboreum* race *sinense* in the eastern half of the country (see Ch. 3). Reliable and up-to-date information on cotton production practices in China is hard to come by, and it is the only country for which the ICAC has no data on staple length (Table 2.1). Between 1949 and 1958 the average fibre length is said to have increased from 22 to 28 mm ($\frac{7}{8}$ in. to $1\frac{3}{32}$ in.) and the area under improved varieties increased from 10 to nearly 28 per cent; in 1958 about 60 per cent of the crop was irrigated, some with a winter or early spring irrigation prior to sowing (Basinski 1960). This indicates an increase in the proportion of *hirsutum* cottons, but some at least of the improved varieties which have been received in other countries through United Nations agencies are diploids. Tregear (1965) writes that growers prefer Chinese to imported varieties because of their quick maturity, pest resistance and resistance to shedding.

South America was the centre of origin of *G. barbadense*, but apart from the kidney cotton of Brazil, var. *braziliense*, there are no well defined local races. The bulk of the commercial production, except for Peruvian Tanguis, is from American Upland varieties, but Paraguay and to some extent Bolivia grow Reba varieties from French West Africa. In Ecuador short staple 'criollo' cotton is grown in the more inaccessible hilly areas as a perennial, inter-cropped with grain and legumes; it is a form of *barbadense*, and growers recognize several different varieties. In the north of Brazil, the Moco variety is derived from *G. hirsutum* race *marie galante* in the West Indies, but is grown as a perennial under conditions of low and uncertain rainfall (Demetriadi 1963). The short-staple cotton of Mexico is *G. hirsutum* race *punctatum*, grown round the coast of the Gulf of Mexico.

Cotton varieties

Lists of the cotton varieties grown in different countries have been published by the ICAC in the cotton production surveys carried out in 1973 and 1979. Such lists are not included here for three reasons.

In the first place varieties are constantly being replaced, and any such list rapidly loses its value; secondly, the names used by breeders are often changed when a new variety is issued, or it may replace an existing variety under the same trade name; and thirdly it is very difficult to decide from published data whether a variety should be included or not, and up-to-date information is often not available outside the country concerned.

In India, for example, the ICAC (1979) list comprises 14 varieties, plus Punjab Americans and Punjab desi; there are at least 10 Punjab American varieties. Three of the varieties on that list are not mentioned by Chokey Singh (1980) in his list of 56 important varieties or hybrids, but in another list giving the counts spun from each variety he names 77 varieties, and some in the first list do not appear in the second. Between 1971 and 1980, 46 new varieties were released.

A new variety usually has very similar fibre characters to those of the one it replaces, so as not to upset established markets for the lint, but shows improvements in other ways such as yield, ginning percentage, disease resistance or vegetative characters. The list in Table 2.1 gives a guide to the staple lengths produced in each country.

Chapter 3

Taxonomy

The order Malvales

The taxonomist tries to classify plants according to a logical system for a variety of reasons: to identify and name an unknown plant; to facilitate the storage and retrieval of information; or to establish genetical and family affinities. The early botanists had to decide the diagnostic characters or combination of characters which gave the most logical grouping of the various taxa, but the key characters, and hence the classification, are continually being revised as new knowledge is acquired from the discovery and description of new specimens; from cytological studies of the number and morphology of the chromosomes; from studies in cross-breeding and the fertility of the crosses; with or without complementary studies of chromosome affinities at meiosis.

Some acquaintance with the genera related to cotton is particularly useful to the cotton grower because many of them are alternate hosts to some of the most important pests of cotton. Sweeney (1960) showed that the initial breeding of stainers (*Dysdercus* spp.) at the start of the rains in Malawi takes place on *Sterculia africana*, one of the local trees, and that the recorded host plants of *Dysdercus* belong almost exclusively to the Malvaceae, Bombacaceae and Sterculiaceae. Similarly in Kenya, La Croix (1966) found that the different species of stainers bred on different wild host plants, mostly Malvaceous species including *Sterculia*, kapok and baobab. The wild hosts of pink bollworm (*Pectinophora*) are mostly species of *Hibiscus: H. esculentus* in India (Husain *et al.* 1931), *H. macrocanthus* in Uganda (Taylor 1936) and *H. dongolensis* in Zimbabwe (Matthews *et al.* 1965). The only alternative host of red bollworm (*Diparopsis*) in southern Africa, apart from the wild and cultivated species of *Gossypium*, is *Cienfugosia hildebrantii* (Parsons and Hutchinson 1939). Spiny bollworm (*Earias*) is found throughout the year in Tanzania on *Waltheria indica* and on species of *Hibiscus* and *Abutilon* (Reed 1974a).

The order Malvales in which cotton is placed includes other important fibre crops, such as the stem fibres of jute (*Corchoris capsularis*) and kenaf (*Hibiscus cannabinus*), and the fibres of kapok (*Ceiba pentandra*) and the silk cotton tree (*Bombax* spp.), borne on the inner wall of the seed capsule. Other well-known members of the order are the baobab (*Adansonia digitata*) and cocoa (*Theobroma cacao*). The main families of the Malvales are shown in Table 3.1.

The family Malvaceae

Classification within the Malvales has seen considerable changes. Bentham and Hooker (1867) described the families shown in Table 3.1 as natural orders, except for the 'Bombacaceae' which they placed as a fourth tribe in the Malvaceae; this classification was used by Watt (1907), but Hutchinson, J. (1959) separated all the groups related to the Malvaceae (as shown in Table 3.1) into the order Tiliales, leaving the Malvales with only one family, the Malvaceae.

Edlin (1935) proposed a revision of the order, placing the Hibisceae, including *Gossypium*, in the Bombacaceae instead of the Malvaceae because of the similarity of the fruits. This was accepted by Hutchinson *et al.* (1947) and even welcomed – 'the species producing seed and capsule hairs are all included in one family, as are also most of the important host plants of pests attacking *Gossypium*'. Mainly on the evidence of chromosome numbers, which shows that the Bombacaceae are highly polyploid compared

Table 3.1   *The family tree of Gossypium*

Rank	*Main line of descent*	*Related groups*
Order	Malvales	
Family	Malvaceae	Dirachmaceae Scyptopetalaceae Tiliaceae Sterculiaceae Peridiscaceae Bombacaceae
Tribe	Gossypieae	Hibisceae Malvinae Ureneae
Genus	*Gossypium*	*Thespesia* *Cienfuegosia* *Kokia* *Gossypioides* *Hampea* *Cephalohibiscus* *Lebronnecia*

with the Malvaceae, Fryxell (1979) disagrees with Edlin and his supporters and reverts to the previous classification.

The Hibisceae have been recognized as a distinct group with capsular fruits since they were first segregated by Reichenbach in 1828. Bentham and Hooker divided them into two series, centred on *Hibiscus* and *Gossypium* respectively. The distinction between the two series has been reinforced by subsequent work, and Fryxell (1968) assembled the data on which he based a separation of the Gossypieae as a tribe distinct from the Hibisceae. He gives Alefeld (1861) the credit for first recognizing the importance of the gossypol glands, which he called 'punctae', in separating the two tribes.

The tribe Gossypieae

Whether it be a series, group or a separate tribe, the genera now included in the Gossypieae form a closely related group: the number of species which have been transferred from one genus to another, and sometimes back again, is evidence of this, even if it is rather confusing to the layman. The central genus *Gossypium* was named by Linnaeus in 1753. *Thespesia* was separated from *Hibiscus* in 1760 by Duhamel; *Cienfuegosia* was described by Cavanilles in 1787; the central American genus *Hampea* was named by Schlechtendal in 1837.

Watt (1907) was not directly concerned with the genera related to *Gossypium*, but suggested a rearrangement should be attempted into four genera – *Gossypium*, *Thurberia*, *Thespesia* and *Fugosia* *(Cienfuegosia)*. The *Thurberia* are lintless wild American diploids, and have been included in *Gossypium* by both Hutchinson *et al.* (1947) and Fryxell (1979). Lewton (1912) considered *G. drynarioides* to be a separate genus, which he named *Kokia*, and this has been accepted by later authors; *Gossypioides* was separated from *Gossypium* by Harland (1932), a conclusion confirmed by Skovsted (1935) on the basis of chromosome numbers. The genera of the cotton tribe are shown in Table 3.2.

Two new genera were described by Lewton (1915) 'to include endemic Australian species that had been placed by various authors in *Gossypium*, *Cienfuegosia* and *Hibiscus*' (Hutchinson *et al.* 1947). These two genera were accepted by Hutchinson *et al.* (1947), but *Notoxylinon* is reincorporated in *Gossypium* by Fryxell (1979); *Alyogyne* is a genus of succulents, which is not included in Fryxell's cotton tribe. On the other hand, Fryxell includes the genus *Hampea*, which is not mentioned by Hutchinson, as well as two single species genera, *Cephalohibiscus peekeli* and *Lebronnecia kokioides*, found in New Guinea in 1935 and the Marquesas Islands in 1966 respectively.

Table 3.2   *The cotton tribe*

Watt	Hutchinson	Fryxell
Gossypium	Gossypium	Gossypium
Thurberia	Gossypium IV*	Gossypium IV & V[†]
Thespesia	Thespesia	Thespesia
Fugosia	Cienfugosia	Cienfuegosia
G. drynarioides	Kokia	Kokia
G. kirkii	Gossypioides	Gossypioides
—	Notoxylinon	Gossypium I
—	Alyogyne	—
—	—	Hampea
—	—	Cephalohibiscus
—	—	Lebronnecia

* Section
[†] Sub-genus
Sources: Watt 1907, Hutchinson *et al.* 1947, Fryxell 1979.

The genus *Gossypium*

The first comprehensive study of the taxonomy of cotton was *The Wild and Cultivated Cotton Plants of the World*, published in 1907 by Sir George Watt. This is still a standard reference for the nineteenth-century literature and herbarium material, although his subdivision of the genus, using the bracteoles, nectaries and seed hairs as diagnostic characters, has been superseded. Current ideas on classification date from a paper by Zaitzev (1928a); these were elaborated by Harland (1932, 1940) and Hutchinson and Ghose (1937b). Cytological studies by Skovsted (1937) and Webber (1939) led to the concept of genomes and their nomenclature by Beasley (1942). In 1947 Hutchinson, Silow and Stephens published *The Evolution of Gossypium*, reviewing the genus as a whole and the related genera of the tribe now called Gossypieae; they discussed the evolution and current status of the genus, with the aim of providing an adequate foundation for the proper planning of breeding work.

This review has been for many years the standard work of reference; it was supplemented by notes on the related genera by Hutchinson (1947) and *The Wild Species of Gossypium* (Saunders 1961), which incorporated subsequent discoveries, further taxonomic and cytogenetic studies of the wild species, and presented a set of botanical drawings of 19 wild species.

Recently Fryxell (1979) has published *The Natural History of the Cotton Tribe* which will probably supersede Hutchinson *et al.* as the taxonomic reference work on the tribe Gossypieae; it does not attempt, however, to deal with classification within the cultivated species of cotton, which is dealt with in great detail by Hutchinson *et al.* (1947), Silow (1944) and Hutchinson (1950, 1962).

Table 3.3 *The species of Gossypium*

Hutchinson	Fryxell
I Sturtiana	I Sturtia
1. *G. sturtii*	1. *G. sturtianum*
2. *G. robinsonii*	2. *G. sturtianum* var. *nandewarense*
G. australe	3. *G. robinsonii*
	II Grandicalyx
	4. *G. costulatum*
	5. *G. pulchellum*
	6. *G. populifolium*
	7. *G. pisolum*
	8. *G. cunninghamii*
	III Hibiscoidea
	9. *G. australe*
	10. *G. nelsonii*
	11. *G. bickii*
	12. *G. triphyllum*
III Klotzschiana	IV Houzingenia
6. *G. klotzschianum* var. *davidsonii*	13. *G. trilobum*
7. *G. raimondii*	14. *G. thurberi*
IV Thurberana	15. *G. klotzschianum*
8. *G. thurberi*	16. *G. davidsonii*
9. *G. trilobum*	17. *G. harknessii*
10. *G. gossypioides*	18. *G. armourianum*
G. lobatum	19. *G. turneri*
II Erioxyla	V Erioxylum
3. *G. aridum*	20. *G. aridum*
4. *G. armourianum*	21. *G. laxum*
5. *G. harknessii*	22. *G. lobatum*
VII Herbacea	23. *G. gossypioides*
16. *G. arboreum*	24. *G. raimondii*
17. *G. herbaceum*	VI Gossypium
V Anomala	25. *G. arboreum*
11. *G. triphyllum*	26. *G. herbaceum*
12. *G. anomalum*	27. *G. anomalum*
13. *G. areysianum*	28. *G. capitis-viridis*
VI Stocksiana	VII Pseudopambak
14. *G. stocksii*	29. *G. stocksii*
15. *G. somalense*	30. *G. somalense*
G. incanum	31. *G. incanum*
G. longicalyx	32. *G. areysianum*
VIII Hirsuta	33. *G. longicalyx*
18. *G. tomentosum*	VIII Karpas
19. *G. hirsutum*	34. *G. tomentosum*
20. *G. barbadense*	35. *G. lanceolatum*
	36. *G. hirsutum*
	37. *G. barbadense*
	38. *G. mustelinum*
	39. *G. darwinii*

Sources: Hutchinson *et al.* 1947, Fryxell 1979.

The following botanical description of the genus is adapted from those given by Hutchinson *et al.* (1947), Fryxell (1979) and Purseglove (1968); Fryxell gives a much fuller description in his book. Hutchinson and Fryxell give keys to the identification both of the genera of the Gossypieae and of the species of *Gossypium*. Table 3.3 compares the grouping of the species of *Gossypium* by the two authors.

Gossypium L.

Annual subshrubs, perennial shrubs or small trees, distributed in the tropical and subtropical regions of Africa, Asia, Australia and America.

Whole *plant* irregularly dotted with black oil glands. *Branches* terete or slightly angled, tomentose, usually with monopodial vegetative branches and sympodial fruiting branches. *Bracteoles* 3, usually foliar and persistent, rarely caducous at anthesis, inserted above nectaries at top of pedicel. *Calyx* cup-shaped, truncate, 5–toothed or deeply 5–lobed. *Petals* 5, imbricate, often large and showy, white, cream-yellow, rose or mauve, usually with darker coloured spot on claw. *Stamens* numerous, the lower parts of the filaments united in a tube, the upper free, bearing unilocular *anthers*. *Style* and *stigma* clavate rarely divided at the tip. *Ovary* 3–5 locular. *Fruit* a dry, brittle, loculicidally dehiscent capsule. *Seeds* several per locule, rarely 2 only, covered with 1 or 2 coats of unicellular hairs, or in some wild species almost naked. Haploid *chromosome* number 13 or 26.

Cytology

The first chromosome counts in cotton were reported by Nikolajewa in 1923 and Zaitzev in 1927. The New World cultivated cottons each had 52 chromosomes, and the Old World Asiatic cottons were found to possess half this number with 26 chromosomes. It was postulated that the basic haploid chromosome number (x) for *Gossypium* was 13; the Asiatics with $2x = 26$ were diploids, and the Upland and Egyptian cottons with $4x = 52$ were tetraploids. A comprehensive survey of the various species was undertaken by Skovsted (1935), but within the genus no other chromosome numbers were discovered.

The next step was to investigate the cross-fertility of the species of *Gossypium*, and where hybrids proved to be viable the pollen mother cells were examined at the first metaphase of meiosis to determine the behaviour of the parental chromosomes during pairing. The hybrid carries one set of chromosomes from each parent, and within a pure species the two sets give as many pairs, or bivalents, as there are chromosomes in the haploid number. But whereas within a species each chromosome must have its strict homologue, this is not necessarily so when different species are crossed. Evolutionary change has created distinct species within the genus; they are usually distinct phenotypically, and the differences may be associated with recognizable changes in the chromosomes. Chromosomal differences may be structural, with exchange of arms or segments of chromosomes, loss of non-essential parts, or inversion of material within the chromosomes; or there may be chemical differences and other incompatibilities which are not fully understood.

When crosses are made between species, the parental chromosomes seek their homologues in pairing at meiosis. Any structural or genetic divergences between chromosomes derived from a common ancestral stock will affect the efficiency of bivalent formation; the greater the divergence, the less chance there will be that the chromosomes will pair. Not all the chromosomes of a particular species will have undergone an equal amount of evolutionary change: the amount of pairing or the number of bivalents formed in an interspecific hybrid is therefore one measure of the degree of relationship that exists between the parent species. A full complement of 13 bivalents in a diploid hybrid would suggest that no great structural changes have occurred, although this would not necessarily preclude what Stephens (1944) refers to as 'cryptic' differences, whereas if as few as five bivalents were found no great affinity is indicated. It is possible in a hybrid between species to find that there is no pairing at all in the meiotic configurations.

If two closely similar forms are hybridized, their F_1 may have a high degree of fertility, with a complete set of bivalents and viable gametes. Chromosome doubling of such a hybrid to give a tetraploid would lead to greatly reduced fertility, if not near-sterility, because the close homology of the four sets of chromosomes creates difficulties at meiosis; there will be a tendency for multivalent groupings of up to four chromosomes, with a number of unmatched chromosomes as univalents. On the other hand, if the two species concerned are already very dissimilar in their chromosome and gene action, but yet have retained sufficient in common with one another to permit a successful cross, the F_1 would most likely be highly sterile. Supposing a spontaneous doubling of the chromosomes of such a hybrid occurred in somatic tissue which could ultimately produce a fruiting branch, every chromosome at meiosis would have its perfect homologue; regular pairing would take place, viable seed would result, and a new species is created which is effectively isolated genetically from its progenitors.

Genomes

Skovsted (1937) was able to distinguish clearly that in the tetraploid cottons there were 13 small and 13 large chromosomes in the haploid complement of 26. He inferred that in the evolutionary past a diploid species with small chromosomes had hybridized with a second diploid species with larger chromosomes, and that a spontaneous doubling had created a 52–chromosome tetraploid species. Triploid hybrids between the cultivated tetraploids and diploid Asiatic cottons showed that the 13 large chromosomes of the tetraploid formed 13 bivalents with the Asiatic chromosomes, while the 13 small chromosomes remained unpaired; a comparable triploid with a small chromosome American wild species as the diploid parent showed that the small chromosomes formed the bivalents, and the larger chromosomes remained unpaired.

Theory was put to the test when Beasley (1940, 1942) synthesized a tetraploid cotton by crossing the American diploid *G. thurberi* with the Asiatic *G. herbaceum*, and induced chromosome doubling by the use of the newly discovered properties of the drug colchicine; the 52–chromosome cotton was male-sterile, but when crossed with the cultivated tetraploid *G. hirsutum*, it gave viable seed and fertile progeny. Thus the evolutionary steps necessary for the formation of the American tetraploid cottons had been retraced.

As a result of his researches, Beasley postulated that there are in the genus *Gossypium* 5 distinct and basic sets of 13 chromosomes each, and that all known species could be associated with one or other of these sets. The basic sets were called 'genomes' and identified by capital letters from A to E; it is not surprising that the genome groups should coincide with geographical areas. The cultivated diploid Asiatic cottons were assigned the A genome. The wild diploids endemic to the African continent became the B genome, and those found in Australia the C genome. The diploids with the smallest chromosomes, found in central America, became the D genome; finally a distinctive group found in an area touching Pakistan, southern Arabia and parts of East Africa became the E genome. The tetraploid cottons, combining an Asiatic and an American ancestor, were symbolized as (AD) genome cottons. The world-wide distribution of the genome groups is shown on the map, Fig. 3.1, and Table 3.4 'lists the currently recognized species with their genome classification.

The cultivated cottons

The cultivated cottons are found in four of the species of *Gossypium*:
    Diploid Old World cottons: *G. arboreum; G. herbaceum*
    Tetraploid New World cottons: *G. hirsutum; G. barbadense*

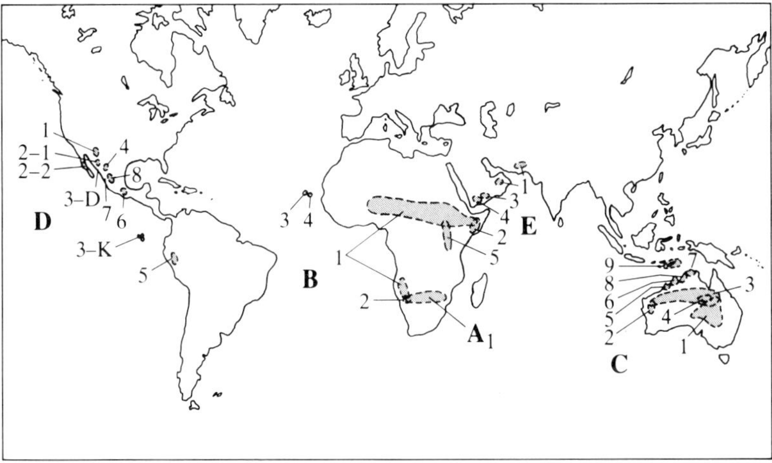

Fig. 3.1 Distribution of diploid genomes of Gossypium (Saunders 1972).

Table 3.4 *Genomes of currently recognized species of Gossypium*

Genome		Species
Asiatic		
A1		*G. herbaceum* L.
A1-W		*G. herbaceum* var. *africanum* (Watt) Hutch. & Ghose
A2		*G. arboreum* L.
African		
B1		*G. anomalum* Wawra *ex* Wawra & Peyr.
B2		*G. triphyllum* (Harvey) Hochreutiner
B3	(1)	*G. barbosanum* Phill. & Clem.
*B4		*G. capitis-viridis* Mauer
Australian		
C1		*G. sturtianum* Willis
C1-N	(2)	*G. sturtianum* var. *nandewarense* (Der.) Fryxell
C2		*G. robinsonii* Mueller
C3		*G. australe* Mueller
*C5		*G. costulatum* Todaro
*C6		*G. populifolium* (Benth.) Mueller
*C7		*G. cunninghamii* Todaro
*C8		*G. pulchellum* (Gardn.) Fryxell
*C9		*G. nelsonii* Fryxell
*C?		*G. pilosum* Fryxell
American		
D1		*G. thurberi* Todaro
D2-1		*G. armourianum* Kearney
D2-2		*G. harknessii* Brandegee
D3-K		*G. klotschianum* Andersson

Table 3.4   *Genomes of currently recognized species of Gossypium (continued)*

Genome	Species
American	
D3-D	*G. davidsonii* Kellogg
D4	*G. aridum* (Rose & Stand.) Skovsted
D5	*G. raimondii* Ulbrich
D6	*G. gossypioides* (Ulb.) Standley
D7	*G. lobatum* Gentry
D8	*G. trilobum* (Mocino & Sesse *ex* DC.) Skovsted
D9	*G. laxum* Phillips
*D?	*G. turneri* Fryxell
Arabian/African	
E1	*G. stocksii* Masters
E2	*G. somalense* (Gurke) Hutch.
E3	*G. areysianum* Deflers
E4	*G. incanum* (Schwartz) Hillcoat
Redesignations	
F1	(3) *G. longicalyx* Hutch. & Lee
G1	(4) *G. bickii* Prokhanov
Tetraploids	
(AD)1	*G. hirsutum* L.
(AD)2	*G. barbadense* L.
(AD)3	*G. tomentosum* Nuttall *ex* Seemann
*(AD)4	*G. mustelinum* Miers *ex* Watt
*(AD)?	*G. lanceolatum* Todaro
*(AD)?	*G. darwinii* Watt

Notes. *Genome affinity not established cytologically
    (1) Fryxell gives *G. barbosanum* as a synonym of *G. capitis-viridis*
    (2) Fryxell numbers this variety as a separate species
    (3) Formerly* E5
    (4) Formerly* C4
Sources: Mirza and Shaikh 1980, Fryxell 1979, J. H. Saunders in Prentice 1972.

A short key to these species (Purseglove 1968) is given in Table 3.5. They have been subdivided into races or varieties; for detailed descriptions and discussion of their centres of origin, migration and distribution the reader is referred to Hutchinson *et al.* (1947), Silow (1944), Hutchinson (1950, 1962) and Purseglove (1968). Notes on the species, races and varieties compiled from these sources are given below, and Fig. 3.2 shows the geographical distribution of the main cultivated races in the Old World.

G. arboreum L

Differentiation in each of the main areas in which the species is distributed has given rise to a series of genetically distinct races, but

Table 3.5 *Key to the cultivated species of cotton*

A. Bracteoles entire or dentate; teeth usually less than 3 times as long as broad.

    B. Bracteoles flaring widely from flower, usually broader than long; upper margin usually with 6–8 teeth; capsule rounded or with prominent shoulders .... *G. herbaceum*

    BB. Bracteoles closely investing flower, longer than broad; entire or with 3–5 teeth near apex; capsule tapering .... *G. arboreum*

AA. Bracteoles deeply laciniate; teeth usually more than 3 times as long as broad.

    B. Leaves deeply laciniate for two-thirds length into 3–5 lobes; petals usually bright yellow with basal reddish spot; anthers compactly arranged; pollen deep yellow; capsule coarsely pitted with black oil glands .... *G. barbadense*

    BB. Leaves less deeply laciniate for half length or less, or rarely 5 lobes; petals usually pale yellow or cream without basal reddish spot; anthers loosely arranged; pollen pale yellow or cream; capsule surface smooth .... *G. hirsutum*

Source: Purseglove 1968.

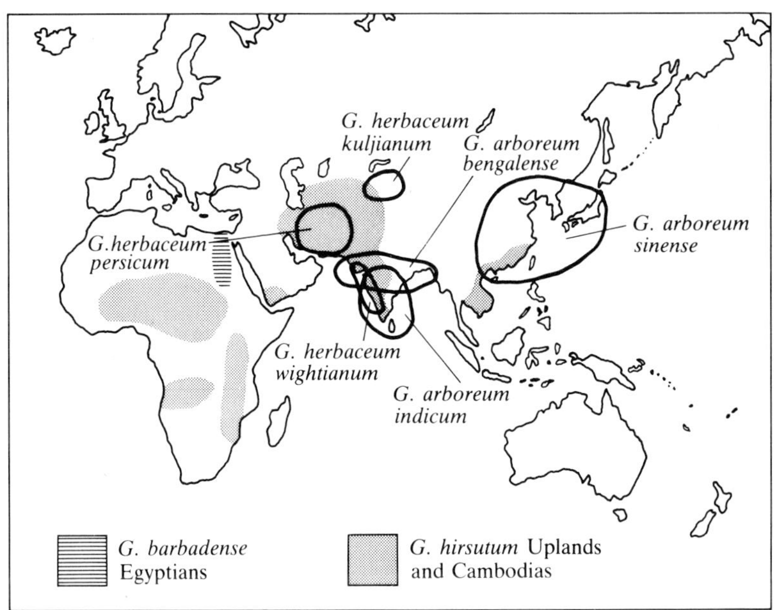

Fig. 3.2 Distribution of the main races of cultivated cotton in the Old World in 1960 (Hutchinson 1962).

there has been sufficient contact between neighbouring races to allow of considerable gene exchange.

1. race *indicum* Silow

The perennial forms are the 'Rozi' cotton of Gujerat in western India and those found in Madagascar and coastal Tanzania. The present annual commercial *arboreums* of southern India have been developed from the perennial forms.

2. race *burmanicum* Silow

A predominantly perennial race found in Burma and parts of Bengal and Assam, Indochina, Malaysia, and the East Indies as far as Timor Laut. Quality ranges from the fine cotton used in producing Dacca muslin to the short coarse types sold as Comillas (race *cernuum*).

3. race *cernuum* (Hutch. and Ghose) Silow

A type developed to suit the high rainfall conditions of the Assam hills, also found in east Bengal. It produces Comilla cotton.

4. race *sinense* Silow

An early fruiting annual which provides the commercial Asiatic cotton of China, Manchuria, Korea and Japan.

5. race *bengalense* Silow

These early, annual, high-ginning and relatively coarse cottons contribute the bulk of the Asiatic crop in Uttar Pradesh, the Punjab and Sind (the Bengals of commerce), and in the central provinces of India and northern Deccan (the Oomras tract).

6. race *soudanense* (Watt) Silow

This race is found in the Sudan region of North Africa and in West Africa. The plants are large perennial shrubs or small trees, the Senaar tree cotton; the plant shows red anthocyanin pigmentation throughout, and the West African types are even more strongly pigmented, with red leaves and flowers.

G. herbaceum

Annual cottons of this species were first developed in Iran and the surrounding countries; they spread southwards into Pakistan and India and north-east as far as China. Perennial forms occur in Africa. The species is divided into five geographical races which are now isolated from each other.

1. race *persicum* Hutch.

This is the typical form of *G. herbaceum*, and is found in Iran, Baluchistan, Afghanistan, Russian Turkestan, Iraq and round the Mediterranean, where it was widely spread by the Muslim invasion in the twelfth century. Watt (1907) considers that 'there is every reason for believing that this was the species first cultivated in Europe', and calls it Levant cotton.

2. race *kuljianum* Hutch.

A very early maturing race which extends further north than any other cultivated cotton in western China (Sinkiang, Kansu) and in Russia.

3. race *acerifolium* (Guill. and Perr.) Chev.

This perennial type spread from Arabia and Ethopia across north Africa as far as the Gambia, but is now of little commercial importance.

4. race *wightianum* (Tod.) Hutch.

These large annual shrubs were developed in western India from seed introduced from Persia in the nineteenth century. They produce some of the better staples among Asiatic cottons, including Coomptas and Broach Desi.

5. race *africanum* (Watt) Hutch. and Ghose

This is a perennial shrub found mainly in south-east Africa (Zimbabwe, Mozambique, Swaziland and the Transvaal lowveld). It is considered to be the most primitive of the Old World linted cottons, and the race in which the mutation for lint production occurred, making it the common ancestor of the diploid cultivated cottons.

G. hirsutum L.

The Upland cottons have spread from a comparatively small area in Mexico until they produce well over half of the world's crop. Seed was taken from Mexico (race *latifolium*) to the United States about AD 1700, where forms were selected which were capable of fruiting irrespective of day length. They have been introduced successfully into practically all the cotton-producing countries of the world. In addition to the species type and the many cultivated varieties, three cultivated races are recognized.

1. race *marie-galante* (Watt) Hutch.

This is a large perennial shrub or small tree, highly photoperiodic, flowering only during short days. It grows wild or cultivated in the islands of the Caribbean from Cuba southwards, and along the coast of South America from Panama to northern Brazil.

2. race *punctatum* (Schum.) Hutch.

The original form is a small perennial bush, found wild and cultivated round the Gulf of Mexico and as far south as Puerto Rico. It has become acclimatized and run wild as an annual in many parts of Africa; in West Africa it is cultivated on the margin of the Sahara Desert for use in the local spinning industry; it occurs as 'Hindi Weed' in Egyptian cotton fields. In India the Nadam crop in Madras is a mixture of *G. arboreum* perennial cotton and *G. hirsutum* race *punctatum*. It has also spread to the Philippines, Polynesia and north-east Australia (syn. *G. taitense* Parl.).

3. race *latifolium* Hutch.

This is an annual form, found in Mexico and Guatemala, which is predominantly photoperiodic. It was seed of this race which was brought to the USA, where it appears to have been the foundation stock of all the annual Upland cottons.

G. barbadense L.

Perennial types occur in western South America from Colombia to Bolivia, either cultivated, semi-cultivated or spontaneous. The dominant commercial variety, Peruvian Tanguis, was originally perennial, but some modern varieties are annuals. Fine-linted types were introduced into the West Indies in the eighteenth century; in 1786 seed from the Bahamas or Jamaica was taken to South Carolina, where the first Sea Island cotton was bred from it as an annual. Perennial types also found their way to West Africa, and gave rise to the Ishan cotton of Nigeria. This was probably the source of the perennial cotton selected by Jumel in Egypt about 1820; it was crossed with the Sea Island cotton introduced about 1850, giving rise to the present annual Egyptian cottons. The Pima varieties of Arizona and New Mexico are derived from the Yuma variety which was selected from Mit Afifi, an introduction from Egypt.

There are no well-defined races, but one variety is distinguished by its seeds:

1. var. *brasiliense* (Macf.) Hutch.

The seeds of this variety are fused into a kidney-shaped mass, from which it gets the name Kidney cotton; the bolls are also larger than the typical *barbadense*. Originating in eastern tropical South America, it is now distributed throughout central America and the Antilles, and is found sporadically in Africa and India.

Chapter 4

The cotton plant

The seedling

When a cotton seed germinates, the first sign of activity is the emergence of the radicle from the pointed hilum end of the seed. This turns downwards, anchoring the seedling in the soil, and resists any downward movement when the hypocotyl develops about two days later. The hypocotyl is bent backwards as it pulls the cotyledons clear of the ground, and then quickly straightens bringing the stem and cotyledons into an upright position (Fig. 4.1). If the soil is too hard or is capped with a hard layer above the seed, the hypocotyl

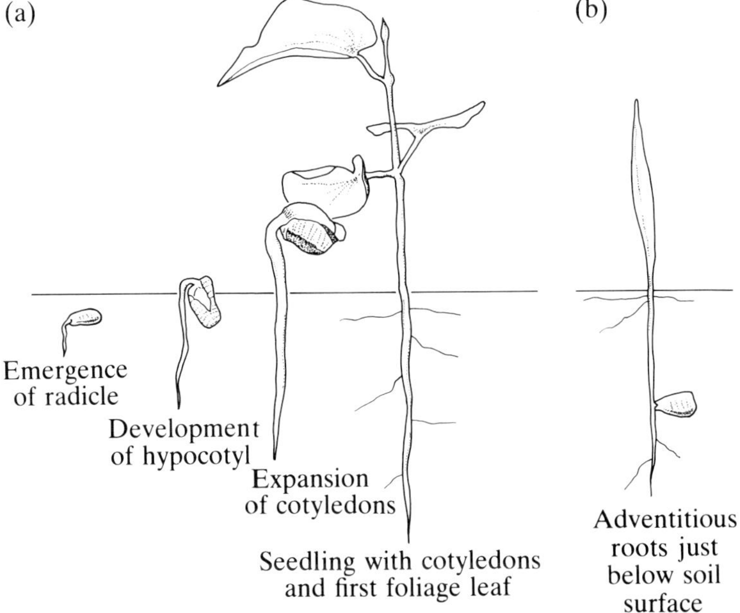

Fig. 4.1 Development of cotton and maize seedlings: (a) cotton; (b) maize.

may not be strong enough to pull the cotyledons free, especially if the seed is sown too deeply; the weakest point is the bend in the hypocotyl, and in these circumstances it may break at this point causing the death of the seedling. In loose soil the seed coat will be pulled out of the soil with the cotyledons and drop off when they expand, but it is usually left behind at or below ground level. The cotton seedling cannot compensate for depth of sowing by producing adventitious roots just below the soil surface like maize (Fig. 4.1); uniform depth of sowing is therefore important in producing a vigorous stand of young cotton. Seed should be sown at a depth of between 2 and 4 cm rather deeper in dry conditions and shallower when the soil is very wet (see Ch. 6).

The root

The cotton plant has a tap-root system. The root starts as a smooth continuation of the stem, of similar thickness, tapers rapidly over its first 15 cm or so and persists as a thin cylindrical root. The depth to which it penetrates is roughly equal to the height of the stem in the adult plant, but is greatly influenced by soil conditions. The tap-root has been measured to a depth of 3 m, and the roots can spread up to 3 m laterally from the tap-root if conditions are favourable. Two-thirds of the root system by weight are commonly found in the top 30 cm of soil. Lateral roots arise freely from the upper, thicker portion of the tap-root, and run for distances of a metre or more, even in closely spaced crops. These laterals run more or less horizontally, and constitute the bulk of the feeding roots; in a crop sown on ridges, the feeding roots are concentrated in the moist soil under the furrows, and root development is often limited by the distribution of soil water. This is clearly illustrated in Fig. 4.2, which shows root development in southern Arabia, where the crop depends on flood water stored in the soil prior to planting (Hearn 1980).

Much less is known about cotton roots than about aerial parts of the plant because of the labour involved in studying them (Hearn 1976). The usual method is to dig a trench alongside the cotton row, and use a jet of water to wash away the soil from the side of the trench, leaving the undamaged roots exposed (Balls 1912, Eaton 1931). The growth of roots can be studied behind a sheet of glass placed against the side of the trench, but there is always the possibility that this upsets the natural conditions of aeration, water supply and light, affecting root growth to an unknown extent. Balls (1915), Rijks (1965) and Farbrother (1972) measured root development and activity indirectly by charting the depletion of soil moisture (see Fig. 10.8) in the soil below the plant, taking samples with soil augers down to a depth of 3 m. Most of the gains and losses occur in

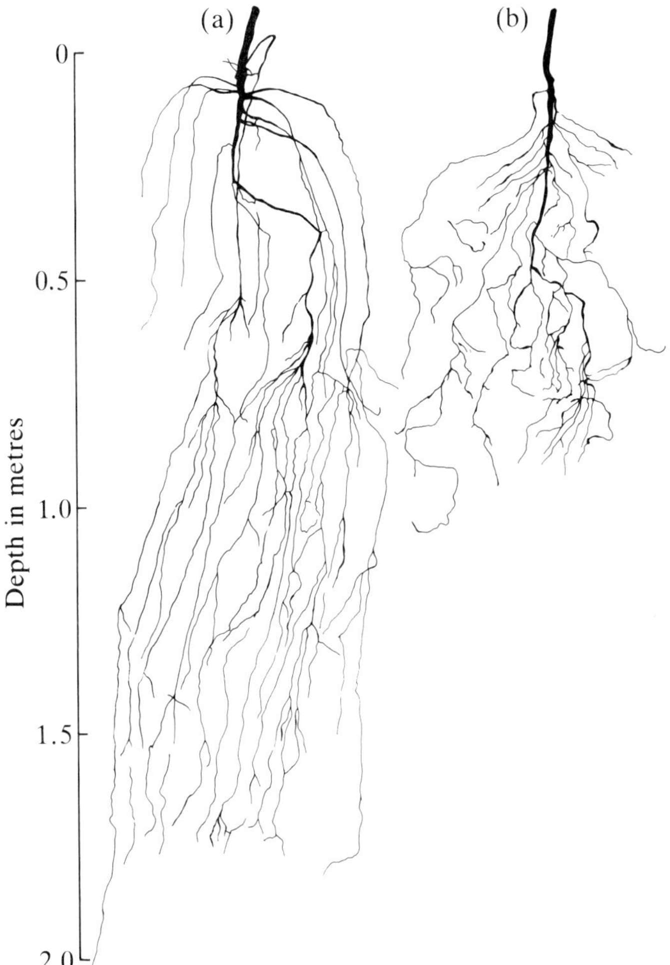

Fig. 4.2   Root systems of cotton plants grown with: (a) 450 mm; and (b) 150 mm of stored water (Hearn 1972b). Compare Fig. 4.10.

the 0–60 cm zone; for control of irrigation, the moisture content of the 0–30 cm zone may be used as an indicator up to flowering, the 0–60 cm zone later in the season (Roe 1950).

The stem and branches

The cotton plant has a vertical stem carrying two kinds of branches: vegetative branches or monopodia are similar to the mainstem in that the primary axis continues to grow from the same growing point, developing successive lateral branches; fruiting branches or sympodia terminate at each node with a flower bud, and growth of the branch is continued by a lateral branch which repeats the process (Fig. 4.3). Thus the mainstem and the monopodia do not bear flowers directly, but produce fruiting branches which do.

The mainstem produces a series of nodes, each with a leaf subtending two or three dormant buds on the side of the stem, which may develop into branches. The first two nodes carry the cotyledons, and are on opposite sides of the stem at the same level; above them, the leaves are arranged in a spiral ⅜ phyllotaxy, that is, they are arranged in a spiral up the stem, each leaf situated ⅜ of the way round the circumference of the stem from its predecessor. The direction of the spiral may be right- or left-handed, about half of the plants in a field belonging to each class. Occasionally, a plant will reverse the direction of its spiral, possibly as a result of injury. The monopodial branches nearly always show the same direction of spiral as the mainstem. The development of the buds at each node largely depends on the environment: under conditions of close spacing, low fertility or drought many of the buds will remain

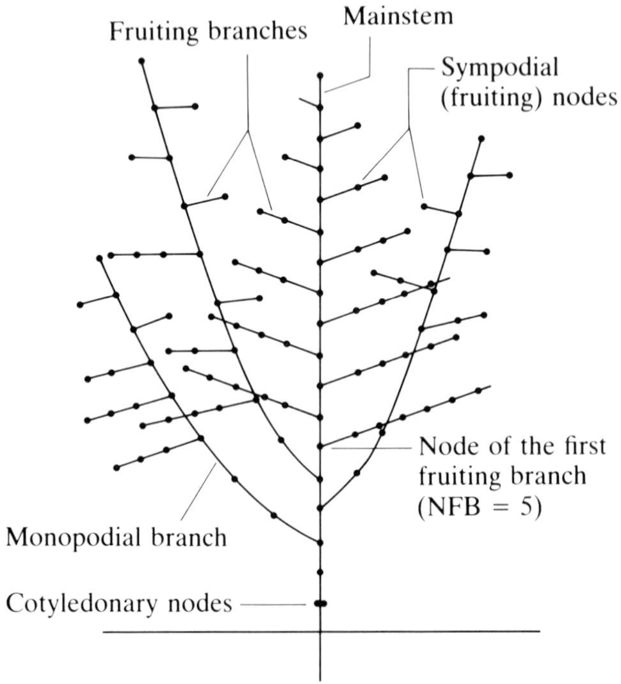

Fig. 4.3   Branching habit of the cotton plant (diagrammatic).

dormant, but given wide spacing and favourable conditions most of them will develop, sometimes producing highly complicated branching patterns.

The first of the buds to develop at each mainstem node is the axillary bud, producing either a sympodial or monopodial branch; which it will be depends on conditions some time before the branch appears (Mauney 1966). Below a certain point the branches are monopodial, above that point they are sympodial; the point at which the change occurs is recorded as the node of the first fruiting branch (NFB), counting the nodes up the stem from the cotyledons, and the significance of this will be discussed later in this chapter. The fruiting branches develop in succession from the NFB upwards, while the vegetative branches develop in succession downwards from the same point, the cotyledonary nodes being the last to develop; usually the lowest nodes on the mainstem and the cotyledonary nodes remain dormant or blind, unless growth is very vigorous indeed. The second and third buds at each node rarely develop below the NFB, but in the fruiting zone, if conditions are favourable, a second bud will develop; this often produces a short sympodium with a single flower, but may be a longer fruiting branch or a vegetative branch. A third branch sometimes develops at the same node.

The fruiting branches are inclined upwards, usually at an angle between 45° and 90° from the vertical, and this angle is a heritable character; selections with horizontal branches (90°) give a more open type of plant. Once a full complement of bolls has been set, the weight of the fruit may bend the branch downwards, but rarely causes it to break. The sympodial type of growth with a flower bud at each node tends to give a zig-zag appearance to these branches, which is more marked in some varieties than in others. Because of their earlier start, the lower fruiting branches are longer than the upper ones, giving the plant its typical pyramid shape, but close spacing restricts the length of the lower branches. In some conditions, notably in *G. barbadense* in the Sudan, the first fruiting branch is retarded and produces no more than one flower bud (H. E. King pers. comm.).

The vegetative branches generally grow upwards at a more acute angle than the fruiting branches; they may reach the same height as the mainstem. The junction of the monopodia with the mainstem is a point of weakness, and heavily laden branches will sometimes break away. The lower internodes of the monopodia are usually long, with one or two blind nodes; this and the straightness of the branch make them easily distinguishable from the fruiting branches. Although the vegetative branch just below the NFB appears first, lower branches develop more strongly, and the lowest of two or three monopodia is usually the strongest.

One of the problems of cotton growing in the tropics is the height of the crop; varieties from the USA, which normally grow to a height of 1.0 to 1.2 m at home, will grow to more than 2.0 m in the tropics. This is the result of increased internode length – the number of branches is much the same in both places. Plant height can be reduced by selection, but the reduction is likely to be small, 5 to 10 cm or so, which is negligible compared to differences of one metre caused by a change in environment.

No adequate definition of the conditions causing excessive vegetative growth has been put forward, although high night temperatures and reduced sunlight probably play a part. The phenomenon is not confined to the tropics, so day length is unlikely to be a factor; it occurs at low altitudes in the subtropics in Malawi and Zimbabwe, in Australia and in the valleys of California and Arizona. In the Pima breeding programmes in Arizona, short internode types (Pima S4) were selected for the low elevations and varieties with longer internodes (Pima S3) for high altitudes (Feaster and Turcotte 1970).

One of the first effects of water stress is a shortening of the internodes, without affecting the timing of node production. In examining a crop in the field, one or two of the mainstem internodes may be found to be much shorter than the others, indicating a dry spell when these nodes were developing. The dry spell can often be dated by reference to the time-scale of plant development; only in extreme drought conditions does growth stop completely.

The 'cluster' and 'semi-cluster' habit, where the flowers are produced very close together, is caused by a hereditary shortening of the internodes of the sympodia; these genes do not affect internode length in the mainstem and vegetative branches.

A tall crop is more liable to 'lodging' than a short one, and a tropical storm of wind and rain can flatten large areas of cotton, especially when it is carrying a heavy crop of green bolls. A lodged crop is difficult to harvest, but in addition the conditions under the leaves of the fallen plants are ideal for fungal growth, and boll rotting sets in. Crop losses can reach 40 per cent of the potential crop, and much of the cotton fit for picking is likely to be of poor quality. Fortunately a reduction in the height of the crop is not the only answer, as marked differences have been noted between varieties and selections in their ability to stand up to these conditions.

The response of the cotton plant to mechanical damage depends on where and when it occurs. When the growing point of the mainstem of a young plant is lost or damaged, additional monopodial branches develop from the last undamaged node to take over the lead (Evenson 1969). When the plant is older (the change-over point has not been defined) no new branches develop, but existing branches grow more vigorously. If the plant is 'topped' at the peak of flowering, the most obvious result is that the highest fruiting branches grow longer and produce more buds. Damage to a fruiting branch usually results in the development of a monopodial branch from the last remaining node, but the new development may be sympodial or no replacement growth may occur. The loss of early buds is discussed in Ch. 12.

The description given here attempts to give a picture of the typical Upland cotton plant, but the structure of the plant is very adaptable and dependent on the environment in which it is growing. Closely spaced plants in a commercial crop tend to have fewer monopodial branches, and the lower sympodia are shorter. Spacing has little effect on plant height, but any loss of early bolls induces additional growth to compensate for the loss, and the plants grow taller. When plants are spaced closely in the row, the stems often incline to left and right of the vertical as if seeking more space to grow.

On infertile soils with a shortage of water, the plant will be short and spindly; many of the mainstem nodes will remain blind, with short internodes between them, the fruiting branches will be short with only one or two flower buds and there will be no monopodial branches. This is usually accompanied by early shedding of the leaves, leaving a thin, short and bare stem perhaps little more than a foot high.

In very fertile conditions with ample water, all the available buds will develop, internodes will be long and nearly all the leaves will be retained, giving a tall, bushy vigorous plant. Each mainstem node will have two or three branches, the fruiting branches will be long with secondary branching, two flower buds will appear at most of the sympodial nodes, and several vigorous monopodia will branch from the base of the stem.

The first fruiting branch

The number of mainstem nodes produced before the first fruiting branch is an important characteristic, affecting the earliness of the crop and the location of the bolls on the plant. It is influenced mainly by three factors: species, variety and ambient temperature, particularly night temperature. In *G. hirsutum* the node of the first fruiting branch (NFB) is typically low, the change from vegetative to fruiting branches occurring about the fourth or fifth node above the cotyledons. *Gossypium barbadense* has a higher NFB, 10 or 12 being typical of Egyptian cotton, while Tanguis goes up to 20 or more. The Asiatic diploid cottons also have a high NFB, from 15 to 20.

Within any species it can be assumed that the rate of node production on the mainstem is fairly uniform, so the lower the NFB the sooner the first flower appears and crop production starts. Selection for low NFB is thus one way of producing an earlier crop, as the character is easily measured and quantified. Counting the mainstem nodes is easier when the plant is young, and still carrying the mainstem leaves; in a mature plant the nodes are not so obvious, and can be masked by the development of vigorous monopodial branches from the lower nodes. Some workers count the two cotyledonary nodes, but it seems simpler to start with the node above the cotyledons as node one.

With a low NFB, the crop can be set lower on the plant, giving a more compact habit; it has the disadvantage that cotton on the lower branches is very near the ground, and may be dirtied by rainsplashed mud or by actual contact with the soil. Wide spacing increases this tendency by allowing long lower branches to develop.

Independent of species and variety, high temperature increases the NFB. This was noted in Egyptian cotton by Balls (1912): 'in certain circumstances, not yet clearly understood, but apparently including excessively high night temperatures, the plant delays its formation of flowering branches.' Using controlled temperature growth chambers, Mauney (1966) has shown this effect on Upland cotton more precisely (see Table 9.1): by controlling the temperature, the NFB of a variety can be raised from node 4 to node 20 (see Ch. 9).

The leaf

The normal adult leaf is palmatifid, but the range in leaf shape and size is considerable (Figs 4.4 and 4.5). The cotyledons are an opposite pair of simple leaves with short stalks; they are rather fleshy and glabrous, and usually expand to a breadth of 4 cm within two or three days. They open about 5 cm above ground level, but this depends on the moisture supply and the depth of sowing; if the cotyledons are very near the ground, the seed has probably been sown too deep. The terminal bud of loosely furled leaves protects the growing point; as this develops from between the cotyledons, the stem elongates and the leaves unfold in succession along the spiral track of the phyllotaxy, one at each node. The leaf arising from the node immediately above the cotyledons is known as the first true or rough leaf; the venation is prominent and a few hairs are present in some varieties. This leaf, however, is rounded or cordate, with only a hint of the lobes typical of the adult leaf; it is not until about the eighth node that the full adult shape with its half-cut lobes is attained. In the same way the hairiness of the under-surface of the leaf becomes more pronounced, reaching its full expression about the twelfth node.

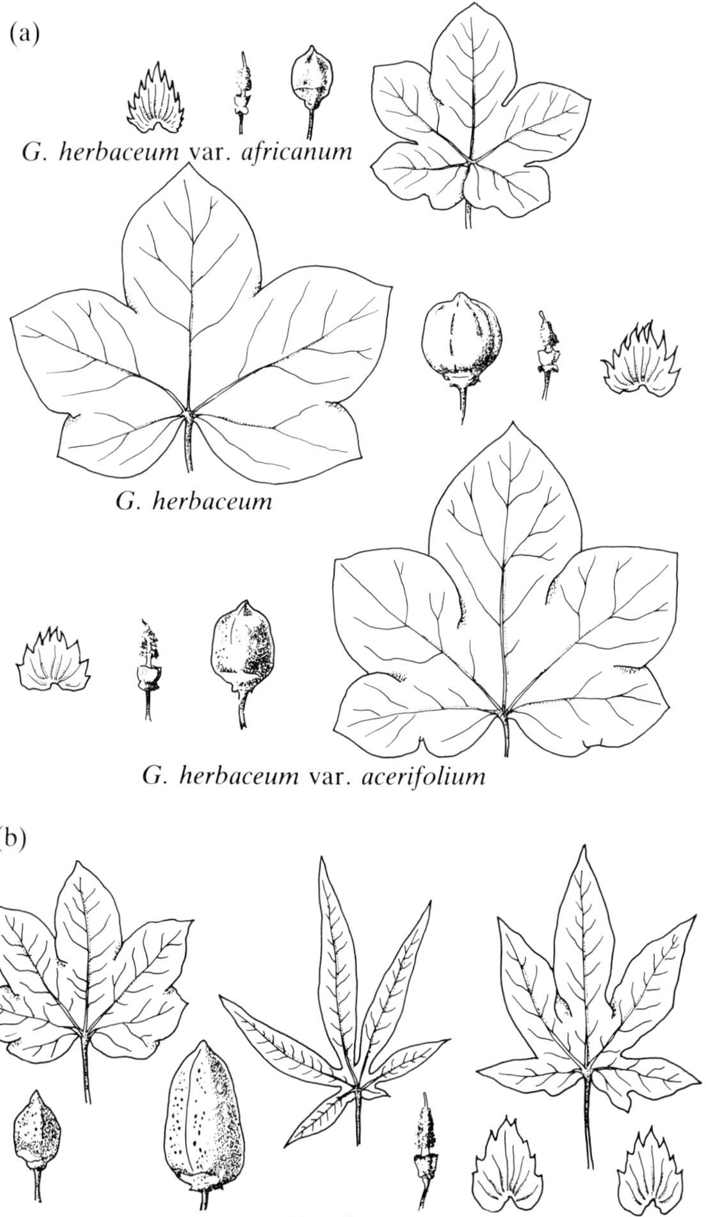

Fig. 4.4 Leaf and boll characters of: (a) *G. herbaceum*; and (b) *G. arboreum* (Hutchinson *et al.* 1947).

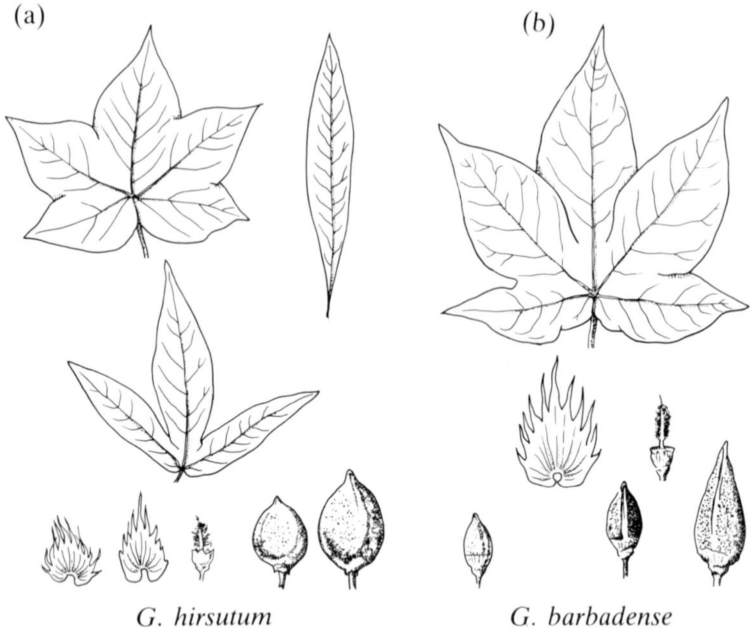

G. hirsutum   G. barbadense

Fig. 4.5 Leaf and boll characters of: (a) *G. hirsutum*; and (b) *G. barbadense* (Hutchinson *et al*. 1947).

The hairiness of the leaf, and of other parts of the plant, is heritable, and is of considerable importance in breeding for resistance to insect pests. The leaf hairs are stellate, and vary both in length and density from very sparse to densely tomatose. There are fewer hairs on the upper surface of the leaf. The concentration of hairs around the growing tip of the stem, before the leaves have expanded, is often sufficient to give the tips a grey, felted look.

The adult leaf is commonly 12–15 cm in length and breadth, but under harder growing conditions, or with the setting of a progressively heavier crop, the average leaf size decreases. Size of leaf, moreover, can vary with the variety of cotton, as can leaf shape and colour. The usual adult leaf has three to five lobes, half-cut as shown in Fig. 4.5, but some varieties show an almost fully cut or laciniate leaf; this is called 'okra-leaf', being similar to the leaves of the okra plant, *Hibiscus esculentus*, and 'super-okra' types have been developed with the leaf often reduced to a single, strap-shaped lobe (Plate 4). In general, the leaves of *G. barbadense* are more deeply divided than those of *G. hirsutum*, and *G. arboreum* more than *G. herbaceum*.

The petiole is 3–4 cm long, with two small green deciduous

stipules. The leaf is shed by the formation of an abscission layer at the base of the petiole.

The lamina is commonly half a millimetre thick, and both surfaces carry stomata, more frequent on the lower surfaces. A nectary is situated on the underside of the leaf, near the base of the main vein.

The leaves of the young plant are of a slightly lighter shade of green than those of an older plant; this is usually noticeable only as a mass effect between adjacent crops of different ages. Most leaves can develop a red pigment, anthocyanin, which is masked in the young leaf by the green colour, except for a small red spot where the petiole joins the lamina; the anthocyanin causes reddening of the leaf under stress – drought, waterlogging, insect damage and senescence; it appears to be a waste product of the cell (Miller 1938). Some plants with an excess of anthocyanin have a dark red colour in all parts of the plants (Plate 3), probably controlled by a single gene and easily transferred; others lack anthocyanin, and turn sear and yellow instead of red under stress.

The flower

The flower bud first becomes visible as a small green cone, closely adpressed to the fruiting branch; it grows steadily as the fruiting branch elongates, its stalk (pedicel) raising it clear of the branch. The stalk has finished growing by the time the flower bud is ready to open; it is usually vertical, carrying the flower in an upright position.

The bud is surrounded by three green bracts or bracteoles, joined at their bases; the bracteoles are heart-shaped, forming a pointed canopy over the bud, often referred to as a 'square' (See Fig. 4.13). At the base of each bracteole, on the outside, is a nectary. The ring or involucre of bracts is pushed open by the expanding petals on the day the flower opens. It has then reached full size, and remains green, investing the developing fruit until the boll is about to ripen, when it becomes dry and brittle but persistent; the dry bracts are an important source of broken pieces of dead leaf which can lower the value of the seed cotton. If a bud or young boll is punctured by bollworm, the bracts flare outwards leaving the bud exposed, and start to lose their green colour; these flared squares are a useful indication of bollworm attack in the crop (Plate 8).

Within the bracts is an inconspicuous green calyx, with a ring of nectaries on the inside; it surrounds the bases of the petals, which are tightly folded in the bud. The day before flowering, the conical bud starts to enlarge rapidly, the petals more than doubling their length in 24 hours. Early on the following day the corolla opens widely, to expose the staminal column and the stigma. The floral diagram is shown in Fig. 4.6, with the formula E 3–4 K (5) C 5

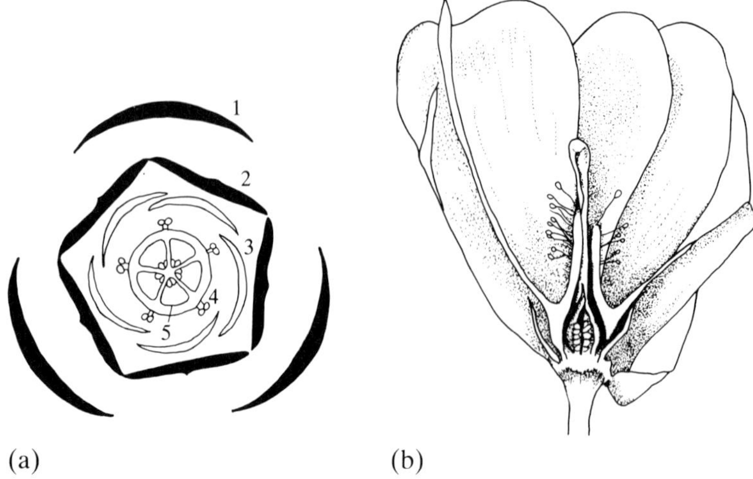

Fig. 4.6 (a) Floral diagram: 1. Epicalyx (3 bracteoles); 2. Calyx (5-undulate); 3. Petals (5); 4. Stamens (indefinite, column); 5. Ovary (3 to 5 locules). (b) Vertical section of cotton flower.

A(100–150) G(3–5). Fertilization takes place on the day of opening, the anthers bursting and the stigma receptive. Although the flower is large and showy, attracting bees and other insects, cross-pollination is not obligatory, and the great majority of seeds are the result of self-pollination. Honey bees are the main agents of cross-pollination, and the pollen grains are comparatively large. The corolla is white or slightly cream-coloured in Upland cotton; brighter yellow in *G. barbadense* and the Asiatic diploids, usually with a red or purple spot on the petals near the base. The day following anthesis, the petals turn pink, and in the next day or two they wither to a purple colour and drop off to reveal the young green boll.

Date of first flower

The date of first flower is an indication of the earliness of the crop, and marks the start of the flowering period with all that it implies for the grower. It is the resultant of a number of factors besides the sowing date, such as temperature, water supply and the NFB (see pp. 47–8); the last of these depends in turn on the species, variety and the day and night temperatures. There is therefore considerable variation in different cotton-growing countries.

At Namulonge in Uganda, with comparatively cool days (max. temp. 28 °C) and a consistent 12-hour day, the first flower on a typical Upland variety appears 65 to 70 days after planting; at Makanga in Malawi (max. temp. 33 °C) the first flower is produced in 45 days. Fifty to fifty-five days may be taken as average for most areas where American Uplands are grown.

Harland (1917) states that the Sea Island cotton in St Vincent 'always flowers within a few days of two months'; Bailey and Trought (1926) report 80 days to flowering of *G. barbadense* in the Sudan; Brown (1953) gives the date of first flower in Egypt as 3 months after sowing; some Punjab/American varieties in Pakistan flower very late, but the M4 variety in Sind takes only 50 days from sowing.

Flowering pattern

Flowering dates provide a very convenient method of studying the sequence of development of the cotton plant. The flowers are conspicuous, and the day of flowering unequivocal, in contrast to other measures such as bud formation, leaf maturity, node development or boll opening; these latter require a precise definition of the state of development at which they are recorded, and it is not too easy to recognize the precise day on which that stage is reached. The flower, on the other hand, opens early in the day and the petals start to turn pink by the evening; the next day they can easily be distinguished by their red colour from newly opened flowers.

The earliest published data on flowering dates are probably those of McClelland (1916) shown in Fig. 4.7; the fruiting branches are shown, alternately on each side of the mainstem, which makes the diagram easy to follow, although in reality they occur in a spiral up the stem. Comprehensive data have been published by Martin *et al.* (1923) from three regions of the USA, and by Megie (1963) from Chad. Other data from *G. hirsutum* have been published in the USA, in Africa and in Australia. Data from *G. barbadense* have been published in the West Indies and Egypt. It should be noted that Balls (1915) described the flowering pattern erroneously as pursuing 'a kind of spiral course, beginning at the innermost flower of the lowest fruiting branch, and ending at the upper and outermost flower of the youngest branch' shown in his Fig. 9, 'Succession of flowers'. The flowers at the same node of each successive sympodium follow the spiral of the phyllotaxy, but there is no way in which the contemporary flowers shown in Fig. 4.7 can be fitted into a continuous spiral. This is one of the very few observations in which he was in error, but his diagram has been copied by some later writers.

All these data confirm that 'the cotton plant develops according to a rather orderly time schedule' (Tharp 1960). Both the vertical flowering interval (VFI) between successive fruiting branches on the mainstem, and the horizontal flowering interval (HFI) between successive flowers on the same fruiting branch, are reasonably

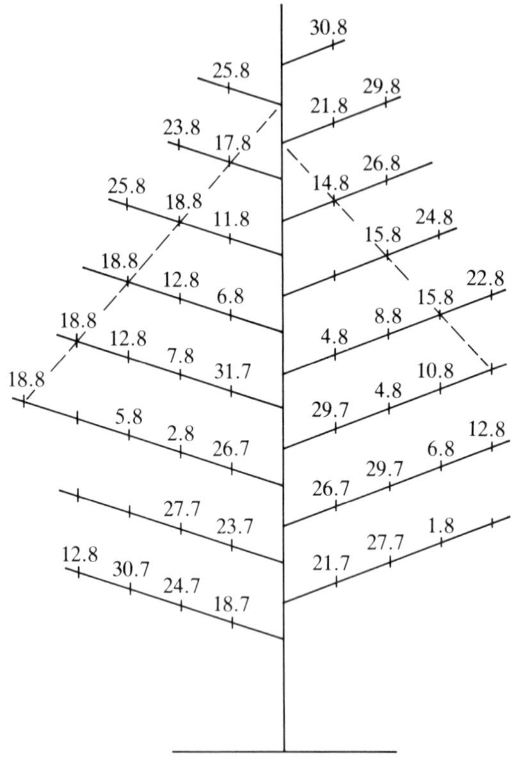

Fig. 4.7 Dates of blooming of a cotton plant (after McClelland 1916; copyright 1916 by the American Association for the Advancement of Science). The dates 18.7 to 30.8 denote 18 July to 30 August. The broken line joins contemporary flowers; buds at undated nodes were shed before opening.

constant throughout the main period of growth of the plant and between plants in the same crop.

Seasonal averages of the HFI range from 5.26 to 8.85 days in the USA, where it is generally accepted that the VFI is about 3 days and the HFI 6 days (Tharp 1960, Thaxton and Bird 1981). The figures published by Megie (1963) for West Africa are between 2.0 and 3.0 days for the VFI and 5.5 to 6.0 days for the HFI. In Uganda, Farbrother (1961) found that the HFI was approximately 11 days, and in Malawi, Munro (1973) found 8.8 days at Makoka (1029 m) and 5.6 days at Makanga (52 m). These differences are caused mainly by differences in temperature although other factors may be involved (see Ch. 9).

The 2 : 1 ratio between HFI and VFI is a useful rule of thumb, but it is not precisely accurate. McClelland and Neely (1931) stated that although the 'flowering ratio' varied greatly, it was in most instances between 2.6 and 2.1. The data published by Megie (1963) suggest that the ratio increases during the season, being 2.1 for the early flowers and 2.6 for the later ones, with an average of 2.27; this is confirmed by unpublished data in the possession of the author. Farbrother (1961) noted that the ratio appeared to be specific for genetically different cottons; he found in Uganda that it was generally 2.0 for American Upland, 2.5 for African Upland and 3.0 for Tanguis (*G. barbadense*).

The importance of the flowering ratio and the flowering interval is that given the date of any flower on the plant, it is possible to predict or estimate the flowering date at any node on the primary fruiting branches. McClelland (Fig. 4.7) could draw a contemporary flowering line on his diagram with a ratio of 2.0, with contemporary flowers appearing two branches below and one node further out along the branch. The flowering zone of the plant may be regarded as a steadily rising cone.

The pattern of flowering on the monopodial branches (Fig. 4.8) is similar to that on the mainstem, but very few data have been published. Flowers on the earliest monopodia are contemporary with mainstem flowers 7 to 10 nodes higher up (counting from the node producing the monopodium), that is to say that the first flower is 20–30 days later on the monopodium than on the mainstem, where VFI = 3 days. While almost every node in the fruiting area of the mainstem produces a fruiting branch, there is usually a high proportion of blank nodes on the vegetative branches, and the secondary fruiting branches are shorter.

The proportion of the total crop produced on the monopodia depends very much on the spacing (Fig. 4.9); at wide spacing the proportion may reach 50 per cent, while at close spacing (over 100,000 plants per hectare) it is less than 10 per cent (see Table 6.2). The proportion is also much less under poor growing conditions.

The proportion of flowers which produce bolls varies throughout the season, and can of course be severely affected by insect attack. Given good control of insects, the earlier flowers set a greater proportion of bolls; we have seen that these early flowers are those close to the mainstem and on the lower fruiting branches. In a normally spaced crop, about 60 per cent of the flowers at the first node of the fruiting branches will produce bolls, 30 per cent at the second and 10 per cent at the third node, with a negligible yield at the fourth and subsequent nodes (Munro 1971). Mauney (1979) gives similar figures for Arizona – 73, 24, 2 and 1 per cent respectively – and similar data from South Arabia by Hearn (1980) is shown graphically in Fig. 4.10.

Complete recording of flowering dates is a time-consuming job, and is only used as a research tool. Once the general flowering

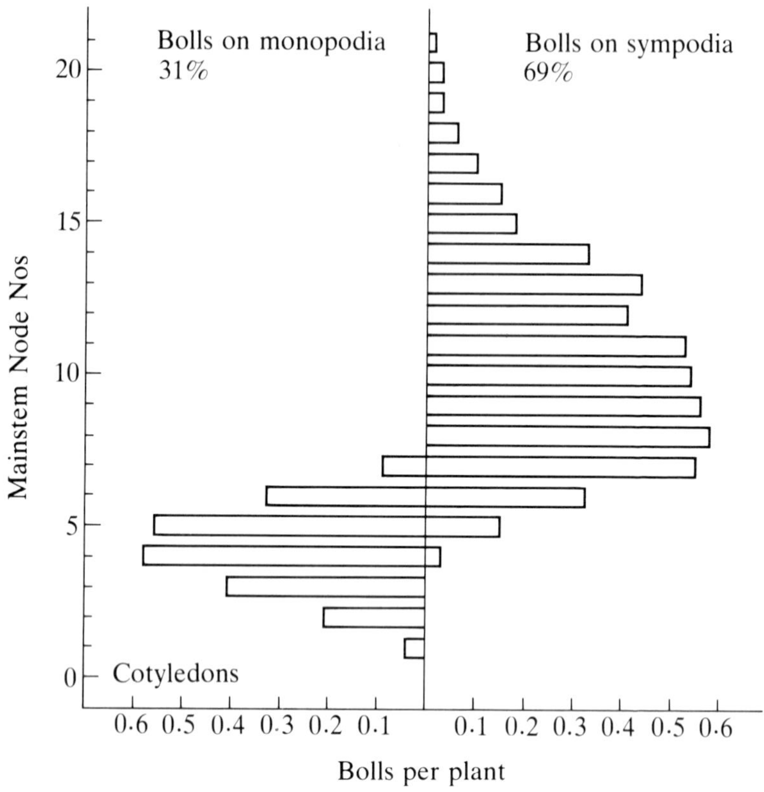

Fig. 4.8 Location of the crop on the plant. Average of 32 plants, spaced 3 ft × 6 in., 1 plant per hill. Makoka, Malawi, 1970–71.

pattern has been established, the number of observations can be drastically reduced, and the intermediate data estimated if required. The reason for such observations was stated long ago by Balls (1914): 'the mere figures for yield mean very little, since the final result may be reached in an infinite number of different ways. If we can not only ascertain exactly how the yield was produced, day by day, but also trace back the fruits to their origin as flowers and the flowers to their origin as buds on the scaffolding of the flowering branches, we have resolved the agricultural problem into components which the botanist can deal with.'

The simplest way of recording crop development is to divide the harvest into a regular series of pickings, and plot a picking curve; this can be supplemented by regular counts of fruiting points, flowers, bolls, etc. Such counts can be used to construct periodic (Fig. 4.9) or cumulative (see Fig. 6.3) graphs showing production at each stage of the fruiting cycle. Distortions caused by insect damage

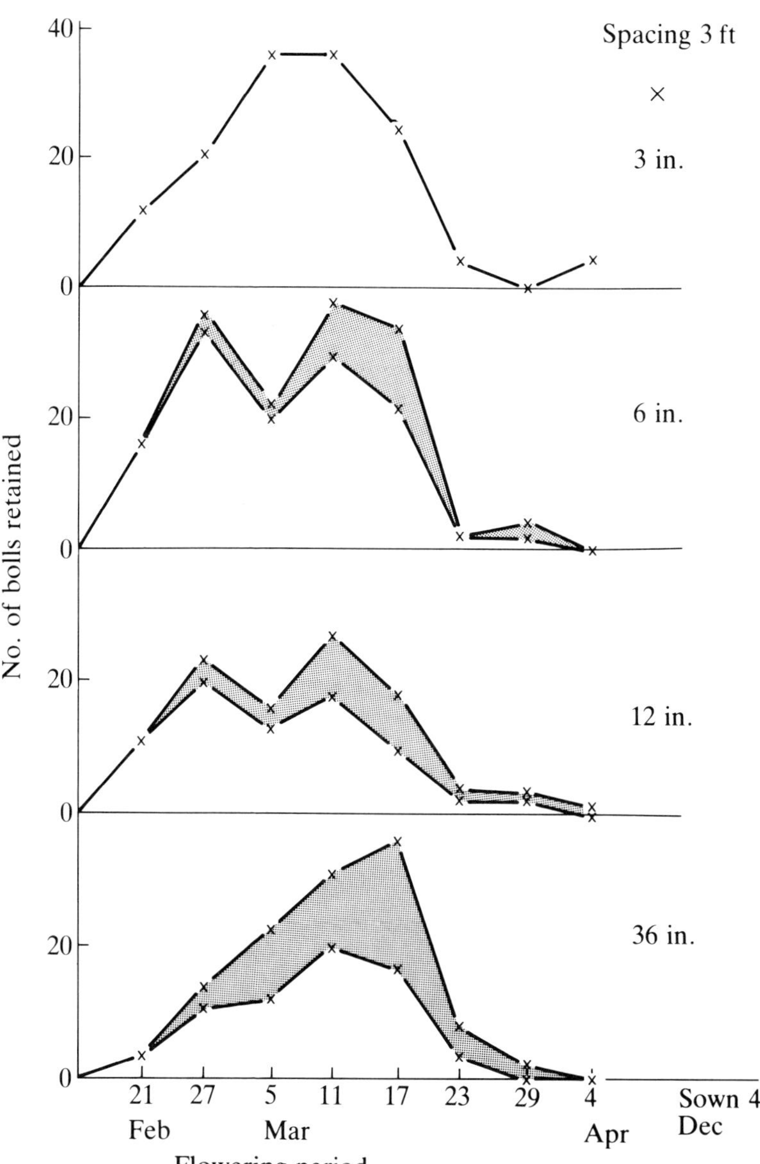

Fig. 4.9 Number of bolls retained on primary and monopodial fruiting branches at different spacings. Bolls in 30 ft² in 6-day flowering periods. Primary branches unshaded, monopodia shaded. Makoka, Malawi, 1969–70.

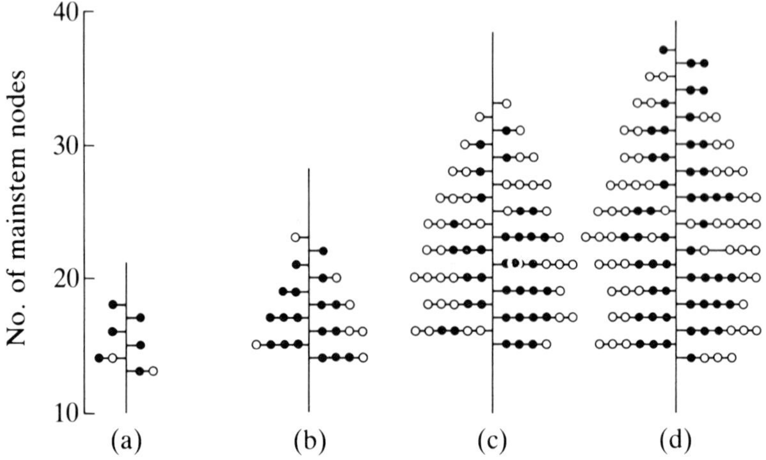

Fig. 4.10   Boll setting on sympodial branches: o = shed fruiting forms; ● = bolls picked. Vegetative branches omitted.
Effect of different amounts of stored water in the soil in the Abyan delta: (a) 140 mm; (b) 340 mm; (c) 460 mm; (d) 670 mm. (Hearn 1980, redrawn from Rijks 1965).

show up clearly on such graphs, which can be used to correct deficiencies in the timing of the programme for insect pest control.

For some purposes, composite plant diagrams (Munro and Farbrother 1969) may be used to reconstruct the pattern of flowering and fruiting throughout the season, and the location of the crop on the plant. A sample of plants is removed from the crop, and each fruiting node is recorded in detail. Contemporary nodes are grouped into zones at intervals equal to the HFI (Fig. 4.11), and the number of buds, flowers, sheds and bolls totalled for each zone. If the HFI is known, flowering dates can be estimated for each group, working back from the group which is flowering at the time of the sample to the date of the first flowers. All these data can be conveniently summarized on a graph (Fig. 4.12), and two or three samples taken at intervals can provide a very complete picture of the flowering and fruiting pattern for the whole season.

Examples of the use of the flowering pattern in selection are given in Chapter 14, and in field experiments in Chapter 16. The cotton farmer, too, will find that a better understanding of how his crop is growing and how it reacts to the weather and to farming practices will help him to grow more profitable crops.

The boll

The cotton boll is a dehiscent schizocarp with 3–5 loculi or locs,

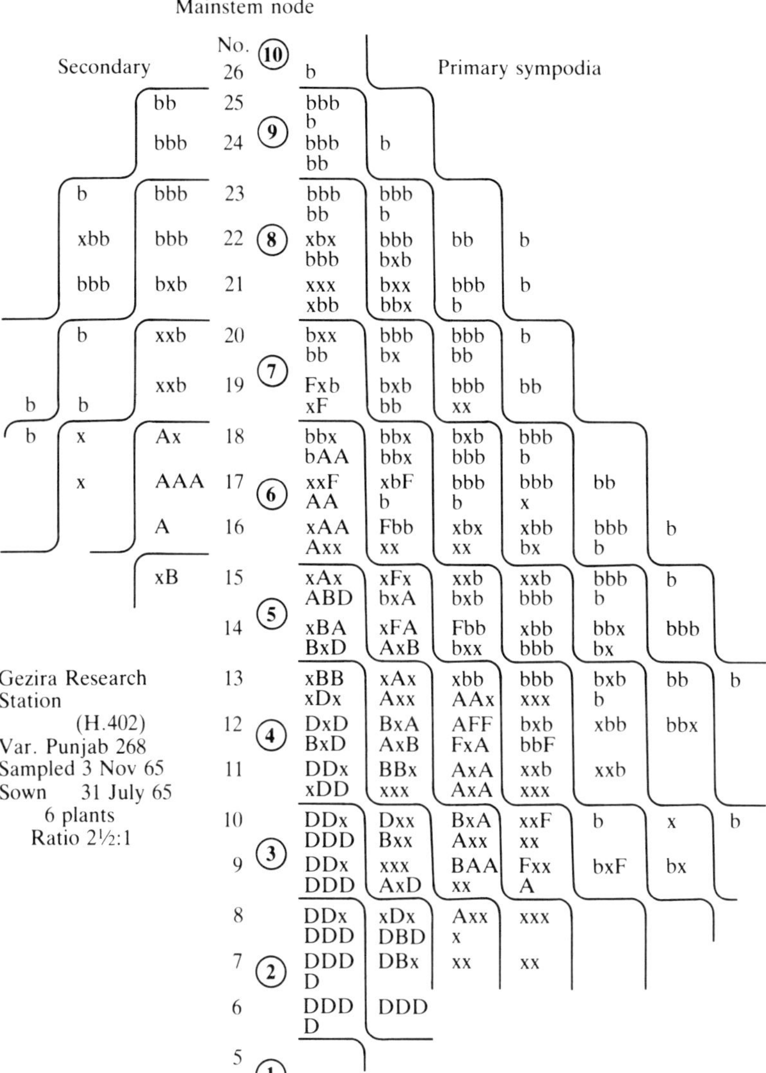

Fig. 4.11 Composite plant diagram – six plants. x = shed; A, B, C, D = bolls of increasing size; b = bud; H = husk of an open or harvested boll; F = flower (Munro and Farbrother 1969).

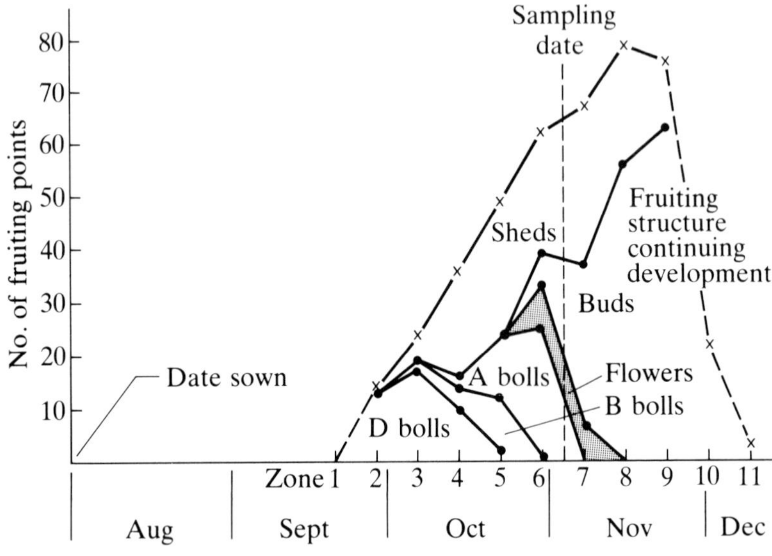

Fig. 4.12 Pattern of crop production derived from the composite plant diagram in Fig. 4.11 (Munro and Farbrother 1969).

each containing about 8 seeds. The number of locs is characteristic of the species and variety, but exceptional bolls can usually be found; the general pattern is as follows:

Species	Loculi per boll	Seeds per loculus
G. arboreum	3, rarely 4	6–17
G. herbaceum	3–4	Not more than 11
G. hirsutum	4, sometimes 5	5–11
G. barbadense	3, sometimes 4	5–8

The fruit usually ripens 50 to 70 days after fertilization (the boll period), the time varying with the variety, the temperature and the humidity; in the field there is a wide variation from plant to plant and boll to boll, which appears to be independent of the date of flowering and the position of the boll on the plant: 100 or more bolls may be needed to give a reliable average. Within a variety, however, under controlled conditions, Gipson and Ray (1970) have shown that temperature accounts for 96 per cent of the variation in the boll period (Table 4.1). Night temperatures have a greater effect than day temperatures.

The boll reaches its full size half-way through the boll period; at about the same time the seeds in it have reached full size, and the lint hairs have grown to their full length and thickening of the fibre

Table 4.1   *Effect of night temperature on boll maturation periods (seasonal means)*

Night temperature °C	Acala 1517 BR-2	Stoneville 7A	Lankart 57	Stripper 31	CA 491
	Number of days of boll period				
27	50.5	53.3	55.3	53.6	50.8
21	66.2	66.0	62.3	64.6	56.3
15	87.3	88.5	84.2	79.5	68.5
11	95.9	94.6	92.6	86.4	81.8
S.E.*	± 1.16	± 1.37	± 1.59	± 1.68	± 1.70

* Minimum of 5 samples per mean.
Source: Gipson and Ray 1970.

walls has started. The boll wall is pitted on the outside with gossypol glands, not very conspicuous in *G. hirsutum*, but black and deeply sunken in the other species. As the boll dries out at maturity the outer wall of each loculus opens outwards exposing the lint and seeds; the mass of cotton expands as it dries, but remains lightly attached to the dry carpels (Fig. 4.13). The bolls of storm-resistant cotton open less fully, and the cotton remains more firmly attached to the boll.

The seed

The cotton seed measures about 10 × 6 mm, and weighs about 80 mg (5–10 g per 100 seeds). It has a hard seed coat covered by a slightly waxy cuticle. Beneath the epidermal cells which produce the seed hairs is a thin outer seed coat with an inner epidermis; this has a different origin from that of the inner seed coat, but the two are fused together. The inner seed coat has a layer of palisade cells, then a layer of protoplasmic or parenchymatous tissue, which is compressed between the palisade layer and the swelling embryo as the seed matures (Balls 1915). Although the endorsperm fills the embryo sac by the third week after anthesis, largely as undifferentiated protoplasm, the rapid growth of the embryo absorbs and digests the endosperm and fills the whole seed cavity.

The seed hairs are unicellular outgrowths of the epidermis of the seed. In the early stages of development they are thin-walled, but cellulose is later deposited within the cuticle. In the ordinary seed hairs, or fuzz, the cellulose deposit builds up until 'at maturity the lumen is practically obliterated, and the dry seed hair on a ripe seed is long, tapered, circular in cross-section and filled with cellulose, except for a small central cavity left by the drying up of the cell contents' (Hutchinson *et al.* 1947).

(a)

(b)

Fig. 4.13 Flowering and fruiting of cotton: (a) buds; (b) flower; (c) green boll; (d) open boll (CSIRO Australia; International Institute for Cotton).

(c)   (d)

Cotton lint, the seed hairs suitable for spinning, develops in a different way; the deposit of cellulose in the cuticle is much less, and leaves quite a large central cavity or lumen at maturity. 'When the capsule opens and the lint hairs dry, the cavity left by the drying up of the cell contents is so large that the hair collapses, forming a ribbon' (Hutchinson *et al.* 1947). The cellulose thickening is deposited in spirally arranged fibrillae (Balls 1924), the spirals reversing direction at intervals so that they are present in the same hair in both clockwise and anticlockwise directions (Clegg 1931). This causes the dry ribbon to twist (Fig. 4.14), and these irregularly twisted ribbons cling together when spun into thread. This cling allows a strong thread to be produced from cotton, even though the individual fibres are much shorter than those of wool or flax.

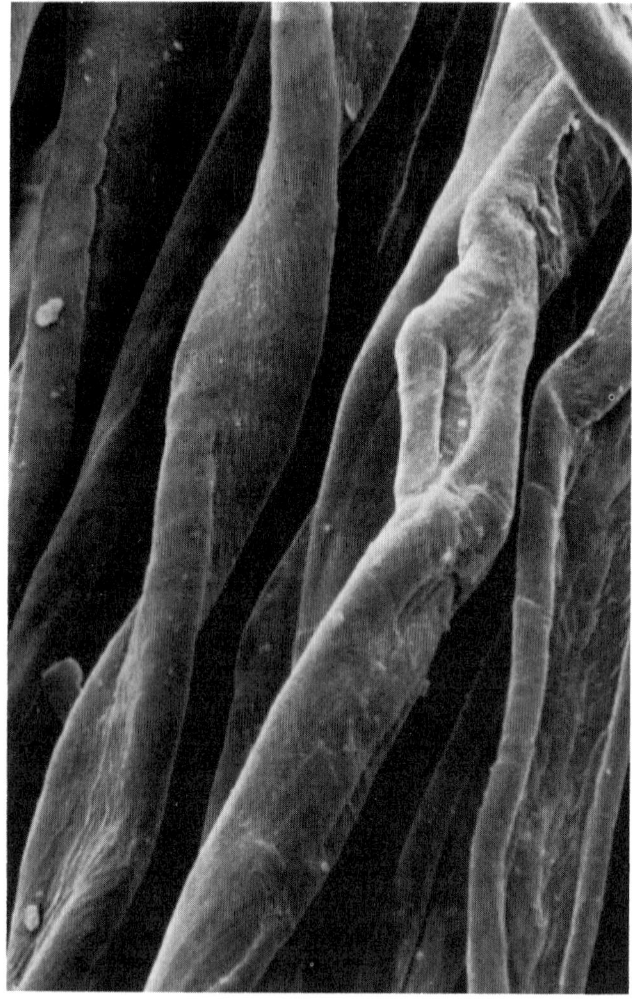

Fig. 4.14   Microphotograph of cotton fibres (Mag. 1100 ×.) (Shirley Institute).

Chapter 5

Land preparation

Soils

Cotton grows well on a wide range of soil types, its main requirement being an adequate root range. Anything which restricts the
depth of rooting reduces growth of the above-ground parts of the
plant, and this limits the cropping potential. The most common
restriction is waterlogging, and good drainage is therefore important.
Shallow soils overlying impermeable rock are unsuitable, but good
growth can be achieved in rocky and stony soils where drainage is
not impeded. Root development can be restricted by a hard pan,
and occasional subsoiling may be advisable to break up the pan; this
may be a plough pan caused by repeated ploughing to the same
depth, or a laterite or ironstone pan formed by the weathering of
minerals in certain types of soil. Soil compaction by the passage of
farm machinery can also restrict root development (Stockton *et al.*
1962, Colwick and Barker 1980).

Cotton grows satisfactorily in sandy soils, such as the hill sands in
Tanzania around Lake Victoria and on the Karonga lake shore in
Malawi; it appears equally at home in the Gezira clay of the Sudan
and in the silty soils of the Shire valley in Malawi. What is called
'black cotton soil' is not necessarily the most suitable soil for cotton;
it is a popular rather than a scientific term, covering a range of soils
with differing characteristics, but is typically a clay soil rich in
organic matter and calcium and magnesium carbonates. If the clay
fraction is montmorillonite as opposed to kaolinite, the soil is
difficult to manage, being too soft when wet and too hard when dry
to produce a good tilth.

Cotton tolerates a fairly wide range of soil acidity and alkalinity;
it stands acidity as well as most major crop plants, and can stand high
pH values better than most. The typical Gezira clay has a pH of 8.5
(Robinson *et al.* 1969), and the *mopani* soils in the Sabi Valley of
Zimbabwe, with a similar pH, can give yields of seed cotton up to
5600 kg per ha. Data from experiments in the Lake Victoria region

of Tanzania show that soil pH was mostly between 5.0 and 6.0, with one zone, Ukerewe Island, as low as 4.5 (Le Mare 1974). In the same region, the application of lime to a long-term fertilizer trial at Ukiriguru increased the yield of the NP treatment from 400 to 1200 kg of seed cotton per ha, while raising the pH from 4.5 to 6.2 (Le Mare 1972). In Indonesia, a cotton estate where the pH is between 4.0 and 4.5 finds it economic to apply 3 tons of limestone per ha (P. T. Kapas Indah, pers. comm.).

Salt concentrations with a specific conductance ($K \times 10^5$) of 400 or less are not deleterious to plant growth; between 400 and 800 (3000 to 6000 ppm) is tolerated by cotton (Roe 1950), but this depends on the kinds of salts and the soil type. Black alkali, containing sodium carbonate as the predominant salt, is much more destructive to vegetation than similar concentrations of white alkali, caused by sodium chloride and the sulphates and chlorides of sodium, calcium, magnesium and potassium (Lyon and Buckman 1949).

Alkaline soils are usually associated with irrigation, and Roe (1950) goes so far as to say that 'alkali occurs, sooner or later, in practically all soils under irrigation'. The accumulation of salts is caused by the use of saline water in conjunction with inadequate drainage; evaporation increases the concentration of salts near the soil surface. The problems of waterlogging and salinity have been studied extensively in the Indus Basin in Pakistan following the Revelle (1964) report (Johnson 1982). When irrigation was developed in the arid parts of the Indus valley, the eventual need for drainage was recognized, but the necessary expenditure was deferred on grounds of political expediency and finance. The cropping intensity catered for was only half of the theoretical potential, so as to obtain the highest returns from the water applied and to benefit the largest possible number of farmers. The Indus water contains 300–400 ppm of dissolved solids, and although there was a build-up of salt in the soil profile, this could be leached out by applying 15 per cent more water than the crop required (Hussain 1963). Where non-saline ground water can be tapped by tube wells, this water may be used to flush out excess salts from the irrigated fields. Gypsum is sometimes added to the irrigation water to replace sodium ions in the soil complex with those of calcium.

In Pakistan, as in most other large-scale irrigation schemes, the intensity of irrigation has been increased over the last 25 years. This has been made possible by advances in the field of irrigation engineering which provide additional water, and has been stimulated by the green revolution, which demands more intensive cultivation for its full development. Existing projects have been intensified to provide more water by increasing the capacity of the canals and pumping stations, and operating them throughout the year instead of a limited period of seasonal operation. As a result, more water is being applied than is being lost by evaporation, and the water table has risen; when it comes within two metres of the soil surface, salinity problems are increased and crop yields decline.

To keep the water table at a safe level, the only alternative to reducing the amount of irrigation water is to drain off the surplus. This can be done in three ways: returning it to the surface by installing tube wells, provided the soil water is not too saline; installing tile drainage in the irrigated fields; or digging open drainage canals. Tile drains are more expensive to install than tube wells, and cost more to maintain; open drains are by far the most expensive both to install and maintain, and occupy a disproportionate amount of the cultivable land area (Carruthers 1985).

Reasons for cultivation

Cotton is grown as an annual crop from seed sown directly in the field, although perennial cottons are still grown in a few places, such as Marie-Galante cotton on Carriacou in the Antilles, Moco cotton in Brazil and Criollo cotton in Ecuador. The annual crop is notoriously intolerant of weed competition in the early stages of growth, so that any operation which favours rapid growth of the crop but checks the growth of weeds is highly desirable.

The functions of land preparation are therefore weed control and the provision of a suitable seed bed. Loosening and aeration of the soil are relatively unimportant, except where a hard pan is present; this is illustrated by the success of zero tillage, where weed control is achieved by herbicide application. Annual weeds which have established themselves prior to land preparation are usually disposed of by burying, but they are mostly prolific seeders; if time and weather permit, weed seeds should be allowed to germinate, and the seedlings destroyed by a light cultivation, before the cotton crop is sown. Perennial weeds are more difficult to control, and the best method will depend on how they propagate themselves (see Ch. 7).

Cotton requires a reasonably smooth seed bed, to facilitate a uniform depth of sowing, with firm soil below the loose surface. Too fine a tilth should be avoided in soils which form a cap or crust when they dry out after heavy rain, which impedes the emergence of the seedlings.

On sloping land, the final preparation should be delayed as long as possible, as a smooth, weed-free sloping surface encourages surface run-off and erosion; a rough surface will accept a much greater amount of rainfall.

Ridging

Most of the cotton in the tropics is grown on the flat, but there are places where cotton and other annual crops have traditionally been grown on ridges for a variety of reasons, which are not always fully understood. In the Sudan Gezira and in Egypt the furrows help to distribute the irrigation water evenly to all parts of the fields, and planting on ridges keeps the cotton plants above the level of the irrigation water. The main reason for ridging in heavy soils is probably the danger of waterlogging, and the disposal of water in times of heavy rainfall, while on lighter soils ridging on the contour or with a graded fall is an accepted method of soil and water conservation (Walton 1962); but the lines of traditional ridges often bear little or no relationship to the contour lines.

In Nigeria, Lawes (1961) showed that tie-ridging or basin listing, in which small bunds are built at intervals to prevent water flowing along the furrows, increased yield in seasons of below average rainfall, but reduced them in wet years; he found that cross-tying every alternate furrow was effective and practical in both dry and wet years.

Weed control is the main reason for ridging the hill sands in Tanzania; the ridges are large, about 1.5 m from crest to crest, and two rows of cotton are planted on each ridge. The land is prepared for sowing by dragging the weeds from the last year's ridges into the furrows, splitting the old ridges and building new ones on top of the weeds (Peat and Prentice 1949).

At the cotton research station in Uganda, straight three-foot ridges with cross-ties were originally used in strips laid out approximately on the contour. After some years of cultivation, the physical condition of the soil had deteriorated, slowing the rate of water acceptance; following heavy rain, the basins formed by the tie-ridges overflowed, the ties were washed away and gullies developed across some of the cultivated strips. From about 1968, the layout of areas showing severe erosion was redesigned, installing graded contour ditches at approximately two-metre vertical intervals with a fall of 1 in 250 (Arnold, Passmore *et al.* 1976). The fields were ploughed consistently downhill with reversible ploughs, gradually forming bench terraces on which crops could be planted on the flat with little or no run-off or erosion.

Hand and ox cultivation

In many parts of the tropics, the heavy hand hoe is still the basic tool of tillage. Hand cultivation ranges from a shallow stirring of the soil (7–10 cm), preceded where necessary by the collection and burning of heavy weeds, to deep cultivation (30 cm or more) with or without ridging. Distinctive methods of cultivation have been developed to suit local conditions. In Sukumaland in Tanzania crops are grown on the large ridges described in the previous section; in parts of the central Sudan and eastern Uganda the weeds are scraped into heaps and burned; the new crop is sown on these heaps where the nutrients from the crop and weed residues have been concentrated. The heavy clay soils in parts of Indonesia are broken up and turned over in the dry season with long-handled chisels.

Hand cultivation limits the amount of land that a family can bring under the cultivation of annual crops to an area of 2–4 ha; the use of the plough can extend these limits. There is a danger, however, that the area sown is then more than the family can weed satisfactorily, and weed competition may reduce the yield to the point where the harvest is little more than it would have been from the smaller, hand-cultivated area. Traditional wooden ox ploughs with a metal share are still used in Pakistan; in the Sudan and India they are being replaced by light metal ploughs drawn by one or two oxen. Other implements have been developed for use with oxen, including toolbar systems (Matthews and Pullen 1976) on which various implements can be mounted as required (Fig. 5.1). In many parts of the tropics, however, there is no tradition of training work oxen, and this has to be developed before ox ploughing is possible. The numbers of oxen which can be maintained all the year round in semi-arid areas is often not sufficient to cope with the demands for land preparation.

Tractor cultivation

Mechanization is not in itself the key to better yields of cotton, as is often thought or implied in development projects; hand-cultivated crops can be just as good or better than those produced with the help of tractors and sophisticated equipment. The principal advantages of mechanized agriculture are that it reduces the demand for labour and allows operations to be carried out quickly, making the most of the sometimes rare periods when conditions are suitable. Mechanization has been forced on industrialized countries by the high cost of labour; where labour is cheap and plentiful, and mechanical knowledge and servicing facilities are poor, mechanization of agriculture is usually uneconomic. These conditions still exist in many underdeveloped countries, although they are becoming rarer. Managerial skill is needed to ensure that mechanical equipment is available and in working order when required, and in deploying it to the best advantage when a suitable opportunity presents itself. An important corollary to the introduction of

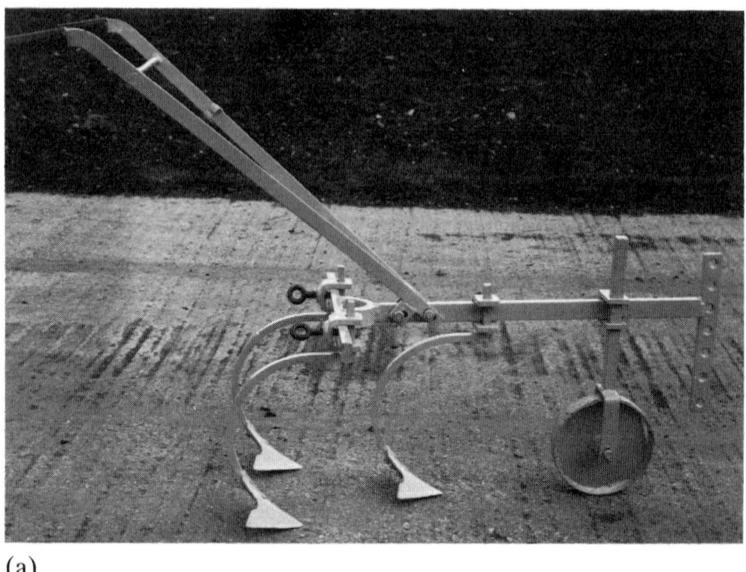

(a)

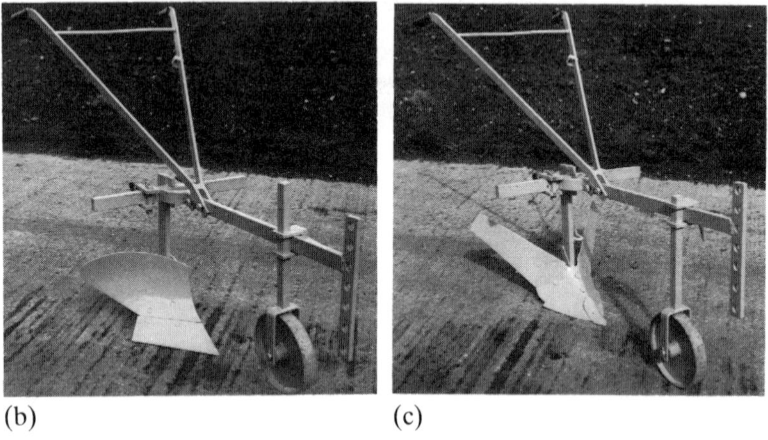

(b)   (c)

Fig. 5.1   Pedestrian controlled multipurpose tool carrier: (a) duckfoot cultivator; (b) mouldboard plough; (c) ridging plough; (d) inter-row cultivation (National Institute of Agricultural Engineering).

mechanization is that the labour released by it should be used productively, and not stand idle while the machines do the work.

In many of the cotton-growing areas of the tropics and subtropics the land is so dry and hard at the end of the dry season that land preparation cannot begin until rain has rendered the soil workable. The use of tractors then allows the land to be prepared much more

(d)

quickly, with the consequent advantage of early sowing, which is nearly always conducive to better yields. It is sometimes practicable to do the initial ploughing by tractor while the land is still dry, but this needs heavier machinery and puts more strain on the equipment; the economic advantages should be carefully assessed before such a policy is advocated. Some heavy clay soils which can grow good crops of cotton may become so soft and wet once the rains start that tractors cannot be used, and cultivation before the rains is the only practicable method of land preparation.

The benefits of mechanized land preparation may be summarized as follows:

1. Reduced labour requirements.
2. Quicker operation, therefore more land can be prepared in time, and
3. Maximum advantage can be taken of suitable weather.
4. Dry season cultivation is possible.
5. Deeper cultivation can be undertaken to break up pan.

Against this, the following disadvantages should be recognized:

1. Dependence on imported oil.
2. Technical skills are necessary.
3. Efficient and regular maintenance must be organized.
4. Repair and replacement services are necessary.
5. Tractors may not be available when conditions are ideal.

6. Small scattered holdings mean excessive road travel.
7. Tractors may be under-utilized during remainder of the year.

Tractor-drawn implements

The implements most commonly used for land preparation are the plough, disc harrow, ridger and rotavator; all except the rotavator may be used with oxen or with tractors. Disc ploughs are more common in the tropics than mouldboard ploughs; they are less vulnerable to damage by rocks, tree roots and other snags left in the ground during land clearance. The discs tend to break up the soil rather than invert it, so that weeds and crop residues are not buried so effectively; the effect on perennial weeds depends on how these regenerate. Disc ploughs depend on their weight to achieve penetration, while the mouldboard plough is drawn into the soil to a depth dependent on its setting: the latter may therefore be of much lighter construction, and is more suitable for use with oxen. Scalloped discs are used where the weed cover is heavy, to cut through the vegetation and prevent it being gathered up and blocking the action of the discs. Reversible ploughs can be used on sloping land to develop contour terraces by turning the soil downhill.

Disc harrows may be used to kill off a proportion of the weeds before ploughing, but their main function is to break down the ploughed land to a tilth suitable for sowing. Both ploughing and harrowing should be done as far as possible when the soil tends to crumble rather than form clods, a condition which depends both on soil type and moisture content; suitable conditions must be judged by experience. The clods produced by ploughing when the soil is wet may be difficult to break down when they dry out; but in some soils they are broken down by the action of even quite light rain, and time should be allowed between ploughing and harrowing to allow this to happen. Other soils produce clods which are not affected by weathering, and additional cultivations are needed to pulverize them. A heavy wooden bar dragged behind the disc harrow is sometimes used to give a more even and level seed bed.

Rotavation may replace either ploughing or disc harrowing or both. This leaves a horizon of loose pulverized soil, and excessive use may destroy the soil structure. Cotton germinates best in a firm seed bed, and the soil should be allowed time to settle and become compacted before sowing. Stoloniferous weeds which can sprout from broken fragments of the stolons (e.g. *Cynodon* sp.) may be spread by the action of the rotavator.

Chisel ploughs have advantages for primary cultivation over disc and mouldboard ploughs under certain conditions. They allow deeper penetration without requiring more power, and can increase the root range of the crop where a hard pan or similar conditions limit root growth. They do not expose the soil to erosion to the same extent as ploughing, but by the same token they have little or no effect on weeds, which must be controlled in some other way.

Other implements sometimes used in seed bed preparation are spike tooth and chain harrows and plain and cambridge rollers.

Ridge cultivation adds an extra operation in preparing the land, and reduces the flexibility in plant spacing. This may be justified by increased yields or by better water conservation and erosion control; probably the most common use of ridges is to aid in the distribution of water under furrow irrigation. Ridging may be carried out with a double mouldboard plough or with a disc ridger; the latter requires a toolbar to hold the discs at the right distance and angle, and like disc ploughs they are dependent on weight for penetration. For ox cultivation, a single furrow formed by a double mouldboard plough is as much as one pair of oxen can manage; a wheeled tractor can make two or three ridges at a time, with the discs or plough bodies mounted on the toolbar.

Chapter 6

Sowing

Time of sowing

Water supply

The optimum sowing date for cotton depends mainly on the quantity and timing of the water supply. In the temperate zone, cotton is a summer crop and temperature limits the choice of sowing date to a short period in spring; but in the tropics and subtropics temperature is much less important. In the Sudan, for example, the mean sowing date is 1 August (Farbrother 1973), and the crop matures during the winter months; in Colombia the central crop is sown in February while the coastal crop is sown in August (Anthony and Bravo 1970).

In areas of low rainfall, less than 1000 mm per annum, the sowing date should be chosen to make the maximum use of the moisture available; where rainfall is more abundant, dry weather at harvest time may be the most important consideration; with irrigation, the period when water is available may decide the sowing date, as in the Abyan Delta in South Arabia (Rijks 1965).

Unimodal rainfall

The rainfall pattern most common in tropical cotton-growing countries is unimodal; that is, the year is divided into a single wet season lasting up to six months and a dry season for the rest of the year. The annual rainfall is typically less than 1000 mm, but both the total rainfall and its distribution tend to vary from year to year; rainfall distribution and probability are discussed in Chapter 10. In these circumstances the best yields are obtained from cotton sown as early as possible in the wet season. Some results of late sowing in east and west Africa were collected by Ruston (1962), and are shown in Table 6.1. The interpretation of the results of sowing date trials is not straightforward, as the arrangement of plots in an orthodox field experiment, even with the use of modern insecticides, allows pests to move from early to late sown treatments; this tends to exaggerate the benefits of early sowing, and confirmation of the

Table 6.1 *Effects of delayed sowing on yield of cotton*

Some typical examples taken from ECGC Progress Reports. Yields are expressed as percentages of the yield from sowing at the optimum time.

Country	Delay in sowing in weeks			
	2	4	6	8
		Yield %		
Uganda – Namulonge – 10-year average	81	70	65	39
Uganda – Serere	—	69	—	42
Uganda – Serere	74	54	52	—
Kenya – 4-year average	—	59	—	48
Kenya – District trials	—	37	—	23
Kenya – District trials	—	60	—	47
Tanganyika – Lake Province	—	90	—	74
– Lake Province	—	83	40	26
– Lake Province	—	—	82	53
– Eastern Province	—	59	—	—
– Eastern Province	—	47	—	—
– Eastern Province	—	53	—	—
– Eastern Province	—	41	—	—
Northern Nigeria	95	56	8	5
– 3-year average	92	67	54	—
Average	86	60	50	38
% loss	14	40	50	62

Source: Ruston 1962.

results should be sought in other ways. Manning (1948) split his trials into overlapping groups of three consecutive sowings, so that the second and third sowing dates in one group were contemporary with the first and second dates in the next; his estimate of the optimum sowing date agreed with independent data from the commercial Uganda cotton crop, in which partial regressions showed that the mean sowing date was almost as important as the planted acreage in determining production (Manning 1952).

Ruston (1962) has commented that 'even in the Sudan, where cotton is grown under irrigation and there is no lack of moisture, delay in sowing still leads to a drop in yield, the figures for one experiment showing a loss of 23 per cent of crop with four weeks' delay and 31 per cent with six weeks' delay'. The loss is not confined to yield alone; quality is also depressed and with it the price. The grade of pickings made in the Sudan at approximately fortnightly intervals over a three-month season showed a very steady drop in grade with advancing date; 'although the pickings were from cotton sown on different dates, the grade of cotton picked on any one date was much the same, whether it came from early-, medium- or late-sown cotton. High production at the earlier picking dates and hence the best average grades can be obtained by early sowing.'

Why then is late sowing so often a feature of small-holder cotton in many countries? One reason is the time required for land preparation (see Ch. 5), which usually takes place after the rains have started, and is limited by the power available. The strong demand for tractor ploughing attests the appreciation by the grower of the importance of early land preparation; but it also highlights the need for working capital or credit to finance the operation. In a subsistence economy where cash crops and food crops compete for priority in planting, the food crops understandably win. When asked why they have planted so late, growers put forward many different reasons; most of these need to be examined critically to see if they are valid, or merely excuses for dilatoriness. Timely sowing will normally depend on how important the cotton crop is in the local agricultural pattern, and on its profitability compared with other crops.

There are various ways in which early sowing can be facilitated. The use of tractors for early land preparation has already been mentioned. A market economy in which the purchase of food supplies can be assured to the cotton growers will allow priority to be given to the cotton crop; such specialization will result in more efficient farming, leading to better yields and a more profitable crop. Inter-cropping is sometimes useful, as in parts of Nigeria where cotton is sown between ripening rows of early sown millet (*Pennisetum*). It is worth mentioning that sowing is sometimes delayed because planting seed was not distributed to the growers in time.

Where cultivation is possible before the start of the rains, it should be encouraged, and may be combined with dry planting. Dry planting means that the cotton seed is sown in dry soil; it rarely comes to any harm in the soil from pests or diseases, and when the rains arrive the cotton gets off to an early start. There is always a risk that resowing may be necessary, but this risk is worth taking if resowing is needed only once in three or four years. Fuzzy seed is preferable for dry planting, as it will not germinate until there is sufficient rain to support the seedlings for some time after germination.

Water planting, a system used for planting out tobacco seedlings in Zimbabwe, has been used successfully to enable cotton to be sown during the optimum planting period, even when there is insufficient rainfall. Small furrows are opened along the crests of the ridges and filled with water, leaving the soil between rows in its naturally dry state. The water is supplied from a tractor drawn bowser, or from the risers of a sprinkler irrigation system. The seed is sown in the wetted soil, and the seedlings will grow without further watering for two or three weeks even if there is no rain. It is essential that the water applied should connect up with residual moisture at depth in the soil, and early ploughing will assist in bringing this residual moisture closer to the surface (Zimbabwe 1980).

Bimodal rainfall

At and near the Equator the rainfall pattern is usually bimodal, with two rainy seasons in the year; this complicates the definition of an optimum sowing date. Where the rainfall in one of the wet seasons is sufficient to mature a crop, it may be treated as a single wet season, as in the western provinces of Kenya. Land preparation should be possible before the start of the rains, as the land does not dry out to the same extent as it does in a long dry season, or the cultivations may be done at the end of the previous rains. It is important that the harvest should be timed to occur between the wet seasons. If both of the wet seasons have sufficient rains for a cotton crop, the climate is probably too wet for cotton and more suited to perennial crops such as rubber or oil palms.

More commonly neither of the wet seasons provides sufficient rain in itself, and the cotton is sown in one season and matures in the other. This is the practice in parts of Uganda and Kenya. In the south west of Uganda (the BP52 area) the first rains in March/June are devoted to annual food crops such as maize and groundnuts; the cotton is sown at the end of these rains (June/July) and harvested in the relatively dry months of December and January (see Fig. 10.4).

In the Central and Eastern Provinces of Kenya, cotton is sown in September/October at the start of the short rains; the annual total is less than 1000 mm, compared with 1000 to 1500 mm in Uganda. The short rains may or may not be sufficient to produce an appreciable bottom crop; the main crop is produced in the long rains, when growth and flowering start again after the dry months December to March. This regime raises problems of pest control – should the pests be controlled during the first rains and the mid-season drought? Usually spraying with insecticides starts in the first rains if the crop is growing well, but is discontinued until the long rains if little or no bottom crop is set. It has been shown that quick-maturing varieties can produce satisfactory yields when they are sown 'late' at the start of the long rains (Kenya 1978 and 1979). Estimates of rainfall reliability are essential for a correct assessment of the optimum sowing date, as it is not unusual for the short rains to be longer and wetter than the long rains (Brown 1975a).

Irrigated crops

The supply of irrigation water is often seasonal, depending on the rainfall in the catchment area, which may be a long way from the cotton-growing area, e.g. the Blue Nile supplying water from Ethiopia to the Sudan and Egypt. As in the case of unimodal rainfall, sowing should be timed to take advantage of the water supply when it is most abundant. Supplementary irrigation will rarely increase cotton yields sufficiently to justify the capital cost of the installation by itself, but the cost may be shared with other crops. If irrigation is available it can give the cotton crop a flying start before the break of the rains; it can also provide an insurance against drought in the middle of the cotton season, and extend the growing season after the rains have finished.

Pest avoidance

Sowing date may be important as a means of pest avoidance. In Ecuador two of the most damaging pests – the leaf miner (*Bucculatrix* spp.) and the boll weevil (*Anthonomus vestitus*) – multiply towards the end of the growing season, and early sown cotton escapes much of the damage by setting its crop before these pests reach critical levels. In the Lower Shire Valley in Malawi, the sowing of cotton before 1952 took place from January to April, and the crop suffered severe losses from red bollworm (*Diparopsis castanea*) and cotton stainers (*Dysdercus* spp.). It was argued by Pearson and Mitchell (1945) that if the cotton crop could be established sufficiently early to produce the greater part of its bolls before mid-May, the red bollworm pupae would be predominantly short-term and would thus be unable to carry the pest over to the next season. The government agreed to enforce a summer-cropping regime, planting between 15 November and 7 January, and uprooting all cotton by 31 July. Reviewing the situation six years later Munro (1958) concluded that there had been considerable control of red bollworm, but that the increase in cotton production expected had not materialized.

In Arizona and southern California the winter carry over of pink bollworm (*Pectinophora gossypiella*) and tobacco budworm (*Heliothis virescens*) depends on diapause larvae, which are only produced late in the season. Watson (1980) concludes that if the cotton crop were mature and senescent by mid-September, there would be virtually no carry over, and these insects would be relegated to the status of minor pests.

Plant population and spacing

Population density

Compared with other field crops, cotton is unusually adaptable to a wide variation in population density because of its indeterminate habit of growth, its ability to produce both vegetative and fruiting branches, and its faculty of shedding buds and young bolls. Closely spaced cotton has short fruiting branches, but few vegetative branches (monopodia) and a high percentage of shedding at the earlier fruiting points (Fig. 6.1). Widely spaced cotton, on the other hand, carries long fruiting branches with a good set of early bolls, highly developed monopodia from the lower mainstem nodes, and additional branches from the nodes already carrying fruiting branches (see Ch. 4). Widely spaced plants carrying a large number of bolls per plant look attractive and give the impression of better yields, but innumerable experiments have shown that this impression is false; yield per unit area, not yield per plant, is the commercial criterion.

Because of the adaptability of individual cotton plants, populations can vary upwards from about 30,000 plants per hectare with little effect on yield, although quite large and significant differences are often found in individual trials. Below 30,000 plants, yield tends to fall fairly sharply, but it is not possible to specify precisely where this occurs; Brown (1971), for example, found that the local variety, 26J, in Nigeria was better able to compensate for wide spacing than was Acala 1517C. It is better to aim for a population well above 30,000 and leave a safe margin in case establishment is poor, and 50,000 to 60,000 plants per hectare are usually recommended. The relationships found by Hearn in Uganda (Fig. 6.2) are probably typical of most cotton-growing countries: 'the curves have very flat peaks so that optimum densities are not clearly defined.' Although an optimum density can be calculated from experimental data, 'to halve or double the optimum population is likely to decrease yield by only five to seven per cent' (Hearn 1972a).

It is a common assumption that bigger plants need more room to develop, and that populations should be less dense on rich soil and denser under poor conditions; experimental evidence, however, suggests the exact opposite. Donald (1963) stated that 'in general, as soil fertility is improved, so the plant density required to give the maximum yield by annual crops will increase'. Gregory *et al.* (1932) found that the response of cotton to nitrogen in the Sudan was greater at close than at wide spacing; Hearn (1972a) in Uganda concluded that: 'the higher the potential of the environment, the denser the optimum population will be, unless drought or bollworm infestation prevent the potential being realized.' This is reflected in a tendency towards higher populations in many countries as the standard of agriculture improves: Uganda, Tanzania, Zimbabwe, Nigeria, Egypt, Sudan, India and Pakistan.

Plant arrangement

A population of plants evenly spaced in all directions gives each one an equal area for expansion, and reduces competition to the minimum; this is important in non-branching tree crops such as the

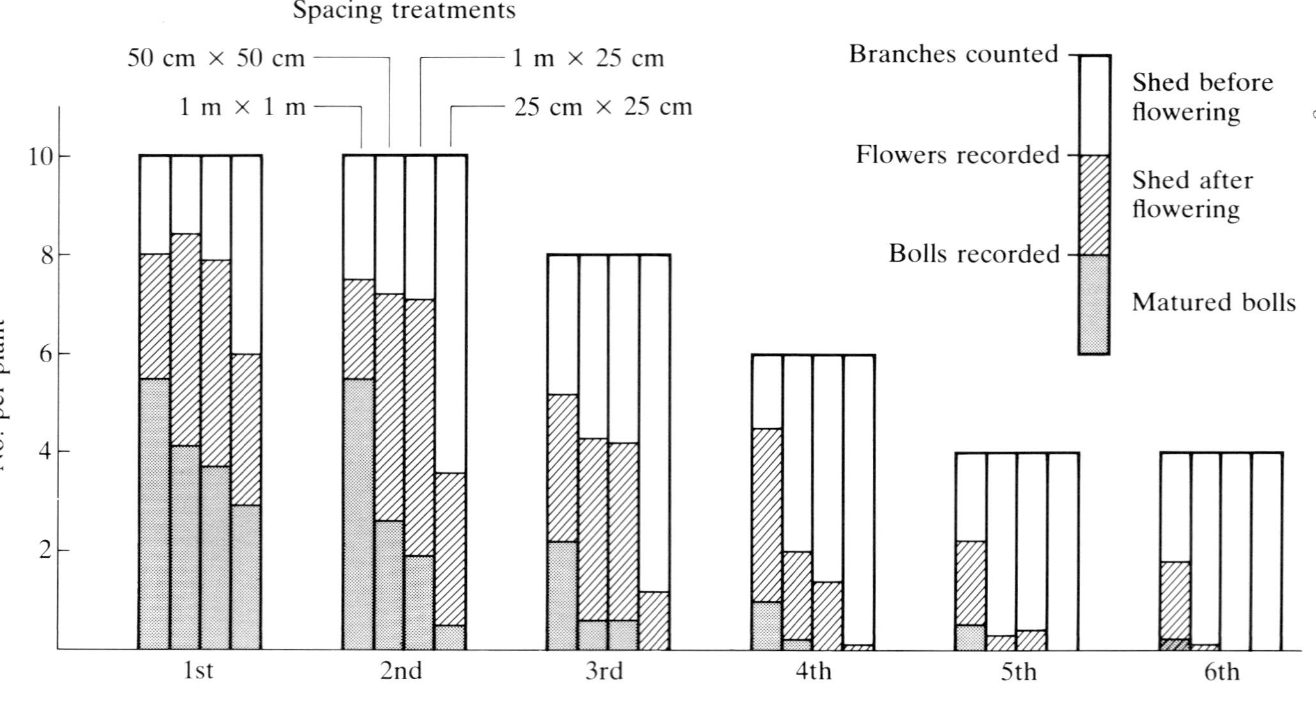

Fig. 6.1 Spacing, flowering and boll set at different nodes. Makanga, Malawi, 1972–73.

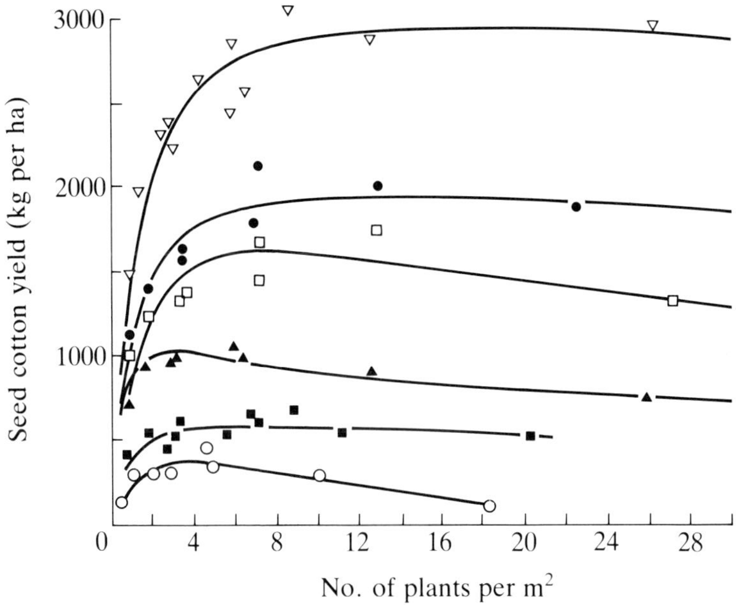

Symbols and experiment numbers:
∇, 1; ■, 6; ▲, 7; ○, 9; ●,11; □, 12.

Fig. 6.2 Relationship between plant population and yield (Hearn 1972a). Actual yields and fitted curves for six representative experiments in Uganda.

oil palm, where triangular planting has been found to give the highest yields (Hartley 1977). With the more adaptable cotton plant, arrangement appears to be much less important than population, especially at fairly low yield levels. In most of the trials reported from tropical countries, population and spacing are confounded, so that it is impossible to apportion the observed effects between them.

Cotton is usually sown in rows with a wider spacing between rows than between plants within the row. When sowing by hand, several seeds are sown together in each planting hole or hill; mechanical seeders can drill a line of seeds or use an attachment for hill dropping as in hand sowing. The seed rate provides an excess of seedlings, which are later thinned to the required number per hill or per metre. One, two or three plants may be left per hill at thinning, and make little or no difference to yield when the stand is good; the extra plants are an insurance against losses between thinning and harvest. Additional plants may be left at thinning adjacent to gaps already present in the stand.

As there is considerable latitude in the choice of population density and row width, the spacing can be chosen to suit the method of sowing, or for convenience in subsequent operations. When sowing is done by hand, narrow rows and closer spacing within the row involve more work and are therefore slower than sowing at wide spacing. When the crop is sown on ridges or beds the row spacing must conform to the size of ridge, which in turn may depend on the equipment used for ridging. Hand-made and machine-made ridges are generally spaced between 0.5 m and 1.0 m from crest to crest; very large ridges, such as the 1.5 m ridges on the hill sand soils of Sukamaland in Tanzania, accommodate two rows of cotton on each ridge. The size of the ridges may be traditional, or may be decided by the infiltration rate, the water-holding capacity of the soil, the intensity of rainfall and whether the ridges are cross-tied or not.

On the flat, the width of the seeder units may determine the minimum distance between rows. For hand planting, the usual spacing is 1 m or 3 ft between rows, and 30 to 40 cm between planting holes, leaving two plants per hole at thinning; this leaves room for a hand hoe to be used for weeding between the plants in the row. Spindle pickers require a spacing of 38–39 in. (1 m) between rows, but strippers are independent of row spacing, and are used for harvesting narrow-row cotton.

Narrow-row cotton

It has been noted earlier in this chapter that there is a tendency towards higher populations as the standard of agriculture improves. In the USA Wilkes and Corley (1968) stated that 110,000 plants per hectare was considered to be the optimum, and small rather than large populations were to be avoided. From about 1967 there has been increasing interest in higher plant populations and closer spacing between the rows. Yield increases of between 10 and 20 per cent have been demonstrated, weeds being controlled by herbicides and the crop harvested by brush strippers. Low and McMahon (1973) report the results of three years' trials in Australia, showing an average increase of about 20 per cent in yield of lint with narrow-row, high-density planting. In Ecuador no significant yield differences were found between row spacings of 1.0 m and 0.5 m, both with the same population of 50,000 plants per hectare (Munro unpublished). In California, with a population of 65,000 plants per hectare, El-Zik and Yamada (1981) report yield increases of 21 per cent from 0.76 m rows and 14 per cent from 1.0 m beds with two rows per bed, compared with one row per bed at 1.0 m; the control yield was 1264 kg of lint per hectare.

As population density increases, the weight of seed cotton per boll decreases; seed weight is reduced most, seed number least and lint per seed is intermediate. Ginning percentage, dependent on seed weight and lint per seed, would be expected to increase, but in fact shows very little change. Fibre quality is only slightly affected, although lint length tends to be marginally shorter with denser populations (Hearn 1972a, El-Zik and Yamada 1981). Low and McMahon (1973) found in Australia that the effect of narrow-row, high-density planting on fibre qualities was very small, much less than the variation from one season to another; if close spacing shortens the growing season, it may be expected to improve the uniformity of the lint.

Earliness

The argument that one boll per plant at a high density will ripen earlier than four bolls per plant at one quarter of the density, and give the same or a better yield, fails to take account of shedding. Very close spacing results in the shedding of early buds and bolls, sometimes in the complete suppression of the first fruiting branches. In trials in Australia 'the effect of establishing six times the usual number of plants was to reduce the number of fruiting forms per plant to about half. Boll retention decreased to only 10 per cent from 25 per cent and boll weights were lower . . . . Earliness did not appear to be a necessary consequence of higher densities.' (Low and McMahon 1973).

In a group of trials carried out in Malawi in 1972–73, all the flowers were tagged on the day of opening at four different sites (Table 6.2); at each site plots of 4 m^2 were planted at each of three different spacings: 1.0 m^2, 0.5 m^2 and 0.25 m^2 – corresponding to 10,000, 40,000 and 160,000 plants per hectare. The cumulative curves of bolls ripened, grouped by date of flowering, are shown in Fig. 6.3. It will be seen that the effect of spacing on both yield and earliness varied widely from site to site: in particular, the closest

Table 6.2   *Spacing and monopodial fruiting*

		Makanga	*Chitala*	*Makoka*	*Chitedze*
Altitude	*m*	*52*	*600*	*1029*	*1149*
Spacing	*Plants per ha*	*% bolls on monopodia*			
1 × 1 m	10,000	44	33	51	44
50 × 50 cm	40,000	21	10	32	21
25 × 25 cm	160,000	6	4	5	1
1 m × 25 cm	40,000	24	8	47	12
Sowing date		21 Jan 73	22 Dec 72	27 Nov 72	2 Dec 72.
Plot size		4 m^2			

Source: Munro unpublished data, Malawi 1972/73.

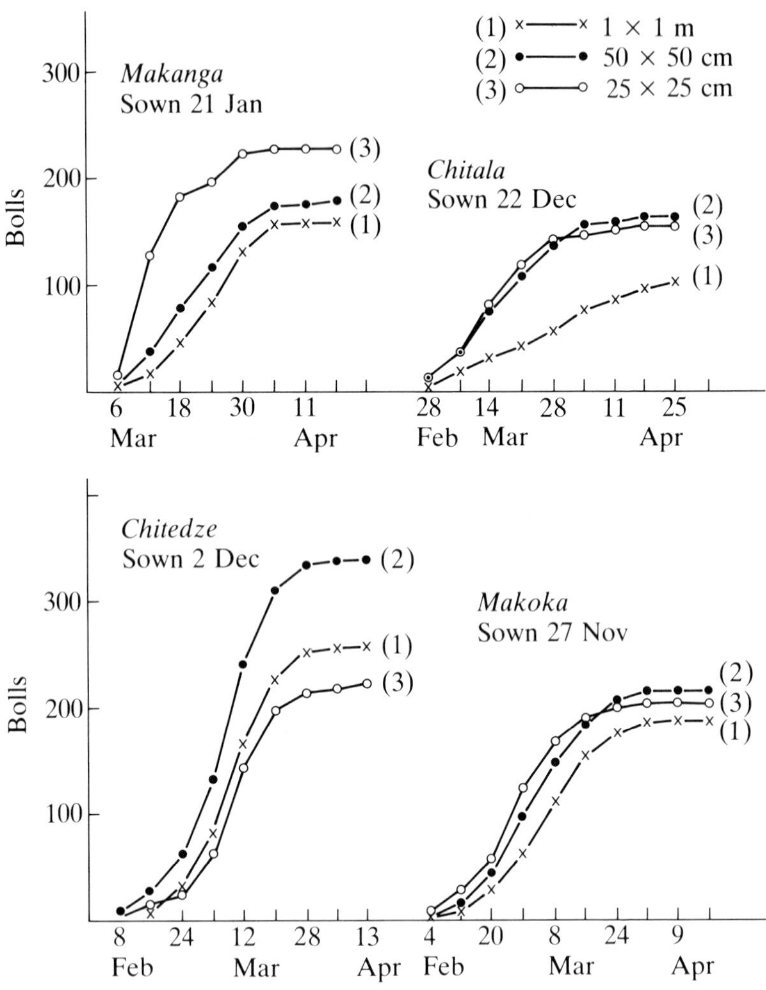

Fig. 6.3   Effect of spacing on yield at different sites. Malawi, 1972–73.

spacing was the earliest and heaviest yielding at Makanga, but the latest and lowest yielding at Chitedze (Munro unpublished). These were exploratory trials which have not been followed up; the plots were small and unreplicated, without guard rows.

In California El-Zik and Yamada (1981) reported that research over the past five years in the San Joaquin Valley showed that 'various narrow row configurations will mature ten or more days earlier than the first pick on conventional 40 inch rows'.

Brown (1953) claimed that in Egypt closer spacing can bring

earlier cropping, with no loss of crop; he recommended about 100,000 plants per hectare as against 35,000 found to be the optimum by Balls (1915), before pink bollworm (*Pectinophora*) became a major pest.

Skip-row planting

Skip-row planting had its origin in the days of acreage restrictions in the USA, when the quota was calculated on the area of cotton actually planted; by leaving every third or fourth row blank, the planted rows benefited from border effects to give higher yields per planted acre. When the unplanted rows are included in the acreage, yields are about 80 per cent of those from a full stand of cotton. In spite of this loss in yield, some growers find skip-row planting is justified by economies in other directions. For example, it allows the passage of tractors along the skip-rows for mechanical cultivation and weeding, reducing hand weeding to the planted rows and the space between them; access for spraying the crop is easier, and causes less damage to the plants; picking by hand is faster in skip-rows than in full stand, and costs less; a saving of water is claimed when irrigation is supplied only in the furrow between the rows. The most common arrangement of skip-rows is 2 : 1, in which every third row is left unsown, but others such as 3 : 1, 4 : 1, 4 : 2, are sometimes preferred.

Method of sowing

Sowing depth

A most important consideration in sowing cotton is the depth at which the seed is placed below the surface. This should be between 2 and 4 cm, and as uniform as possible to produce an even stand of seedlings all at the same stage of development. Deeper sowing leads to irregular and sometimes poor emergence of seedlings; the deep-sown seedling has to struggle to push the seed to the surface, and it will certainly emerge late or possibly not at all. A good stand of cotton should carry the cotyledons 4–7 cm above ground level; cotyledons which open just above ground level indicate that the seed has been sown too deep.

The seed must be in contact with moist soil from the start of germination until the radicle can tap the moisture in the soil below the seed. If the surface soil is dry it may be necessary to sow the seed rather deeper, but the wetter the soil the shallower the planting depth should be. The lower temperature of wet soil slows up germination, rendering the seedling liable to fungal attack, and the upper layers of the soil will be the first to warm up as it dries out.

The seed bed

The soil surface must be smooth with a fairly fine tilth to ensure a uniform depth of sowing. The soil below the seed should be firm and moist, that above the seed should be fairly loose and easily penetrated. If a surface crust forms after the seed is sown, the seedlings may have difficulty in emerging, and the hypocotyl can break at the U-bend in attempting to push through the crust. This is one of the reasons for sowing several seeds in the same hole, called 'hill dropping' in America, as the combined force of three or four seedlings germinating together can break or move a surface cap which would defeat a single seedling.

Before sowing starts, except for dry planting, there must be adequate moisture in the soil for germination of the seed and survival of the seedlings until the next fall of rain. As a rule of thumb, about 50 mm of effective rainfall should have been recorded, ignoring light showers that evaporate without contributing to the soil moisture reserve. If the surface soil has dried out before sowing, the seed should be sown deep enough to make contact with moist soil, but it is sometimes possible to clear aside the dry surface soil, and sow the seed at the recommended depth in the moist soil below. This is the practice under irrigation in the USA, where ridges are formed before the preplanting irrigation; the dry soil on top of the ridge is swept aside into the furrows by a special attachment to the seeder before the seed is dropped. In sandy soil on the shores of Lake Malawi in the Karonga district, holes sometimes 10 to 15 cm deep are excavated to reach moisture and the seeds sown at the bottom of the hole which is left open until the plants are well established.

Hand sowing

Much of the cotton grown in the tropics is still sown by hand. In Africa the normal method is to scrape a small hole, drop in the seeds and cover them with soil; the hole may be made with a stick, the fingers or the toes, and the soil covering the seed is pressed down with the foot. Four or five seeds are usually recommended, but a dozen seeds per hole is quite usual and thirty to forty can occasionally be found. It is difficult to count out and separate fuzzy seeds, and the more lint and fuzz left on the seed the more extravagant the number of seeds per hill. Not only is this wasteful of seed, but thinning is more difficult where the best of a bunch of spindly seedlings has to be selected, and left undisturbed while the other seedlings are uprooted.

The alignment of cotton rows may be done by eye, or by following a string pegged out in the field. In South America up to half a dozen strings are pegged out at the required distance apart, and the planting holes made by jabbing a long stick into the ground at intervals of about 40 cm. Each string has one labourer making the holes, and another following him dropping the seed by hand and covering it with his foot. As one set of rows is completed, the strings are moved on to the next.

In Egypt, the irrigation ridges make the use of strings unnecessary and each labourer, often a child, is provided with a stick about 20 cm long. 'This is pushed into the soil and the seed dropped in behind it, when withdrawal of the stick leaves the seed covered by soil, the stick serving at the same time as a rough measure of distance between holes. Seed is sown dry, and the soil watered immediately afterwards. The usual number of seeds sown per hole is fifteen to twenty, the seedlings being later thinned to two per hole.' (Brown 1953).

The least efficient method of sowing is to broadcast the seed on ploughed land. Most of the seed goes too deep, and germination is poor and uneven; weeding and spraying are more difficult than in cotton planted in rows.

Machine sowing

Hill drop planting with several seeds to a hill is typical of hand planting, while drill or ribbon sowing is typical of machine planting. The cotton planter has been developed from the traditional maize or corn planter, which drops a stream of seed into a small furrow opened by a share, and the seed is covered by loose soil dropping back into the furrow. The depth of sowing is controlled by a trailing wheel, which acts both as a press wheel to consolidate the loose soil over the seed and as a drive wheel to operate the metering mechanism in the hopper through a chain drive (Fig. 6.4).

The metering mechanism for large-seeded crops is usually of the plate type, in which a revolving horizontal disc in the base of the hopper has a series of holes near its outer edge, each hole carrying a single seed from the bulk in the hopper to the top of the seed spout. The discs are interchangeable, providing holes of different sizes to suit the size of the seeds being sown – maize, cotton, groundnuts, beans, etc. Other types of metering system employ fingers or cups to select and deliver individual seeds; naked or acid-delinted cotton seed can be sown by any of these types, with no modification other than that for size of seed, but machine-delinted seed is slightly fuzzy and does not flow easily. It requires the addition of a stirring rod to ensure a continuous flow, which can easily be fitted to a revolving disc or plate; types of seed delivery to which a stirrer cannot be fitted are not satisfactory for sowing machine-delinted cotton seed. The plate type can be further modified to give hill drop planting (Fig. 6.5).

Fig. 6.4 Unit cotton planter: opening runner, seed firming wheel and press wheel driving seed metering system (Deere & Co.).

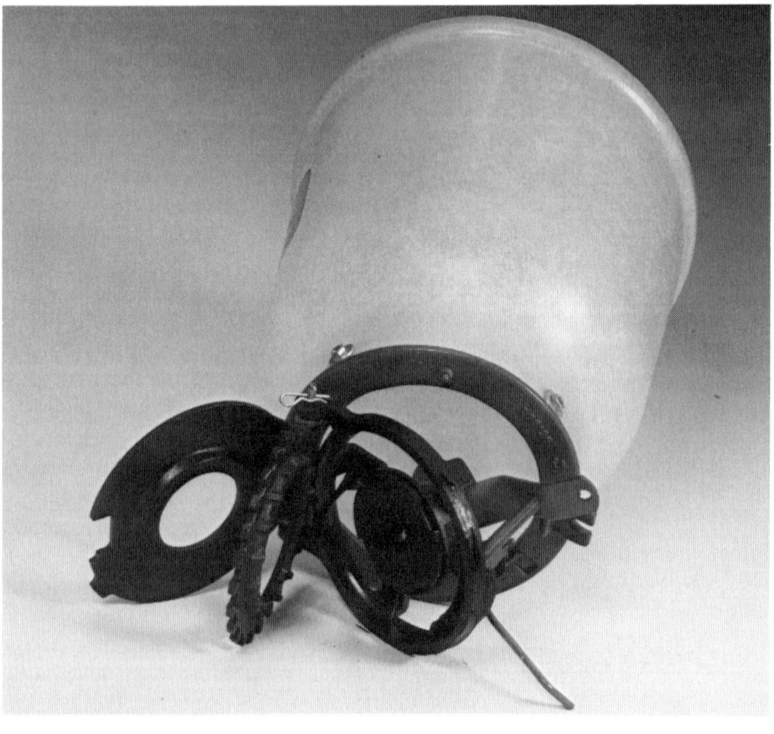

Fig. 6.5 Hopper and plate metering system for hill drop planting of machine-delinted seed (Deere & Co.).

Uniform depth of sowing is essential to obtaining an even stand of cotton; a smooth, even seed bed presents no problems in this respect, but where the surface is uneven, as it very often is, the control of depth is not always satisfactory. Individual seeder units are often supported by the tractor toolbar in front and by the press wheel behind; the tractor will tilt, rise and fall as the tractor passes over uneven ground, and the depth of sowing will vary correspondingly. A more satisfactory system is one in which each seeder unit has its own depth control, situated as near as possible to the seed delivery, with a floating attachment to the toolbar through spring-loaded parallel arms. Depth control is most precise when the depth control wheels are mounted on either side of the seed spout and opener (Fig. 6.6).

The seed furrow may be opened by a runner shaped like a miniature plough (Fig. 6.4), or by a pair of hollow discs set at an angle to the line of travel (Fig. 6.7); the seed is dropped just before

Fig. 6.6 Spring-loaded parallel arm floating attachment to toolbar: drive wheels on either side of opener (Deere & Co.).

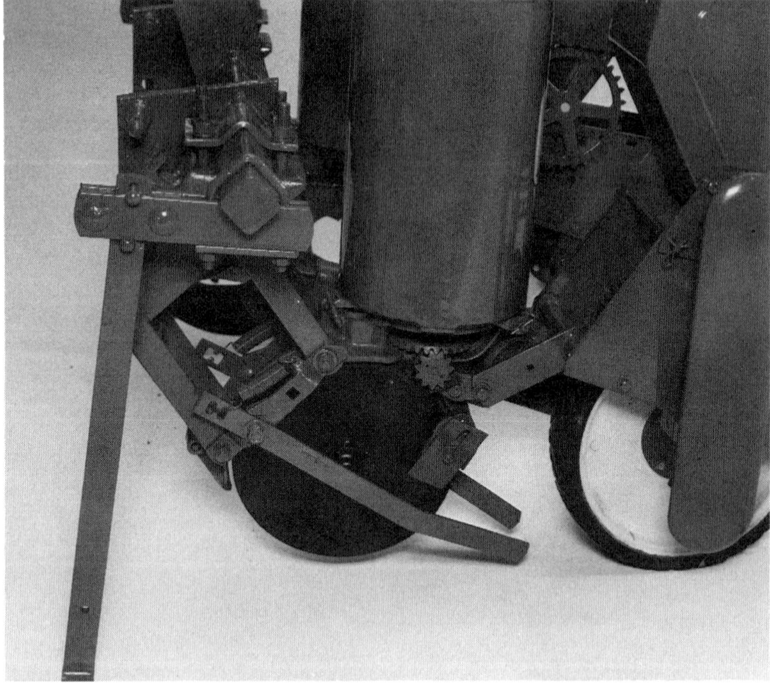

Fig. 6.7 Disc opener with scrapers and spring-loaded knife coverers, followed by press and drive wheel (Deere & Co.).

the pressure on the sides of the furrow is released. If there is much trash left in the seed bed, a disc opener is preferable as it is less likely to get clogged up with accumulated trash. Coverage of the seed may be assisted by small coulters pushing additional soil over the planted row (Fig. 6.7). The press and driving wheel at the rear of the planter consolidates the soil round the seed; it should be designed with a hollow tread to press hardest on both sides of the seed row, leaving loose soil immediately above the seed (Fig. 6.8).

Other attachments available include a V-wing coulter in front of the seeder which sweeps clods out of the way, and skims the dry soil off the top of the ridge or bed (Fig. 6.9); a firming wheel which presses the seed into moist soil at the bottom of the furrow before covering (Fig. 6.4); an adjustable bracket at the rear for mounting a nozzle to apply pre-emergent herbicides (Fig. 6.10); angled firming wheels to close the seed trench leaving the soil relatively loose over the seed (Fig. 6.9); and a choice of tyre treads to suit different soil conditions (Fig. 6.8). Separate hoppers can be mounted on the planter for applying dry fertilizers and granular insecticides and herbicides (Fig. 6.9).

Fig. 6.8   Choice of tyre types; the three in the centre are concave (Deere & Co.).

Fig. 6.9   V-wing attachment to sweep aside dry soil, double driving wheels, angled firming wheels, seed hopper and divided hopper for granular insecticide and herbicide (Deere & Co.).

Fig. 6.10   Pre-emergent herbicide spray nozzle fitted to adjustable bracket (Deere & Co.).

Manually operated seeders of the market garden type have been recommended for use by small-holder cotton growers, but have not proved very popular. Most of them will only work with naked seed, but the main reason probably is that they require capital expenditure for an implement only used for a limited season; the saving in labour is comparatively small, and hand sowing can often be carried out by the family. Only when the scale of operations is such that oxen or tractors are used to prepare the land is mechanical seeding worth while, and then the larger farm seeders are more suitable and cost little more.

Chapter 7

Weed control, thinning and gap filling

Weed competition

Cotton is especially sensitive to weed competition because it grows relatively slowly in the early stages, and does not reach full ground cover until eight or more weeks after germination; often under poor conditions full ground cover is never achieved. To keep the crop free of weeds until the cotton plants meet across the rows is an expensive business, and it is important to know whether and when weed control can be relaxed without seriously affecting yields. Many experiments have been carried out to determine the period in which weed control is most critical, but the results of these experiments have often been contradictory.

Crowther (1943) in the Sudan found that the seventh and eighth week after sowing were the most critical, but Thomas (1969) considered that the first six weeks were the most important. Arnold, Passmore *et al.* (1976) compared two successive seasons in Uganda, and found that the trial results were markedly affected by seasonal conditions; the effects of delayed weeding were much greater in the drier season, when the young cotton was adversely affected by water stress. Thomas and Schwerzel (1968) found in Zimbabwe that the critical period for weed competition was between two and four weeks after crop emergence, but they thought that the critical period would be longer in wetter seasons. On the Kafue flats in Zambia, Kerkhoven (1964) found that the competition between weeds and cotton was for nitrogen; while weeding up to six weeks was important, subsequent weedings had an even greater effect on yield until the cotton plants met across the rows. Lawes (1964) pointed out that the process of weed eradication in Nigeria may be confounded with the effect of cultivation on the soil surface, and hence with infiltration of water into the soil. In the eastern region of Tanzania, the drop in yield associated with delay in weeding and thinning is large, of the order of 50 per cent in potentially high-yielding cotton, when these operations were delayed for 10 weeks after sowing instead of being carried out when the crop was about a month old (Smith and Brown 1963). In unweeded cotton, faster growing weeds can completely suppress the growth of cotton if they are left unchecked; the young cotton seedlings will become drawn, spindly and susceptible to leaf and stem diseases and ground-level pests which are not normally of much importance. Shading by the weeds in these circumstances is at least as damaging as competition for nutrients and moisture.

It is obvious that the nature and severity of weed competition is by no means straightforward and simple to measure. The effect of weeds on the cotton crop can be caused by competition for light, water and nutrients, and will depend on the type of weed and the density of weed growth as well as on the soil type, rainfall and the level of fertility. Reviewing numerous studies on the effects of weed competition on the growth and yield of cotton in various parts of the world, Hawtree (1980) concluded that 'the weight of evidence indicates that weed competition during the first two months of the crop's life is more injurious than during the second two months'.

Once full ground cover has been achieved, cotton can compete satisfactorily with most common species of weeds, but some tall and climbing weeds can still present a problem, not only because of their effect on the yield of cotton, but also because they interfere with picking and other field operations. Climbing weeds hinder access to the crop for spraying and picking; weed foliage can impede the full impact of a spray intended to give full coverage to the leaves of the cotton plants; late weed growth can interfere with mechanical harvesting and cause staining of the lint; grasses and other species which shed their seeds on the open bolls add to the labour of clearing trash from the seed cotton; weed seeds produced at this time can add to the weed problems in the subsequent crop.

Preplanting operations

Weed control starts with land preparation, one of the principal functions of which is to kill and bury existing weed growth and weed seeds; methods of doing this have already been discussed in Chapter 5. Some weeds may require special treatment before the crop is sown; in the Sudan, nut grass or purple nutsedge (*Cyperus rotundus*) was brought under control by a specially designed implement used between successive crops. Research work had shown that by cutting the roots of the sedge at a depth of one foot, the tubers were denied access to soil moisture during the long dry season and died; the practical problem of achieving a cut at this considerable depth has been solved by using a knife blade drawn by a heavy caterpillar tractor (Toms 1957, Little 1965). It was noted at Namulonge in Uganda that rotavation as a primary cultivation was far more effective than mouldboard ploughing in reducing nut grass infestation, but in the case of *lumbugu* (*Digitaria scalarum*) the reverse was true.

Fallow weeding was practised commercially in the Sudan Gezira from 1941 to 1964, when it was abandoned because the introduction of nitrogenous fertilizers applied to the cotton crop had rendered it uneconomic (Thomas 1969). It can be useful in controlling specific weeds such as *Cyperus* and *Digitaria*, either mechanically or with the aid of herbicides, when their control in the cotton crop is impracticable.

Postplanting operations

The control of weeds in cotton during the growing season is similar to that in most annual crops, with the emphasis on the first six weeks after germination. Where herbicides are not used, the number of weedings will vary according to the circumstances, but cotton will usually require:

1 A weeding about a fortnight after emergence, when the weed seedlings are quite small; this may be done by hand or with a rolling cultivator.

2 A major weeding at the time of thinning by hand hoe or inter-row cultivator.

3 An inter-row cultivation before the plants meet across the rows.

4 A late weeding by hand to deal with weeds likely to be troublesome at harvest time.

Close spacing between the rows of cotton will mean that the foliage closes earlier to give a full canopy and shade out the weeds, and 3. may not be necessary.

Weeding tools

Traditional tools are used for hand weeding, the short-handled heavy hoe in Africa, long-handled lighter hoes in Asia, and the machete in Latin America. Lighter swan-necked hoes have been introduced from Europe, but have not displaced the traditional implements. The hoe should be used to cut off the weeds just below ground level to prevent regrowth, and their effectiveness depends on soil and weather conditions as well as on how they are used. Conditions should be such that the severed weeds dry out and die; in wet weather they quickly re-establish themselves unless they are heaped up or carried off the field. The machete usually cuts off the weeds at or above ground level; it is thus less effective, and weedings have to be carried out more frequently; there is less chance of the cut weeds re-establishing themselves in wet weather, more of regrowth from the roots remaining in the ground. For hand weeding, the cotton plants should be spaced sufficiently widely in the row to leave room for the weeding tool to pass between them without damaging the crop.

For inter-row cultivation with oxen and tractors, tined cultivators are commonly used; the toolbar must have sufficient clearance to pass over the cotton rows without damaging the plants. Straight and regular rows are essential if the tines are to work close to the cotton, reducing the need for hand labour. Where multirow cultivators are used, the number of units should as far as possible be the same for sowing as for weeding. Rolling cultivators are efficient against small seedling weeds; they can operate at high speeds in young cotton, and can be adjusted to throw loose soil towards or away from the cotton rows; the soil must be moist but not sticky.

Flame control of weeds has been used in areas where natural gas is cheap and readily available, but is rarely used in the tropics.

Herbicides

Chemical weed control

Chemicals such as naphtha-based oils were used to control weeds in cotton in the United States before the discovery of selective herbicides in 1942, but it is only since chemicals were identified which could safely be applied to the cotton crop that herbicide use in the tropics began to develop. The first selective herbicides for use in cotton were monuron and diuron, and the latter is still one of the most widely used. Not only do such herbicides reduce the labour and power required for weeding, but their persistence can maintain a weed-free crop during periods of wet weather when other methods of weed control are impracticable.

Herbicides are now widely used on cotton in the United States and Latin America, South Africa, Egypt and the Sudan. In the rest of Africa and in Asia experimental stations are giving much time to their study in relation to local weeds and crops. The way in which small-scale cotton growers have adopted insecticides for pest control suggests that it will not be long before they appreciate the benefits of herbicides in dealing with one of their most important problems.

The appropriate herbicide depends on the weed flora, and the following references containing lists of cotton weed species in different countries are quoted from Kasasian (1969):

Mozambique	Balsinhas (1963)
Uganda	Riggs (1967), Tiley (1970)
Kenya	Bradley (1968)
Sudan Gezira	Thomas (1968)
Hyderabad	Bederker (1957)
Bombay State	Satyanarayan and Ranadive (1960)
USA	Holstun and Wooten (1968)

A list of the principal weeds of cotton world-wide is given in Table 7.1.

Pre-sowing herbicides

Herbicides which are poorly mobile in plant and soil, such as trifluralin, are normally incorporated in the soil by cultivation prior to planting the crop; this prolongs the life of the herbicide and ensures close contact with germinating weed seeds. The growth of lateral cotton roots is inhibited in the zone of incorporation of trifluralin, but this does not apparently affect yield (Hawtree 1980). With other herbicides there is a choice between pre-sowing (pre-plant) and pre-emergence applications. Even though some herbicides may lose part of their potency within a week of being applied to the soil surface, Whitwell and Buchanan (1981) in Alabama found that pre-emergence application was generally more consistent in weed control than preplant incorporated application.

Pre-emergence herbicides

Pre-emergent application has proved to be one of the easiest and most efficient methods of using many herbicides. They are applied after the cotton is sown but before it appears above ground four or five days later. The seed must be adequately covered with soil to prevent direct contact with the chemicals. The herbicide can be applied to the whole area or limited to bands covering the planted rows, where weed growth is most difficult and expensive to deal with once the crop is established. Band application reduces the amount of chemical required, e.g. a 25 cm band on cotton sown in 1 m rows will reduce the chemicals used by 75 per cent; weeds in the untreated area can be controlled by cultivation.

Often a mixture of chemicals is used in the pre-emergent spray; such a mixture can include a herbicide effective against grasses, another against broadleaved weeds, paraquat to kill off any weeds already germinated and an insecticide to control seedling pests such as cutworms. Pre-emergent applications can be made with hand-

Table 7.1   *Principal weeds of cotton*

	USA	Mexico	C. America	S. America	Europe	Africa	Asia	Australia
Grass weeds								
Cynodon dactylon	X		X	X		X		
Cyperus rotundus	X	X	X	X	X	X		X
Cyperus esculentus	X					X		
Dactyloctenium aegyptium	X				X	X	X	
Digitaria spp.	X		X	X	X	X		
Echinochloa colonum		X			X	X	X	X
Echinochloa crus-galli	X	X			X	X	X	X
Eleusine spp.	X		X			X	X	
Leptochloa panicea		X						
Panicum maximum					X		X	
Panicum spp. (annuals)	X							
Rottboellia exaltata					X		X	
Sorghum halepense	X	X	X	X	X		X	
Broadleaf weeds								
Amaranthus spp.	X							
Convolvulus arvensis						X		
Ipomoea spp.	X							
Physalis angulata		X		X				
Portulaca oleracea	X	X	X	X	X		X	X
Sida spinosa	X							
Solanum nigrum						X		X
Trianthema portulacastrum			X				X	X
Tribulus terrestris	X					X		X
Xanthium pennsylvanicum	X							

Source: Hawtree 1980.

operated knapsack sprayers fitted with a hand-held lance, with a tractor-mounted spray boom, from nozzles fitted behind a mechanical seeder, or from the air. Spinning disc sprayers have been used experimentally for the application of herbicides, but appropriate dosage and volume rates have yet to be established (Matthews 1982); Garnett (1980) reported the development of a low-volume herbicide applicator for tropical small-holder farmers, with a pump and spinning disc mounted on a wheelbarrow and driven by the groundwheel.

Under suitable conditions pre-emergence herbicides should be effective for about six weeks, and at least one mechanical weeding or post-emergent treatment will be needed before full canopy is reached. Many of the herbicides depend on the establishment of a film of the chemical on the soil surface, and mechanical cultivation is not recommended until their effect is exhausted; they are more effective when applied to moist soil (Tollervey 1974). Extra selec-

Table 7.2   *Classification of herbicides*

Group	Examples
Dinitrophenol	Nitrofen
Dinitroaniline	Trifluralin, Profluralin
Phenoxyacids (acetic)	2, 4-D, MCPA, 2, 4, 5-T
(butyric)	2, 4-DB, MCPB
(propionic)	Fenoprop, Diclofop-methyl
Benzoic acid	Chloramben, Dicamba
Pyridine acid	Picloram
Chlorinated aliphatic acid	TCA, Dalapon
Carbamate	EPTC, Butylate
Amide	Propanil, Alachlor
Nitrile	Dichlobenil
Urea	Diuron, Linuron, Fluometuron
Triazine	Atrazine, Prometryne, Cyanazine
Uracil	Terbacil
Pyridazine	Norflurazon, Pyrazone
Bipyridylium	Paraquat, Diquat
Organophosphorus	Glyphosate
Arsenical	DSMA, MSMA

Source: Adapted from Green *et al.* 1979.

tivity is achieved because the weed seeds are accessible while the crop seedlings are still below the soil; the cotton radicle is growing away from the herbicide while small-seeded weeds are germinating near the surface where they receive a lethal dose (Green *et al.* 1979).

Post-emergence herbicides

Post-emergence herbicides have to be used with more caution than pre-emergence applications, as they range from a few which can safely be applied 'over the top' – that is, as a spray covering the crop plants as well as the weeds – to others which are toxic to cotton and must be kept well away from the cotton plants. Some may be allowed in contact with the woody stems of mature cotton without affecting yield, even though the leaves show visible symptoms of toxicity.

Most post-emergent herbicides are applied as a directed spray, covering the inter-row space but not reaching the cotton rows. Fan jet nozzles are used, with or without metal or fibreglass shields to prevent accidental contact with the cotton; rigid plastic sheets can be pulled along on either side of the operator of a knapsack sprayer when the cotton is still comparatively small. Many farmers in the USA construct their own equipment to suit their particular needs, but a wide range is now available commercially for the application of post-emergence sprays (Williford 1981). The post-emergent

Table 7.3   *Herbicides used most widely in the USA, 1979 and 1980*

Preplant	Pre-emergent	Post-emergent	Contact
Treflan	**Cotoran/Lanex**	**MSMA/DSMA**	**Roundup**
(trifluralin)	(fluometuron)		(glyphosate)
	Karmex/Dynex	**Cotoran/Lanex**	Gramoxone
Prowl	(diuron)	(fluometuron)	(paraquat)
(penoxalin)*	Zorial	Karmex/Dynex	
Tolban	(norflurazon)	(diuron)	
(profluralin)	Bladex	Lanex	
Basalin	(cyanazine)	†	
†	Caporal	Bladex	
	†	(cynazine)	
	Sancap	Probe	
	†	(methazole)	
		Dinitro	
		†	
		Lorox	
		(linuron)	

Common name in UK in brackets.
Herbicides in bold type applied to over 1 million acres per annum.
* No common name approved.
† Not listed in Martin and Worthing 1977.
Sources: Buchanan *et al.* 1980, Whitwell *et al.* 1981.

herbicides should preferably be chemically different from those used pre-emergent, as there is a danger that a second application of the same chemical could raise the concentration to a level toxic to the cotton crop.

Where weed growth is sporadic, spot application of the herbicide will reduce the amount required per hectare; this is usually done with knapsack sprayers and a hand-held lance, but an operator can travel with tractor-mounted equipment, turning the spray on and off as required.

The last application of herbicide before the canopy closes is called a layby spray; usually a persistent chemical is used, ensuring as far as possible that the crop remains weed-free until harvest. High-clearance tractors have been designed to allow late applications to be made without damaging the tops of the plants, and are provided with wheel guards to push the branches out of the way.

Herbicides injurious to cotton

Some weeds have proved to be resistant to the range of selective herbicides which do not injure the cotton crop, and a different approach is called for. The choice of chemical is greatly enlarged if it can be applied during a fallow period before the crop is sown, or if it can be applied to another crop grown in rotation with cotton.

The most troublesome weeds are perennials like nut grass (*Cyperus* spp.), Bermuda grass (*Cynodon dactylon*) and Johnsongrass (*Sorghum halepense*), and other species are of local importance. Not only are they resistant to selective herbicides, but mechanical cultivation breaks them up into fragments, each of which can develop into a new plant.

Dalapon has been used successfully in Uganda to control *Digitaria scalarum* by applying it during the fallow period; the herbicide is more effective when the *Digitaria* is growing actively, and a nitrogen fertilizer applied a few weeks before the dalapon spray was found to enhance the control of the weed (Arnold, Passmore *et al.* 1976). Dalapon has little residual effect on cotton, which can safely be sown a few weeks after the herbicide has been applied; but other herbicides, such as fluometuron in the Sudan, must be applied four months before the cotton sowing date (Idris, 1969). This residual effect has always been a problem, and may extend beyond the cotton crop to other crops in the rotation, but there are some general herbicides which have no residual effects.

One of these is paraquat, which is quickly inactivated by absorption on to the clay minerals in the soil, but it only kills the aerial parts of the plant with which it makes contact, without affecting the ability of the underground stolons or tubers to regenerate. Another is glyphosate, which was introduced in the 1970s; it is a general herbicide in the organophosphorus group, which is translocated systemically to all parts of the plant from the surface of the leaves, but is rendered biologically inert on contact with the soil. It can be sprayed on persistent weeds in the fallow period in the same way as other herbicides, and has been used successfully alone or in combination to control *Cyperus* (Pandey 1984, Reeves *et al.* 1982), *Cynodon* (Chandler 1981) and Johnsongrass (Weaver 1981) in this way, but special types of applicator are being developed to reduce the quantity of herbicide required. These work on the principle of brushing the leaves with a pad or wick kept moist with the herbicide solution, and some include a system of recovery and recirculation of surplus liquid. Although glyphosate is lethal to cotton, applicators have been designed for using it in the growing crop as a post-emergence herbicide: the pad may be carried above the crop to contact weeds that grow taller than the cotton, or may be drawn along about an inch above the soil between the cotton rows to contact low growing weeds (Williford 1981).

Where cotton is grown in rotation with other crops to which herbicides are applied, the weeds are exposed to a different range of products, and control of persistent weeds in these other crops can supplement the weed control programme in cotton. For example, some herbicides useful against *Cyperus* can safely be applied to the maize crop, but are not tolerated by cotton (Hawtree 1980).

Fig. 7.1　Damage caused by 2,4-D: normal plants on left, damaged plants on right (Perry 1962b).

Cotton is extremely sensitive to 2,4-D. The symptoms of damage by 2,4-D (Fig. 7.1) have been described by Dunlap (1948) in the USA and by Lawes (1955) in Nigeria; the ester is more volatile than the sodium salt, and can affect cotton more than three miles away. Contamination has also been reported round a building in which 2, 4-D was last stored three years before, and from a spray tank which had lain in the open unused for several years. Cotton can recover from the distortion caused to the leaves, but no data are available which relate the severity and duration of the symptoms to yield loss.

Naming of herbicides

Herbicides (and insecticides) have at least three names each, the chemical name, the common name and the trade name. The chemical name means little to anyone but a chemist, e.g. N'-(3,4-dichlorophenyl)-NN-dimethylurea is diuron; it is one of a group called substituted ureas. A new product probably is identified by a code number while it is undergoing tests. Once it is recognized, it receives a common name, e.g. diuron, by which it can be referred to in experimental work; the name is agreed by the International Standards Organization Technical Committee 81 for most countries except the USA and USSR; the USA have their own American National Standards Institute Committee K62, which decides the names to be used there, although most of the common names are accepted by both organizations (Green *et al.* 1979). In Britain, the common names are published by the British Standards Institution (BSI 1969, etc).

The trade name is given by the manufacturer or distributer to his particular formulation of the active ingredient, and different names are often used in different countries for the same formulation. A directory relating the trade names of a country to the common name of the active ingredient is needed if the farmer and the extension worker are to be able to choose the most effective product, as well as for treatment in case of accidents. The Thailand Pesticide Handbook is such a guide (Rushtapakornchai and Vattanatungum 1981), showing that diuron, for example, is sold under eight different trade names there.

The names of the chemical groups to which the herbicides belong are often mentioned in the literature, and a list of the main groups is given in Table 7.2. The herbicides most widely used in the USA in 1979 and 1980 are shown in Table 7.3.

Thinning and gap filling

Planting to a stand

The ideal seed rate should use just sufficient seed to provide the required population and spacing of plants in the field after germination. This is called planting to a stand, but even with irrigation and the best of conditions success can rarely be guaranteed, still less when dependent on rainfall. Consequently the seed rate for cotton is almost invariably in excess of the ideal, and unless conditions turn out to be very bad the excess of seedlings has to be thinned out after the crop is established. Thinning also allows local variations in stand to be evened out, and the more vigorous seedlings to be selected; Walker (1968) has shown that the selection pressure exerted at thinning by choosing the most vigorous seedlings can increase the proportion of heterozygotes in the crop.

Thinning

With hill drop planting, thinning consists of reducing the number of plants per hill to one or two according to the spacing desired; in the case of drill sowing, single plants are left at the required spacing, which may be measured as so many per metre of row or so many inches or centimetres between plants. The distance between plants may be estimated by eye or checked by measurement with the hand or a short stick. As far as possible the weaker plants are removed and the stronger ones left. Where a gap occurs which is longer than the required distance, extra plants may be left on either side of the gap to utilize the vacant space. The excess seedlings are uprooted most easily and with the least disturbance of the remaining plants when the soil is moist.

Thinning is nearly always done by hand, even if most other operations are mechanized, as the exercise of judgement and selection is important in achieving a uniform crop. The excess seedlings are pulled sideways away from those which are to be left, and disturbed soil should be pressed back round the remaining seedlings. Singling can be done with a light hoe, but the African hoe of the heavy digging type is not suited to the light touch needed; there is some conservatism in the use of hoes other than those of the traditional weight.

The simplest method of mechanical thinning is to run a tined cultivator across the cotton rows at right angles, with the tines set to leave seedlings untouched at the desired interval; more sophisticated machines are available and others are being developed for use when labour costs become prohibitive. These machines may be used to give a partial reduction in stand, reducing the amount of labour necessary to complete the operation by hand.

Thinning takes place after the danger of loss from seedling disease and pests is over, and before inter-plant competition acts to the detriment of the crop. Timing is not very critical within a week or so, and thinning usually takes place four to six weeks after germination, when the plants are about 15 cm tall. Growing conditions, however, notably temperature, affect the rate of growth and development, and a better criterion is when the seedlings are showing four true leaves, not counting the cotyledons.

Gap filling

The value of resowing gaps in the stand – variously referred to as refilling, gapping, infilling or gap filling – is doubtful. In experiments it has shown no worthwhile benefit, and may in fact reduce yield per hectare, besides involving an extra field operation (MacDonald *et al.* 1947); but it is not easy to reproduce experimentally the plant arrangement that actually occurs in a commercial stand. These gaps do not occur in a regular or uniform pattern, but are likely to be patchy, with areas of good stand interspersed with patches where the stand is poor. Matthews (1972) achieved a natural appearance by locating the gaps at random, using tables of random numbers to select the row and hill to be eliminated and refilled; his results confirmed those of MacDonald, showing that a loss of 60 per cent of the recommended stand of 72,000 plants per hectare did not reduce yields significantly, and that yields were reduced on average by 5 per cent (not significant) as a result of refilling.

Refilling results in plants at different stages of growth being present in the same field, which makes insect control and other operations more complicated. The sort of gap which is probably worth refilling is a length of row which has been left unsown or failed to germinate, or a noticeably large empty space; the refilling should be done as early as possible, not more than 10 days or so from the original sowing. Even if the stand is only 20 per cent of the intended population, it is not worth while uprooting and resowing the crop; the reduction in yield resulting from delayed planting will likely be greater than the loss resulting from a poor stand. In experimental plots the requirements are rather different, and such operations may be worth while (see Ch. 16).

Transplanting of cotton seedlings to give a more uniform stand has never been found to be successful. In experimental work, spare seedlings can be grown in plastic bags for refilling gaps.

Chapter 8

Rotation, soil fertility and fertilizers

Continuous cotton

Cotton is much less sensitive to crop rotation than other annual crops such as maize, and in many parts of the world it is grown year after year on the same land with no apparent ill effects. Continuous cotton plots have been maintained at research stations in the Sudan since 1918, at Ukiriguru in Tanzania since 1939 and at Namulonge in Uganda since 1950; the Uganda plot grows two crops a year, and cotton has been rotated each year with beans and maize. Where no nutrients are added, the plots settle down to a low but fairly stable level of yield, which can be raised to normal commercial levels when desired by the use of fertilizers and insecticides (Lee *et al.* 1974, Arnold, Passmore *et al.* 1976). These results have been obtained on land free of nematodes and *Fusarium* wilt, which could change the picture considerably.

Interaction of cotton with other crops

A few crops have been shown to have a depressing effect on the yield of the following cotton crop: sorghum in Tanzania (Prentice 1972) and in the Sudan Gezira (Burhan 1969), and *Stylosanthes* at Serere in Uganda (A. Low pers. comm.). On the other hand, rotation with cotton is known to benefit other crops. Tobacco in Zimbabwe is rotated with cotton as well as other crops to limit the eelworm population in the soil; Munro (unpublished) found that maize yields fell rapidly when grown year after year in Malawi, but could be maintained with a two-year rotation with cotton. Well-weeded cotton is a good cleaning crop in a rotation, and its deep root system may help to improve soil structure.

Legumes

The early improvers of tropical agriculture usually stressed the value of including a legume in the rotation, and green manuring with sunhemp (*Crotalaria juncea*) or beans such as *Stizolobium* sp. was recommended; but the economic value of green manuring is debatable (Webster and Wilson 1966), and in Zimbabwe and South Africa it has been largely replaced by applications of nitrogenous fertilizers or a leguminous crop which yields a return in its own right. Similarly in the Sudan the inclusion of lubia (*Dolichos lablab*) in the rotation increases the yield of cotton (Burhan 1969), and Simpson (1969) estimates its value as a little over 40 kg of nitrogen per hectare. In the standard two-year rotation in Egypt cotton follows berseem (*Trifolium alexandrinum*), but there have been no long-term rotation experiments to measure the effect of the legume crop (Brown 1953).

Rotations

The planned succession of crops on a given piece of land is followed more in irrigated than in raingrown cotton, because irrigation schemes have to make the maximum use of the available water, and the water supply cannot easily be adjusted to the needs of the individual farmer. Little importance is attached to rotation by the growers of raingrown cotton, and any regular crop sequences are followed more for convenience than on theoretical grounds.

Where two crops a year are planted, and cotton is restricted to one season, a simple rotation is achieved automatically. Some African tribes have evolved quite elaborate rotations, for example, that practiced by the Wakara on an island in Lake Victoria where population pressure is strong (Rounce and Thornton 1956), or that of the northern Zambians who adhere to a rigid crop succession on land which has been fired in a traditional manner (Prentice 1972). Cotton is generally considered as a good opening crop following bush clearing; in northern Nigeria it often follows, and is interplanted with, maize or finger millet (*Eleusine coracana*) (Norman *et al.* 1974); but over much of Africa the crop sequence is haphazard.

Agricultural departments in east Africa have long recommended the 3 : 3 rotation, three years cropping followed by three years grass, which has some claim to be indigenous to the area (Stephens 1970), but the recommendation has never been widely adopted and no particular sequence has been recommended for the cropping phase of the rotation.

Rotation trials were started at the Gezira Research Farm when it was inaugurated in 1918, and further trials were started in 1925 and 1931. These trials were carried on for at least 30 years without fertilizer, while the Continuous Cotton Trial started in 1918 is still running. More recent trials have included fertilizers and more intensified and diversified cropping (Burhan 1969). The Gezira eight-course rotation was first adopted in 1933, with 25 per cent of the area devoted to cotton and about 55 per cent to fallows; while the rotation has remained basically unchanged, the fallow area has been reduced to below 40 per cent since 1960 by the introduction of wheat, groundnuts and vegetable crops to replace some of the fallows. The standard eight-course rotation of the main Gezira is now cotton – wheat – fallow – cotton – legume – sorghum – fallow, or phillipesara (raingrown) – fallow (Simpson 1969, Farbrother 1973). The long-term trials without fertilizer showed the value of fallows preceding cotton, the beneficial effects of legumes and the depressing effect of sorghum; more recent trials with fertilizer show that these effects are significantly reduced but not eliminated. More intensive rotations, where the fallows are replaced by crops and hence receive more frequent irrigation, have not reduced the average yields of cotton (Burhan 1969).

In the former French colonies of West Africa, an inventory was taken of the local arable land and the annual crops were arranged in a four-year rotation giving proportions similar to the inventory: cotton – millet – fallow – fallow (Carbon 1957).

Mixed cropping

Mixed cropping is widely practised in the tropics, and fulfils some of the functions of a rotation; research tends to show that the returns under mixed cropping are as good as, or better than, those when the component crops are grown in pure stand. It is quite possible that inter-planting of cotton with other crops provides some protection against insect pests. The fact that mixed cropping is so widely practised suggests that it is eminently suited to small-scale agriculture cultivated by hand, but difficulties arise when improved husbandry methods are introduced. It is not compatible with efficient pest control, mechanization and above all with the use of herbicides. For this reason it is usually recommended that cotton should be grown as a pure stand, opening the way to more efficient and labour-saving methods.

Land resting

Under tropical conditions of temperature and rainfall, soil weathering proceeds rapidly, leaving a more or less inert residue of minerals in mature soils; this residue depends more on the climate than the parent rock from which it is derived. Overlying this the natural vegetation maintains a thin topsoil 15–45 cm deep, enriched by the biological accumulation of nutrients; when this natural vegetation is cleared to allow arable cropping, the topsoil is exposed to the climate, and the nutrients are lost by weathering and leaching.

The traditional answer to this decline in fertility in less densely populated areas is shifting cultivation: when yields on the original land decline, virgin land is opened for cultivation and the natural vegetation allowed to regenerate on the abandoned land and restore the *status quo*. The periods of cropping and regeneration vary according to the climate and cropping; in the Congo forests, a minimum of 12 years is reckoned to be necessary to restore fertility (Demol and Banninck 1957), while in East Africa the Departments of Agriculture recommended the 3 : 3 rotation. At Serere in eastern Uganda, planted grass, grazed or not, for two or three years out of five, has barely maintained fertility; at Kawanda near Kampala a 3 : 3 rotation with *Chloris guayana* (Rhodes grass) as the grass rest was equivalent to continuous cropping with 10 tons of farmyard manure per acre applied every third year (Stephens 1970). In the Lake Province of Tanzania Peat and Brown (1962) found that a 'tumbledown fallow', allowing the natural regeneration of weeds and grass for three seasons, is generally as effective in restoring fertility as planted grasses or reasonable dressings of cattle manure or phosphates. Because of pressure of population such a fallow was not in general practicable for the cultivators of hill sand soils, but by planting cassava, weeded only in the first season, similar results were achieved with the added profit from the cassava crop.

The recycling of nutrients that occurs within the soil profile under shifting cultivation has been illustrated diagrammatically by Jones (1976) whose diagram is reproduced in Fig. 8.1. Between 1961 and

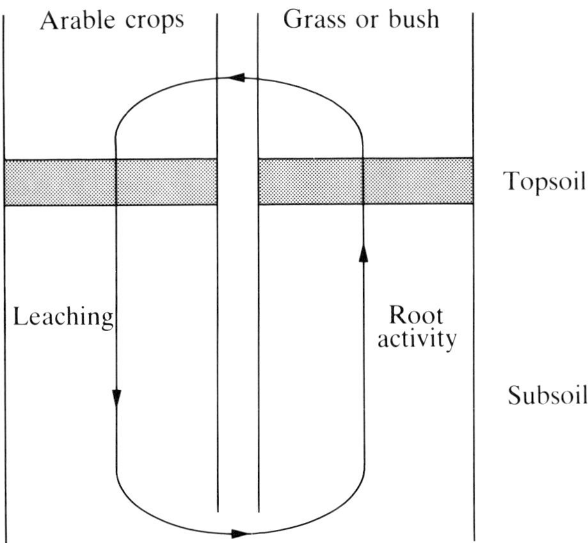

Fig. 8.1 Nutrient recycling within a soil profile through a ley-arable rotation (after Jones 1976).

1969 Jones measured the changes in nutrient content of the soil profile under a 3 : 3 rotation in Uganda, in a project financed jointly by the Nuffield Foundation, the Overseas Development Administration and the Cotton Research Corporation. He showed that the fall in yields under arable cropping was accompanied by the leaching downwards beyond the root zone of a proportion of all the major soil nutrients and organic carbon, together with a lowering of the pH. The resting phase was planted with either Rhodes grass or elephant grass (*Pennisetum purpureum*); the improvement in crop yields following the grass rest was the result of a gain in all the major nutrients in the topsoil plus vegetation, and a higher organic carbon and pH value in the soil (Table 8.1). Jones (1976) argues that uptake by the grass roots cannot fully explain the movement of nutrients during the resting phase, and concludes that 'the major mechanism indicated for the gains and losses is the mass movement of soil water up the profile in response to a hydraulic gradient, or down the profile in response to gravity…this mechanism is dependent on the weather pattern'. He goes on to discuss how this weather pattern could affect the useful length of the arable period.

Conclusions

Dependence on a single crop is dangerous both agriculturally and economically, and a suitable choice of crops grown in conjunction with cotton will spread the risk of crop failure and will help to even out labour and machinery requirements throughout the year. There

Table 8.1 *Change in the nutrient content and pH of the 0–45 cm depth of soil following a resting phase and an arable phase on the same area of land*

	Measured after resting phase 1963–66 (kg per ha)		Measured after arable phase 1966–69 (kg per ha)	
Organic carbon	+ 15,950	± 2,850	− 19,700	± 1,151
Total nitrogen	+ 769	± 149	− 968	± 220
Total phosphorus	+ 85	± 53	− 88	± 40
Total sulphur	+ 45	± 39	− 86	± 25
pH (CaCl$_2$)	+ 0.26	± 0.03	− 0.31	± 0.06
Exchangeable K	+ 471	± 129	− 461	± 165
Exchangeable Ca	+ 971	± 360	− 1,897	± 270
Exchangeable Mg	+ 465	± 175	− 420	± 84

Each figure is the average for 10 plots (5 elephant grass undisturbed and 5 elephant grass self-mulched).

The figures for N,P,S,K,Ca and Mg after rest include the amounts of those nutrients in the standing vegetation after 3 years. All vegetation was incorporated into the soil by rotavation.

Source: Jones 1972.

is no conclusive evidence that cotton benefits from rotation with other crops, but such a rotation may benefit the other crops and does no harm to the cotton; it is only prudent therefore to grow cotton in a simple rotation, which can be left fairly flexible. The order of cropping does not seem to affect cotton to any great extent, especially when fertilizers are applied, and will depend more on the needs of the other crops than on those of cotton.

Fertilizer requirements

The quantities of nutrients removed from the field as seed cotton have little value as a guide to the practical and economic use of fertilizers. They differ widely from the amount of nutrients actually drawn from the soil by the crop (nutrient uptake) where these have been measured; Hearn (1981) has summarized the published data. They take no account of the losses of nitrogen, phosphorus and sulphur which occur when crop residues are burned, nor of the varying percentages of the nutrient applied which find their way into the plant. The process of determining the nutrient uptake is too elaborate to carry out on a large scale, so how then should fertilizer requirements be determined?

The obvious answer is to lay down fertilizer trials, but this is not as simple an answer as it might appear. Fertilizer requirements vary from one part of a field to another, and even more so over a whole district or region; a trial can only record the effect of fertilizers at the site of the trial in the season in which it was carried out, and trial results often differ from one season to the next on the same land, over and above the experimental error. A large number of trials are therefore necessary before any firm recommendations can be made. Trials in new areas are usually planned in two stages, preliminary trials to determine the elements most deficient, followed by factorial trials to establish response curves for these elements, any interactions between them and hence the most economic rate of application.

Soil analysis is another way of assessing fertilizer requirements. This depends on adequate sampling, as only a very small quantity of soil is actually analysed: a number of points and the profile at each point must be sampled. The analysis may be purely chemical, or chemical and physical, with or without a biological assay such as the 'minus-one' technique (Le Mare 1963). Chemical analysis has not proved very reliable as a direct guide to fertilizer requirements, except in particular areas where local research has been intensive enough to relate the level of nutrients in the soil to fertilizer responses (Hearn 1981). This situation is common in the USA, but sufficient background information is rarely to be found in less well-developed countries. Relationships found in one area are unlikely to be valid elsewhere; Zimbabwe recommendations for the application of nitrogen, based on the soil analysis of individual fields, did not work when applied in Malawi. Birch (1952), using all the data from the East African Agriculture and Forestry Research Organization (EAAFRO), was unable to establish a correlation between phosphate response and any of the standard chemical tests for soil phosphate, total or available; the best correlation was with base saturation.

Soil analysis can, however, show up major deficiencies, and give an indication of the nutrients which must be included in fertilizer trials. Table 8.2 shows the critical soil nutrient levels which various workers have given as the minimum below which deficiency symptoms are likely to appear, and Hearn (1981) has given the relative frequency of occurrence of such deficiencies:

very common	N					
common	P	K				
occasional	Mg	S	Zn	B	Mn	
rare	Ca	Fe	Cu			
unknown	Mo	Cl	Na			

Table 8.2   *Critical soil nutrient levels*

Element	Method	Level
N	NO_3 initial	10 ppm
	NO_3 incubated	40 ppm
P	0.025 N H_2SO_4/0.05 N HCl extract	16 ppm
	0.5 M $NaHCO_3$ extract	5 ppm
K	Exchangeable, sands	40–80 ppm
	other loams	60–100 ppm
	clays	75–125 ppm
Mg	Exchangeable	4–6% CEC*
S	SO_4 ions	10–15 ppm
	Total	50–100 ppm
Zn	Dithizane extract	0.5 ppm
B	Hot water extract	0.15 ppm
Mn	Water extract	3 ppm

* Cation exchange capacity.
Source: Hearn 1981, Appendix 4.

Soil analysis, climate and vegetation have been successfully combined in Nigeria, where at least 300 major soil series in the northern region were grouped into about 30 vegetation-soil units by Higgins (1961). Experimental farms and district trials were located to give adequate representation to each of these vegetation-soil units. Goldsworthy (1967) was able to group these units into areas giving similar responses to N and P – five response areas for sorghum and three for maize – based on the results of a series of factorial N × P trials covering all the units in which these crops are grown.

Tissue tests, although mainly used for perennial crops, have been used successfully in the USA to signal the need for fertilizers in cotton. Joham (1951) showed that the leaf petioles are a more sensitive index of plant nitrogen (N) than other parts of the plant; samples of 25 to 30 petioles from the youngest fully expanded leaves of the mainstem are taken at two-week intervals, beginning at the time of the first square. The analysis measures the flow of nitrate to the leaf, rather than the N status of the plant; the level therefore indicates when fertilizer is required, but not how much (Hearn 1981). In the Sudan, Burhan and Babiker (1968) showed that there was a significant positive correlation between final yield and petiole nitrogen at the early flowering stage, and that yield reduction took place when the level fell below 2000 ppm. Braud (1974) used the lamina of the mainstem leaf subtending the fruiting branch carrying the most recently opened flower, and showed an interaction between N and P critical levels. Levels of nutrients in a well-grown crop vary according to the age of the crop and the variety, as do the critical levels for N deficiency, and the success of this technique depends on the delivery of samples to the laboratory, analysis and reporting of results being completed in a matter of days.

Nitrogen

Nitrogen (N) in general stimulates vegetative growth, and its effect on the cotton crop, derived as it is from the fruit, is indirect. The plant must produce an adequate number of nodes on the fruiting branches to set the crop, but must not be so vigorous that it induces excessive shedding of buds and young bolls. Under tropical conditions the internodes both on the mainstem and on the branches are longer than those growing in more temperate climates (see Ch. 4); the application of nitrogenous fertilizers can increase internode length, and thus increase the tendency towards rank growth.

Burton (1972) was concerned with controlling rank growth in the cotton grown on the Triangle Sugar Estate in Zimbabwe; he determined the N status of each field by soil analysis, and applied no nitrogen if it was high. If the N status was low, half the estimated requirement was applied at sowing, and the remainder only if visual symptoms of N deficiency (yellowing of leaves) were observed between 6 and 12 weeks after sowing.

Birch (1964) has shown that alternate wetting and drying cycles in

Table 8.3 *Nutrient content of common fertilizers*

	N	P (P_2O_5)	K (K_2O)	Ca (CaO)	S	Mg (MgO)
			% by weight			
Ammonium sulphate	20–21	—	—	—	24	—
Ammonium nitrate	33–35	—	—	—	—	—
Ammonium sulphate/nitrate	26	—	—	—	12	—
Calcium ammonium nitrate	20–26	—	—	10	—	7
Calcium nitrate	13–15	—	—	25–30	—	2
Calcium cyanamide	21–22	—	—	53	—	—
Urea	44–46	—	—	—	—	—
Anhydrous ammonia	77–82	—	—	—	—	—
Single superphosphate	—	18–22	—	27	12	—
Triple superphosphate	—	45–48	—	20	1	—
Mono ammonium phosphate	11	48–50	—	2	3	—
Diammonium phosphate	21	53–54	—	—	—	—
Ammoniated supers	3–4	16–18	—	23	10	—
Basic slag	—	10–20	—	45	—	6
Rock phosphate	—	25–30	—	46	—	—
Bone meal	2–4	23–30	—	42	—	—
Potassium sulphate	—	—	48–50	—	18	—
Potassium chloride	—	—	50–62	—	—	—
Potassium metaphosphate	—	27	35–40	—	—	—
Potassium nitrate	13	—	44	—	—	—
Sulphur potash magnesia	—	—	22–23	—	24	18
*	—	—	26–30	—	24	9–12
Gypsum	—	—	—	31	18	—
Kieserite	—	—	—	—	23	29
Wood ash	—	1–2	3–7	—	—	—
Flowers of sulphur	—	—	—	—	95	—
To convert oxide to element, divide by:	1	2.3	1.2	1.4	1	1.7

* Sulphate of potash-magnesia in Geus (1973).
Sources: Tisdale and Nelson 1966, Geus 1973, Halley 1982.

tropical soils cause the release of mineral nitrogen from organic matter. This may be adequate for the cotton crop, but Jones (1976) has shown that in a wet season the mineral N may be leached down beyond the root zone before the plant can take it up and use it; as a result, the response of raingrown cotton to added nitrogen is usually greater in wet years than in dry ones.

Equal quantities of nitrogen applied in different forms, urea, sulphate of ammonia (S/A), ammonium nitrate, etc., have much the same effect, so the cheapest form per unit of N is usually chosen (Table 8.3). Urea, prepared commercially from ammonia and carbon dioxide, has a high proportion of N and is easily stored; the high proportion of N means that transport costs, especially for imported fertilizers, are low. On the other hand, the use of urea instead of sulphate of ammonia can result in sulphur deficiency (see below).

Nitrogen deficiency is indicated by a pallid or yellowish green colour in the leaves, compared with a vivid green where nitrogen is plentiful. The plants respond within a few days to an application of a soluble form of nitrogen, but the improved colour of the leaves is not always accompanied by higher yields. In fact, once the crop is set the yield response to nitrogen is small, if any, and the best results are obtained from applications made not later than six weeks from germination; but responses have been demonstrated from applications made up to 12 weeks from sowing (Burton 1972, Smithson and Heathcote 1977).

Split applications of nitrogen are often recommended. It seems logical that some of the N should be applied about the period of maximum vegetative growth, but experimental results are conflicting – most of the evidence indicates that it is just as effective to apply all the N at or soon after sowing. A split application means that an extra field operation is required, adding to production costs; on the other hand, the second application may be omitted if the season turns out to be a dry one, when responses are likely to be small, and only applied if the weather and/or the appearance of the crop suggest that a worthwhile response is likely (Brown 1962, Jones 1976).

The placement of N fertilizers is not critical, except that soluble compounds should never be placed in contact with the cotton seed at sowing, and urea should always be lightly covered with soil to prevent the vaporisation of the nitrogen in sunlight. At sowing, the N fertilizer will probably be mixed beforehand with any other fertilizers needed, or a compound fertilizer will be used. In mixing, S/A should not be mixed with an alkaline product, or much of the N will be released as ammonia gas and lost. With hand-sown cotton the fertilizer is usually applied in holes between the plants in a row, or in a band alongside the planting row. Where cotton is sown on ridges the fertilizer may be buried in the ridge by broadcasting it before ridging or placing it along the old furrow before the old ridges are split. The second half of a split application may be applied in a band alongside the cotton row, or if the leaf canopy is well developed, the N may be broadcast along the rows under the canopy, without covering it with soil. Mechanical seeders often have a fertilizer attachment which allows very precise placement of the fertilizer, usually 5 cm below and 5 cm to the side of the seed. Without the fertilizer attachment, the fertilizer may be applied by hand in the same way as for hand-sown seed. Even if a fertilizer attachment is available, keeping the fertilizer hoppers supplied will slow down the operation of sowing, and the benefit from timely sowing may be greater than that from doing the two operations together.

The continuous use of S/A can lead to an increase in soil acidity (Table 8.4). At Ukiriguru in Tanzania, pH fell from 4.9 to 4.1 over a period of 14 years in an experiment fertilized with sulphate of ammonia every year; a single application of 3.5 tons of lime per acre (8.6 tons per hectare) raised the pH to 6.0, and subsequently the N treatment was changed to nitrochalk, supplying both N and Ca (Le Mare 1972).

Phosphorus

Mineral phosphates are barely soluble in water and depend on weathering and bacterial action in the soil to convert them to a form available to the plant; this is a slow process, but proceeds rather faster in acid soils. To obtain a quick-acting fertilizer, rock phosphate is treated with sulphuric acid to produce superphosphate (supers) and gypsum, or with phosphoric acid to give triple superphosphate. Soda phosphate was prepared from locally available materials in Kenya when sulphuric acid was in short supply, but has now been replaced by superphosphate, although it gave comparable responses in field trials. Single supers contains less phosphorus (P) but more calcium and sulphur than triple supers, and has a lower percentage of soluble phosphate. Solubility in citric acid used to be considered as a measure of the immediate availability to the plant, but water solubility is now usually quoted.

Various analytical processes claim to measure the amount of P in the soil which is in a form readily available for absorption by plant roots, but none of them gives a fully satisfactory measure, and they are continually being modified to suit local conditions. When a

Table 8.4   *Acidity of nitrogenous fertilizers*

Fertilizer	*% N*	*Equivalent in pounds of calcium carbonate per ton of product*	
		Acidity	*Basicity*
Ammonium chloride	24	2560	
Ammonium sulphate	21	2200	
Ammonium sulphate/nitrate	26	1700	
Ammonium nitrate	33–34.5	1200	
Urea	45–46	1500	
Calcium ammonium nitrate	20.5–26	neutral*	
Calcium nitrate	15.5		400
Sodium nitrate	16		580
Calcium cyanamide	21		1245

* At 21 % N

Source: Geus 1973.

soluble form of phosphate is applied to the soil, there is no guarantee that it will remain soluble for long enough for the plant to make use of it; recovery rates are particularly low in acid and alkaline soils, below 6 and above 8 pH respectively. This does not mean that the P is lost, as residual effects have been recorded for at least two seasons after the year of application and reserves can build up to such an extent that crops no longer respond to further applications of P (Peat and Brown 1962, Hearn 1981, Le Mare 1974).

Ferrallitic soils have a high capacity for the fixation of phosphates; the added P combines with soil minerals to form insoluble compounds which are not available to the plant. Until the capacity for fixation is satisfied, added P does not benefit the crop, but beyond that capacity the responses follow the usual pattern. This was investigated by Le Mare (1968) in experiments on red clay loam soil in Uganda. He showed that applications of 1 cwt per acre of supers (10 kg P per hectare) reduced the yield of cotton, but applications greater than 2 cwt showed a progressive increase. Kabaara (1965) showed that in pot experiments the yield depression could be prevented by raising the pH of the soil above 5.5, but other factors are probably involved. The same term 'fixation' is sometimes used to describe the adsorption of phosphates in the soil complex, resistant to leaching but available to the plant.

In other soils responses to P can be very different. In Ecuador Mestanza (1978) found that small applications of supers, up to 40 kg P_2O_5 per hectare gave significant yield increases in seed cotton, but higher dosages gave no further increase and sometimes reduced yields. This type of response is the more usual one in soils with pH in the medium range, between 6 and 8.

Phosphate is invariably applied before, at, or soon after sowing, as it is important for early growth and is not readily leached. It may be broadcast before the final harrowing, but more usually it is placed in a band alongside the planting row, especially if there is any danger of fixation in the soil; the band is in contact with a relatively small volume of soil, whose capacity for fixation is soon satisfied, leaving a high proportion of the fertilizer available to the plant. The more common P fertilizers are listed in Table 8.3.

Potassium

Potassium (K) is the least likely of the three major nutrients to need attention, as it is a constituent of many soil-forming minerals. Only a small proportion, 2 to 10 per cent, of the total K in the soil is readily available to plants (Tisdale and Nelson 1966), but deficiencies are not likely to develop until after a number of years of arable cropping (Stephens 1969, Foster 1972). The K in crop residues is held in the soil in a readily available form after decomposition or burning; Hearn (1981) quotes figures ranging from 3 to 6 kg per bale of lint (5 to 9 kg per ton of seed cotton) for the K lost by the removal of seed cotton at harvest.

Deficiency symptoms are a bronzing of the margins of the older leaves, which later shrivel and are shed, but this is not a very reliable guide as similar symptoms are often seen in the tropics where K is known to be adequate. Presley and Bird (1968) state that cotton diseases are more likely to be serious when K levels are low.

Common potash fertilizers are listed in Table 8.3.

Minor elements

Calcium

Calcium (Ca) has occasionally been shown to be deficient as a nutrient, but it is usually applied to correct soil acidity, when doses of one ton or more of lime per hectare may be necessary; the lime may be applied in the form of ground limestone or slaked lime. More often a Ca-rich form of nitrogenous fertilizer (Table 8.4) is used to correct minor deficiencies and prevent any increase in soil acidity; most phosphatic fertilizers contain a considerable amount of Ca.

Calcium can be displaced from the base exchange complex of the soil by the addition of magnesium, and the Ca : Mg ratio should not be allowed to fall below 1.0. 'Application of limestone can temporarily raise the supply of all nutrients by increasing the rate of mineralization of organic matter, and over a longer term it will improve the supply of phosphorus by raising the pH' (Jones 1972).

Sulphur

Sulphur (S) deficiency has become more common since urea has been used in place of sulphate of ammonia as a nitrogen fertilizer. It is recognized by the fact that increasing doses of urea increase the yellowing of the leaves instead of intensifying their green colour. The deficiency is quickly alleviated by the application of sulphur itself or of a fertilizer containing sulphur; it is commonly avoided by using sulphate of ammonia, or a compound fertilizer containing sulphur, to supply nitrogen at sowing, followed by urea as a second application later in the season.

Magnesium

Cotton and other oil-seed crops require magnesium (Mg) to stimulate the production of organic phosphorus compounds in the formation of oil. Magnesium ions are readily leached from acid sandy soils, and deficiencies have been reported in the south-eastern USA, Tanzania, Zambia and Brazil (Hearn 1981). In the soil a suitable ratio must be maintained between exchangeable Mg ions and those of K and Ca; heavy applications of K can increase the K : Mg ratio above 1.5 and cause Mg deficiency, and too much Mg can cause leaching of Ca ions (see above). On acid soils Mg may be applied in the form of dolomite or dolomitic limestone to reduce the acidity, while kieserite is recommended for other types of soil.

Trace elements or micronutrients

Deficiency of boron (B) is probably the most easily identified of the trace element deficiencies, as the symptoms are quite characteristic. A full description of the symptoms in Zambia, where it was found to be unexpectedly widespread, was published by Rothwell *et al.* (1967); the earliest symptoms are necrotic areas in the pith of the leaf petioles (Plate 19), corresponding with dark and often swollen bands round the outside of the petiole. The petals of the flowers are shortened and do not open fully; early bolls are abnormally shaped, and as the season advances bolls fail to set and terminal growth is distorted. The symptoms have been noted in many other tropical countries and in the USA, but rarely in a form extreme enough to cause crop failure. The deficiency can be corrected with quite small applications of 1–2 kg of B per hectare, usually as borax (10–20 kg per hectare), either as a leaf spray or applied to the soil. Only sufficient B to correct the deficiency should be applied, as in some soils the phytotoxic level is not much more than that needed to correct the deficiency; this level is lower in sandy than in clay soils. Boronated fertilizers are available in some countries.

Manganese (Mn) deficiency is rarely encountered in cotton, but an excess of soluble Mn can be phytotoxic. Manganese is released to the plant in acid soils, but is precipitated in insoluble compounds when the pH is high, so that an excess can be corrected by liming. Manganese poisoning causes one form of 'crinkle leaf', and has been reported in the USA, Chad, Congo (Brazzaville) and Uganda.

High pH also reduces the availability of zinc (Zn), and deficiency is usually associated with heavy neutral to alkaline soils. Such deficiencies have been reported from the USSR, California, Texas and Australia. In Zambia deficiency symptoms were widespread on the Mount Makulu research station in 1973–74, where Zn deficiency had long been recognized in maize. Dust deposited from a nearby cement factory had given high levels of Ca and Mg in the topsoil, and raised the pH value; the severest symptoms were associated with continuous arable cultivation and soil erosion.

Deficiencies have been reported of iron (Fe) in Greece and California, and of copper (Cu) in Georgia, USA, but are very rare; deficiency of molybdenum (Mo) has not been reported on cotton in the field. If a trace element deficiency is suspected, a mixture containing the elements mentioned can be included as a treatment in local fertilizer trials or applied as a foliar spray; only if this gives a positive response is it necessary to investigate further. The concentrations are not critical, and the mixtures used by Singh *et al.* (1970) are given in Table 8.5 as an example.

Plant growth regulators

Biochemists have been studying the chemical processes associated with the growth and development of plants for many years. This has led naturally to the proposition that if a certain chemical is associated in the plant with a specific development, it can be applied to the plant from outside to produce the same effect. Considerable progress in identifying such chemicals has been made in the laboratory, but apart from some of the growth-promoting substances they have not been very effective when applied to the plant in the field. The most successful growth regulators have little or no connection with the internal chemicals which have been identified.

Progress has been hampered by the difficulty of defining the targets for a screening programme, and by the fact that results obtained in the laboratory or glasshouse are not always repeated in the field (Green *et al.* 1979). In cotton, the main targets have been the abscission of buds and leaves and promoting and retarding growth. Commercial products have been developed to control two of these processes in the field: the use of harvest aid chemicals to induce leaf shedding is well established, while growth retardents are just emerging from the experimental stage.

Table 8.5   *Trace element mixtures*

	Foliar spray kg per ha	Soil application kg per ha
Borax	0.5	10
Manganese sulphate	2.0	20
Zinc sulphate	2.0	20
Ferrous sulphate	5.0	20
Copper sulphate	1.0	10
Ammonium molybdate	0.5	1
Hydrated lime (50% of sulphates)	5.0	
Water	650 litres	
Teepol	650 ml	

Source: Adapted from Sing *et al.* 1970.

Considering these four processes in turn, the proportion of flower buds which produce mature bolls is very small (see Fig. 4.12), and it was thought that if shedding could be prevented or reduced there would be an increase in yield. Suitable substances have been found, but there appears to be a limit to the number of bolls which the plant can bring to maturity, and any reduction in shedding at one stage is nullified by an increase at another. With respect to leaves, the aim is to induce shedding so that the bolls are exposed for picking. Harvest aid chemicals include defoliants, leaf desiccants and others which promote boll opening and senescence to terminate the harvest as early as possible without loss of yield. Defoliants are mainly used with spindle pickers, when the object is to shed the leaves while they are still green so that they fall free of the lint in open bolls. There are four main groups of defoliants – organophosphorus compounds, sodium chlorate mixtures, magnesium chlorate and cacodylic acid. Stripper harvesting requires more complete removal and drying of the leaves, and desiccants are commonly used instead of defoliants; a desiccant may also be used to remove any leaves left by the defoliant, and some farmers add a small quantity of desiccant to the defoliant because they find that it enhances its effectiveness. The main chemical desiccants are arsenic acid and paraquat (Cathey 1980).

Several growth-promoting chemicals have been identified, but have not found any practical application to cotton; increased plant size is generally a disadvantage, but there is a possibility that they might be used to increase seedling vigour, and some of them reduce bud and boll shedding. Growth retardants are much more useful, as excessive vegetative growth is frequently a problem, encouraging bollrot and lodging and making harvesting more difficult. The danger is that they may reduce yield or lint quality as well as growth. Growth retardants are now used extensively to control the growth of ornamental pot plants, but different species react to different chemicals. Cyclocel (CCC) has been tested experimentally on cotton for many years, and is used to control the growth of irrigated cotton in Greece (G. Wrigley pers. comm.). Mepiquat chloride is being tested extensively on cotton in the USA.

Chapter 9

Climate, temperature and light

Climate

Like any other green plant cotton needs light, heat, water and nutrients for growth; the soil supports the plant and acts as the medium for the supply of water and nutrients. Soil and nutrients have been discussed in previous chapters, and the other three factors – light, heat and water – go to make up the climate. An ecological balance exists between the natural vegetation and the climate, to the extent that a study of the vegetation can often give as accurate a picture of the climate as years of meteorological measurements. The climate plays a decisive role in determining the farming systems which develop under the influence of many natural and economic circumstances. Engledow (1947) considered that the rainfall component, which will be discussed in the next chapter, was certainly 'the chief climatic determinant of the usability of the land'; yet so far as cotton is concerned, temperature sets a limit to the regions where it can be grown as a commercial crop.

Temperature

Mean annual temperature is of limited value when considering annual crops; the daily and seasonal variations have to be taken into account. The range of diurnal variation (Fig. 9.2) depends largely on the amount of water in the vicinity, as water has a greater specific heat than dry soil, and therefore changes in temperature occur more slowly when a significant volume of water is present. The water may be in the soil, so that the daily range is less in the wet than in the dry season; or it may be present as open water in lakes or the sea, so that coastal areas vary in temperature much less than inland or continental climates. Seasonal variation depends more on latitude, being less near the equator than at the tropics (22°N and S), and becomes more extreme at higher latitudes. Some examples from Africa are shown in Fig. 9.1.

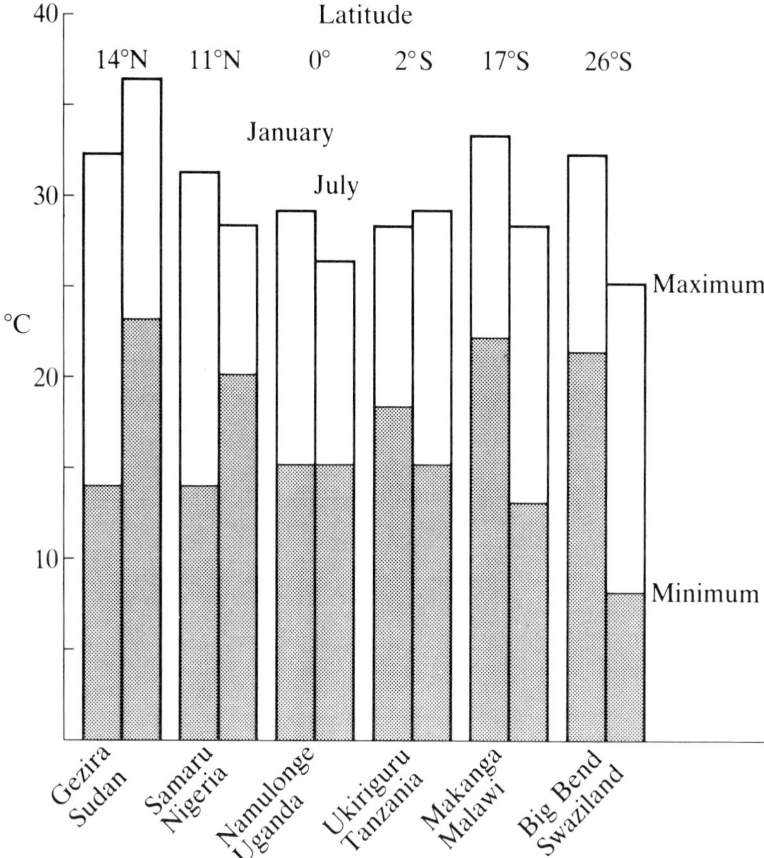

Fig. 9.1 Maximum and minimum temperatures in January and July at different latitudes (Cott. Res. Reps.).

Cotton evolved in the hotter and drier parts of the world, and it is not surprising to find that the modern plant retains some drought resistance and needs sunshine and high temperatures for optimum development. Johnson (1926), writing of Upland cotton in the USA, has said that 'other conditions being satisfactory, the most favourable temperature for the optimum growth of cotton is approximately 32 °C, with a correspondingly warm night'. Tharp (1960) considers that 'the optimum, or the soil temperature at which earliest germination and most rapid seedling growth may be expected, is near 34 °C'. For Egyptian cotton Balls (1953) puts the optimum temperature for a week's exposure, as in a germination test, at about 32 °C. Sikka and Dastur (1960) give the optimum range for vegetative growth of Asiatic cotton in India as 21–27 °C; 'during the period of fruiting day temperatures of 27–32 °C and cool nights are needed for the best results'. According to Purseglove (1968) 'the optimum temperature for germination is 94 °F (34 °C), for the growth of seedlings 75–85 °F (24–29 °C) and for later continuous growth 90 °F (32 °C)'.

It is in fact difficult if not impossible to define the optimum temperature. The cotton plant in the field is not exposed to continuous high temperatures, and the cool nights enable it to withstand high daytime temperatures without injury; Powell (1969) has shown that continuous exposure to 29.4 °C caused more damage than 32.2 °C for 13 hours and 21.1 °C for 11 hours. Nobody knows whether artificial high/low regimes have the same effect as the more gradual natural cycle (Fig. 9.2). In many experiments of this kind it has been found that the minimum rather than the maximum temperature has a greater effect on plant behaviour. The effect of extreme temperatures also differs according to the stage of growth; in general, younger plants are more susceptible, but also recover more quickly. The duration of cold, the ambient humidity and the age of the tissues are all important factors in determining the extent of cold injury (Amin and Powell 1966, McMichael and Powell 1971).

Zaitzev (1928b) postulated that all stages of growth of the cotton plant were equally influenced by temperature, and that the time-scale could be measured in units which were independent of temperature which he called 'isophases'; one isophase is the time interval between flowers at the same node on successive fruiting branches, or the vertical flowering interval (VFI) (Fig. 9.3). Later work has shown that this concept was only a useful approximation. Mauney (1979) found that the rate of production of new sympodia

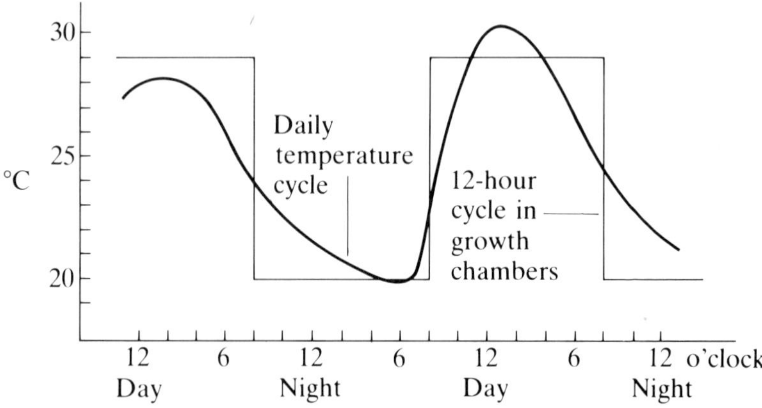

Fig. 9.2 Typical daily temperature cycle and temperatures used in growth chambers (not actual data).

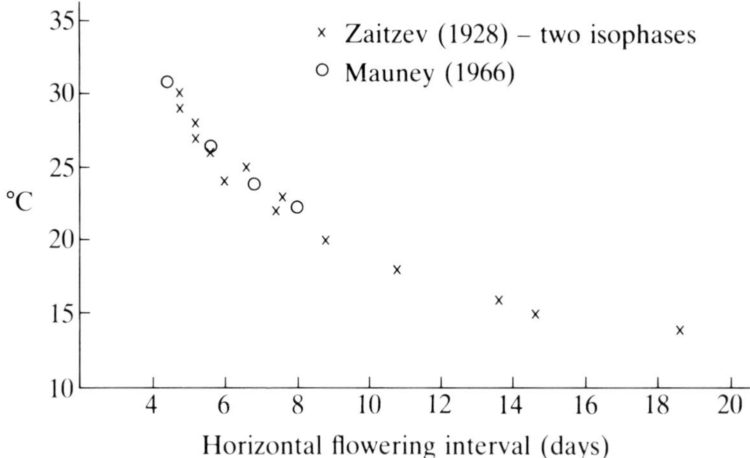

Fig. 9.3 Temperature and flowering interval (Munro 1971).

decreases with the onset of squaring when temperature is held constant; the rate of flowering also slows down as the season progresses, but not to the same extent. 'The inevitable result of this excess of flower opening over new leaf and sympodia formation was that the flowers opened at nodes closer and closer to the apex of the stem. In time, the plants expended the positions available and flowering stopped.'

On Zaitzev's scale the interval between flowers opening on the same branch, the horizontal flowering interval (HFI), was two isophases, but the ratio between HFI and VFI has been found to vary between 2 : 1 and 3 : 1 (see Ch. 4); it is affected by variety, temperature and probably other factors as well. Omitting temperatures high enough to cause damage, it is now well established that lower temperatures slow down the development of the cotton plant at all stages of growth. Erickson and Michelini (1957) introduced the term plastochron for the number of days between the appearance of successive organs of the same type, such as buds, branches, flowers and leaves, and Hesketh *et al.* (1972) have calculated a curve showing the average relationship between temperature and the plastochron for cotton.

Temperature must therefore take its place as one of the factors determining the length of the growing season for cotton, alongside water supply, insect pest damage and varietal differences. Temperature is the limiting factor in temperate climates and at high altitudes in the tropics and subtropics; in Russia, particularly, the limits of latitude for cotton production are being extended northwards by the development of quick-maturing varieties which can complete their growing cycle in the short, hot summers; in Zambia and Zimbabwe at altitudes over about 1200 m, the summer is considered to be too short for a profitable cotton crop. At similar altitudes near the equator, the length of the wet season is the limiting factor, as growth can go on, albeit slowly, throughout the year; in the Kenya Highlands coffee replaces cotton as the main cash crop at altitudes above 1500 m. In such marginal areas early maturity is important, and ratooning of cotton can sometimes be justified because the ratooned plants get a quicker start than cotton grown from seed (see Ch. 17).

According to Tharp (1960) it is usually considered that air temperatures below 16 °C contribute little if anything to the growth of the cotton plant in the USA, where the sowing date for Upland cotton moves north with advancing spring, making due allowance for altitude, and follows roughly the movement of the 16 °C isotherm. Cotton planted in conditions appreciably colder than this is very slow to germinate and grow, and is susceptible to an undue amout of seedling damage from soil fungi such as *Rhizoctonia*. Bird (1975) in Texas has used the ability to produce disease-free seedlings when germinated at low temperature as a criterion for selection in a breeding programme for multi-adversity resistance.

The minimum temperature below which growth does not take place is put at approximately 12 °C by Balls (1953) for *G. barbadense* cotton in Egypt. In estimating growing degree days in Australia, McMahon and Low (1972) used a mean daily temperature of 10 °C as the base temperature, and summed the number of degrees of daily mean temperature above this base for a given time to find the degree days needed to produce a crop of 1500 kg of seed cotton per hectare. The concept of growing degree days is of little value in the tropics, except at high altitudes, as the mean daily temperature rarely falls below the base value, but it provides a useful criterion for the economic production of cotton in higher latitudes. Wallach *et al.* (1978) calculated a 'physiological day' as 14 day degrees above 12 °C.

Both these base temperatures seem to be too low for cotton, and the curves relating growth to temperature are probably a more reliable guide (Fig. 9.3); they show that growth becomes infinitely slow at about 14 °C. In a 25-year study in New Mexico, Malm and Kerby (1981) used a base of 55 °F (12.8 °C), but found practically identical correlations with yield using a base of 59 °F (15 °C).

At the other end of the scale there is a limit to the temperature that cotton can stand. Balls (1953) gives around 37.5 °C as the short-term maximum for growth of *G. barbadense* in Egypt. In India, Sikka and Dastur (1960) state that Asiatic cottons 'given good moisture conditions... can stand temperatures even as high as 43–46 °C'. In Pakistan cotton is normally sown in June; if it is sown earlier, high temperatures in July/August cause shedding of fruiting points in the current varieties because the anthers fail to dehisce. New heat-tolerant varieties are now being tested, which can set their crop during the hot months (Taha *et al.* 1981). Data on *G. hirsutum* (Ehlig and Le Mert 1973) showed a reduction in the number of flowers per metre of row three weeks after periods when the maximum temperature exceeded 42 °C.

The adverse effects of high night temperatures on boll set has been noted in Arizona (Fisher 1973). If this occurs when the first flowers are due to open, the absence of early bolls allows vegetative development further up the plant, resulting in taller plants. Thus a tendency to rank growth is found in areas or seasons where the minimum temperature exceeds 75 °F (24 °C), as in the Sabi Valley in Zimbabwe, the Lower Shire Valley in Malawi, and the Imperial and Death Valleys in the USA. Temperature is, however, only one of several factors which may be responsible for rank growth (see Ch. 4).

Lower temperatures are associated with longer boll maturation periods (see Table 4.1), and this can affect fibre properties and seed quality in various ways. Gipson and his co-workers concentrated on night temperatures, which are a limiting factor in the Texas high plains (Gipson and Joham 1968, Gipson and Ray 1970); using field growth chambers at night, the plants were exposed to ambient temperatures during the day (about 16 hours). They found that lower night temperatures reduced micronaire values, but the relationship was not very close, and the effect on fibre strength was minimal. The effect on fibre length was curvilinear, with the optimum night temperature between 15 and 20 °C (Gipson 1980). Hesketh and Low (1968), working with similar varieties in controlled temperature glasshouses, found that boll production was best under day/night temperature regimes of 33/28 and 30/25 °C; at higher temperatures vegetative growth was stronger, but there was more shedding of fruiting bodies. Among fibre characters the greatest effect was an increase in fibre strength with increased temperature, along with reduced ginning percentage; changes in fibre length and micronaire were less consistent. In both these studies the authors commented on the marked differences in response to temperature between varieties. Morris (1962) in Uganda records only a small drop in lint length of the order of 1 mm on a 30 mm cotton between two seasons, when the average maximum temperatures were 28.3 and 25.9 °C respectively, but it may be significant that the minimum temperatures differed by only 0.15 °C. Frost will cause premature opening of bolls, in which the lint may still be immature.

We have seen in Chapter 4 that the lower mainstem nodes tend to produce vegetative branches, but once the first fruiting branch has developed, all subsequent nodes are likely to produce fruiting branches. The point at which the change-over occurs, the node of the first fruiting branch (NFB), is identified by its mainstem node number, counting up the stem from the cotyledons. Using controlled day and night temperatures, Mauney (1966) showed that both maximum and minimum temperatures affect the NFB: higher night temperatures progressively delay the appearance of the first fruiting branch (Table 9.1). At a night temperature of 25 °C, day temperatures have little effect; but at higher night temperatures, high day temperature raises the NFB, and lowers it when the night temperature is lower. Similar results have been reported in Australia by Low *et al.* (1969).

Light

The traditional measurement of light in agrometeorology has been the number of hours of bright sunlight recorded by the Campbell-Stokes solarimeter. This may be reported directly as hours per day, but in some countries it is related to day length and expressed as a percentage of the possible hours of bright sun. In some parts of the tropics, however, local conditions cause a persistent cloud cover during the day, as on the Pacific coast of South America and in parts of Uganda; this results in low readings of sunshine hours that would cause extensive shedding of fruiting points in other cotton-growing areas, and yet excellent cotton crops are produced. This has led to the use of more discriminating measures of light, the commonest being radiation in calories per cm^2, with a distinction between long- and short-wave radiation. The instruments for measuring radiation usually depend on a photoelectric cell, in which the cell itself, the cell wall or the protective cover limit the cell's response to certain

Table 9.1   *Interaction of day and night temperature on node of floral initiation*

Night temperature[*] *(°C)*	*Day temperature*[†] *(°C)*			
	22	*25*	*28*	*32*
	Node of first floral branch			
32	13.0	13.0	16.6	24.5
28	8.7	9.5	8.8	11.7
25	9.1	8.7	8.7	8.5
22	9.0	8.5	7.6	7.3
20	8.9	8.7	7.3	7.2
Significant difference	($P = 0.05$) 0.9			
	($P = 0.01$) 1.2			

[*] Night temperature 10 hours per day.
[†] Day temperature 14 hours per day.
Source: Mauney 1966.

wavelengths of light, and the question then arises whether these wavebands are the same as those which the plant uses in growth and photosynthesis. Sunlight comprises a wide spectrum of light waves from short-wave ultraviolet light to long-wave heat rays.

As with all green plants, cotton requires light for photosynthesis, but when grown outdoors light rarely if ever limits growth. Even in the temperate zone, the intensity of sunlight at midday is four to five times the requirement for optimum growth, and most plants will grow fairly well and produce seed at levels one-fourth of the optimum (Withrow 1936).

Cloudy days appear to increase the percentage of shedding of young bolls in cotton (Dunlap 1945), the loss occurring two to eight days later. It is difficult to be sure that reduced light intensity is the direct cause, as it is practically impossible in the field to separate it from the effects of other factors such as temperature and moisture. Goodman (1955) has shown at Tokar in the Sudan that shedding was significantly correlated with cloud cover two to four days earlier. In Malawi it was thought that less than four hours of sunshine per day was the cause of shedding of practically all young bolls in the early months of 1957, but in Ecuador four hours per day is quite normal in the growing season and does not affect fruit setting.

Photoperiodicity is a mechanism by which the production of fruiting branches in perennial cotton occurs at the most suitable time of year for flowering and fruiting. In the development of the annual habit, selection is for early fruiting in the first season; behaviour in later seasons is immaterial, and although the photo-periodic response may persist, as it has done among the tropical types of *G. hirsutum*, the establishment of really early annual cottons has involved the loss of the periodicity controls (Hutchinson *et al.* 1947).

The intensity of light reaching the cotton crop may be reduced by shading from trees, tall weeds or even by delayed thinning, and this causes etiolation of the stems and an increase in the size of the leaves, which further reduce the intensity of light reaching the lower leaves. On the other hand, strong sunlight will slow or even stop growth; it appears to act indirectly by increasing the water loss from growing tissues (Balls 1912). In a sunny climate most of the growth occurs at night. At tropical temperatures direct sunlight cancels out the normal effect of high temperature on growth during the day, and prevents excessive vegetative growth; this may explain why minimum or night temperatures are often more closely correlated with growth than maximum temperatures.

Chapter 10

Water

Water requirements

A knowledge of crop water requirements is important for both irrigated and rainfed crops. In irrigation, water conservation is becoming increasingly important as demand is outrunning supply in many existing projects, and when new developments are being planned a crucial question is whether the water supply is sufficient to support the project. Once their pattern of crop water use is known, rainfed crops can be timed to make the best use of the available rainfall; the effect of excesses and deficiencies in rainfall also plays an important part in crop forecasting, as well as being of basic scientific interest.

The water required to produce a high-yielding crop depends on a large number of factors. There is the water transpired through the crop itself, water evaporated from the soil between the plants, water lost by surface run-off, water lost through deep drainage and differences in the amount of water stored in the soil at the start and finish of the growing season; in irrigated crops water is also lost by evaporation and seepage between the supply point and the irrigated site. None of these factors can be calculated with absolute accuracy, and field measurements are subject to a variety of errors.

Nevertheless, a large number of estimates have been published, particularly in connection with irrigation. A distinction can be drawn between the first two factors mentioned above, usually combined in the term evapotranspiration, and the remainder: evapotranspiration depends largely on the evaporative potential of the atmosphere surrounding the crop, while the remainder are either partially controllable or depend on very local conditions. Attention has therefore been concentrated on the measurement of evapotranspiration, and methods of estimating it by correlation with measurable climatic and crop characteristics.

Measurement of evapotranspiration

The changing balance of water in the soil, taken in conjunction with rainfall, provides an estimate of water lost to the atmosphere, provided that soil moisture is not lost by deep drainage out of the rooting zone. Rainfall can be measured with reasonable accuracy using simple and standardized equipment. Soil moisture measurement is more difficult, and various methods have been used: gravimetrically, by tensiometers, by gypsum or nylon resistance units and by neutron probes; each of these has its advantages and disadvantages (see below). In studying soil moisture under annual crops, frequent sampling is necessary, there must be an adequate number of replicates and the soil should not be disturbed excessively by the sampling.

In order to apply these methods, soil characters such as wilting point, field capacity, bulk density and the depth of the rooting zone must be established. To cope with losses by deep drainage, it may be assumed that any water reaching the soil in excess of its field capacity is lost in this way; alternatively, only those periods when a layer of dry soil exists below the root zone may be used in estimating water use.

Deep drainage and water use can be measured directly in a lysimeter. This isolates a block of soil in the field in a metal, concrete or plastic box; water applied to the surface of the soil is measured as well as any seepage from the bottom of the box. A refinement of this is the weighing lysimeter, where the whole box can be weighed. Lysimeters also have their disadvantages, the main one being the difficulty of ensuring that the soil in the box is reasonably similar in profile, texture and compaction to the original soil, so that the crop grown in the lysimeter is typical of the rest of the field.

Physical principles

In planning new irrigation schemes an estimate is often required of the seasonal and maximum demands for water before any crops have been grown in the area. Harry F. Blaney, who had been engaged in such planning since the early 1920s, recognized that evapotranspiration accounted for much of the difference in irrigation practice between sites. He coined the term consumptive use of water (U) for the water consumed in the complete evapotranspiration process by a crop, and postulated that it varied directly with a factor (f) calculated from monthly temperature, per cent of daytime hours in each month, and the number of months in the growing season. He correlated the sum of these monthly factors (F) with existing consumptive use data from lysimeters, and determined a crop co-efficient (K) such that:

$U = KF$

'The co-efficients so developed for different crops are used to transfer consumptive use data from one section to another where only climatological data are available' (Blaney and Criddle 1945).

Thornthwaite and Mather (1955) also estimated potential evapotranspiration from mean monthly temperatures and day length in order to calculate the water balance in the soil. In both cases a crop co-efficient took into account the length of the growing season and the crop. H. L. Penman took this concept further, maintaining that evapotranspiration is a purely physical process, and that potential crop water use (E_t) is determined by the same meteorological factors as those which determine evaporation from an open water surface (E_o). He concluded that E_o and E_t can be calculated precisely from meteorological data (Penman 1948), and that the result was more accurate than direct measurements of open pan evaporation. Direct measurements suffer from a number of disadvantages (Rijks 1976): the pans are seldom exactly similar in design, almost never exposed in a uniform way, suffer from an oasis effect in a dry environment, readings are affected by diurnal variation in the temperature of water in the pan, and rainfall is underestimated because of losses by splash. Penman (1948) proposed a formula using meteorological data for temperature, net radiation or sunshine hours, wind speed and vapour pressure deficit; an account of the principles underlying the Penman formula is given by Monteith (1973), and the calculations together with the necessary tables are given by Doorenbos and Pruitt (1977). Differences in water use between crops are explained by differences in the co-efficient of reflection of light and in turbulence caused by wind passing over an uneven crop surface.

Pattern of crop water use

Much of the work on evapotranspiration has been done on perennial crops and grass, and when applied to an annual crop like cotton allowance must be made for the changes in size of plant and ground cover from sowing to harvest. Changes in the water requirements of the cotton crop (E_t) were studied in Uganda by Hutchinson *et al.* (1958b); soil moisture was recorded by nylon electrical resistance units placed at 1 ft depth, crop development was measured by leaf area index (LAI), evaporation from open water (E_o) was recorded in an open pan evaporimeter, and periods of moisture stress in the plant were identified by measurement of leaf turgidity (Weatherley 1950). The seasonal pattern of crop water use was expressed at each stage of growth as a ratio E_t/E_o, and ranged from 0.4 to 1.8; the starting figure of 0.4 is an average figure for evaporation from bare soil under Uganda conditions at the time of sowing. The maximum ratio was later revised downwards to between 1.2 and 1.4, giving the pattern of crop water use shown in Fig. 10.1 (Farbrother and Munro 1970).

While the magnitude of the quantitative differences...thus derived were criticized (and this criticism was later shown to be largely valid), the main conclusion of the work on the dominant influence of crop development on the seasonal pattern of water use was ultimately accepted, and has become a recognized feature of almost all later work on crop water use on seasonal crops throughout the world. (Rijks 1976).

This later work has further reduced the maximum value of the ratio E_t/E_o, as shown in Fig. 10.1 by the values used by Thorp (1973) and Doorenbos and Pruitt (1977).

The terminology used by different authors in discussing crop water use can be confusing, but the main thing to watch is the reference point. Blaney and Criddle (1945) used a monthly factor (F), calculated from mean temperature and daylight hours; Penman (1948) and Hutchinson *et al.* (1958b) used the evaporation from an open water surface (E, E_o); others have used a reference crop: Doorenbos and Pruitt (1977) define their reference (ET_o) as 'the rate of evapotranspiration from an extensive surface of 8 to 15 cm tall, green grass cover of uniform height, actively growing, completely shading the ground and not short of water'. This is the

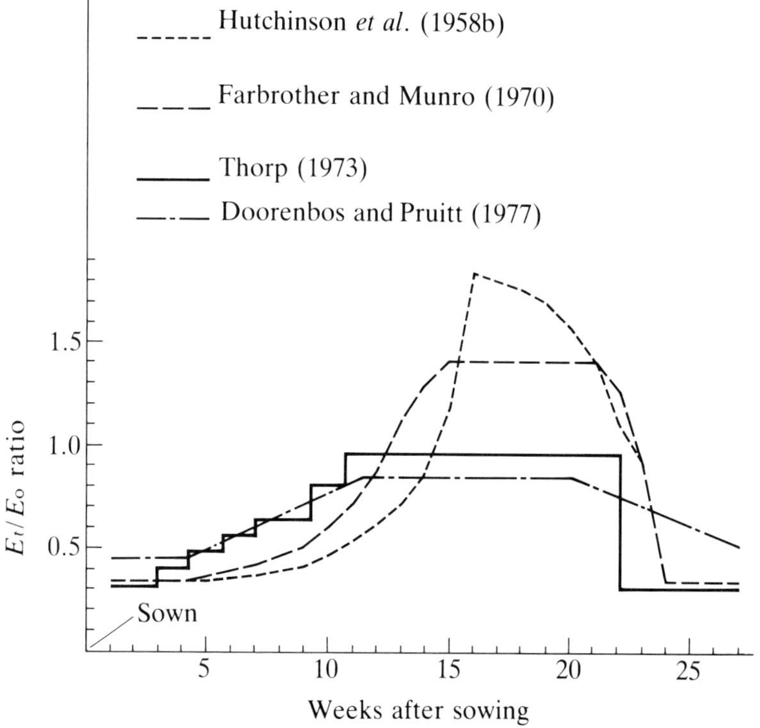

Fig. 10.1   Water requirements of the cotton crop at Namulonge in Uganda.

same as Penman's E_t, except that he specifies 'a short green crop', and this was the reference point used by Thorp (1973).

Water stress

The effects of water stress, a general term which includes the practical effect of periods of drought on the cotton crop, are quite complicated, and the highest yields are not necessarily produced by uninterrupted vegetative growth.

The most easily recognized symptom of water stress is wilting of the leaves; some wilting of raingrown cotton during the hottest part of the day is acceptable, provided the leaves recover before the evening; wilting of irrigated cotton can be used as an indicator of water need. More precise work on turgidity suggests that the water content of the leaves can fall to 83 per cent of the content at full turgidity before growth of the plant is affected. This percentage is called 'relative water content' or 'relative turgidity'. Hearn (1972b) found that the plant behaved as though growth stopped at a certain critical soil water deficit, which increased with age; the relative water content of the leaves associated with this deficit did not change with crop age, and was estimated to be 92 per cent at dawn and 82 per cent at 2 p.m. Growth was freely resumed when the deficit was removed, demonstrating a mechanism for survival during periods of drought.

The stage of growth at which moisture stress has the greatest effect on final yield is the early flowering period (Marani and Horwitz 1963). Periods of stress sufficient to cause bud and boll shedding can be tolerated without affecting yield (Stockton *et al.* 1961). Water stress does not cause shedding directly, but acts through a mechanism common to other physiological processes which allows the abscission layer to develop (Hearn 1980).

Considerable work has been done on the opening and closing of the stomata, and their function in controlling water loss from the plant. Little of practical value has come out of this, and Hearn (1972b) concluded that 'stomata will not be completely closed until all available water (in the root zone) is consumed'.

Water supply

The best cotton yields are generally achieved in a dry desert climate with irrigation, and in the major producing countries this method of production is predominant. Pakistan, Egypt, the Sudan, Peru, Russia and the western United States all depend on large-scale irrigation schemes for the bulk of their cotton crops. On the other hand, most of the smaller producers in South America, Africa and

Asia depend directly on rainfall. Water supply to rainfed crops is usually less reliable, and is certainly less controllable, than that for irrigation, and irrigation schemes have problems of soil salinity and control of the water table (see Ch. 5). Nevertheless, the principles underlying crop water requirements are the same for all, and all are ultimately dependent on rainfall; a better understanding of rainfall occurrence is therefore of value in both rainfed and irrigated cotton production.

Rainfall

The main factor determining the pattern of rainfall is latitude. At the equator rainfall typically occurs all the year round, with two minor peaks in March–May and September–November; in the northern hemisphere these peaks coalesce to form a single peak lasting from May to October, while in the southern hemisphere the pattern moves the other way forming a single peak from November to April (Fig. 10.2).

The amount of rain also tends to diminish from the equator to the subtropics, where desert conditions are found: southwards in Australia, the Kalahari in Africa and the Gran Chaco in Argentina and Paraguay, northwards in Pakistan, the Middle East, North Africa, USSR, China, Texas and Mexico.

Both pattern and quantity of rainfall are affected by other factors – topography, trade winds, and ocean currents. In South America, for example, the cold Humboldt or Peruvian current travelling north brings desert conditions along the Peruvian coast within 5° S of the equator and a one-peak rainfall to the coastal regions of Ecuador. In Indonesia the high central spine of the western leg of Sulawesi combined with the change in direction of the trade winds give a single-peak rainfall, but the peak changes from January in the west to May/June in the east within a few miles. The fact that the only available rainfall station in the Kongwa area lay on the rainy side of a ridge misled the planners of the East African groundnut scheme.

Rainfall records are usually started many years before other meteorological instruments are installed, because of the simplicity and durability of the equipment needed, but the data should be treated with caution where the recorders are comparatively untrained and unsupervised. Fully equipped meteorological stations are usually confined to centres where the standard of supervision and staffing is higher.

Reliability of rainfall

The application of statistical methods has enabled the expectation of rainfall for a locality to be adduced with some confidence. Manning

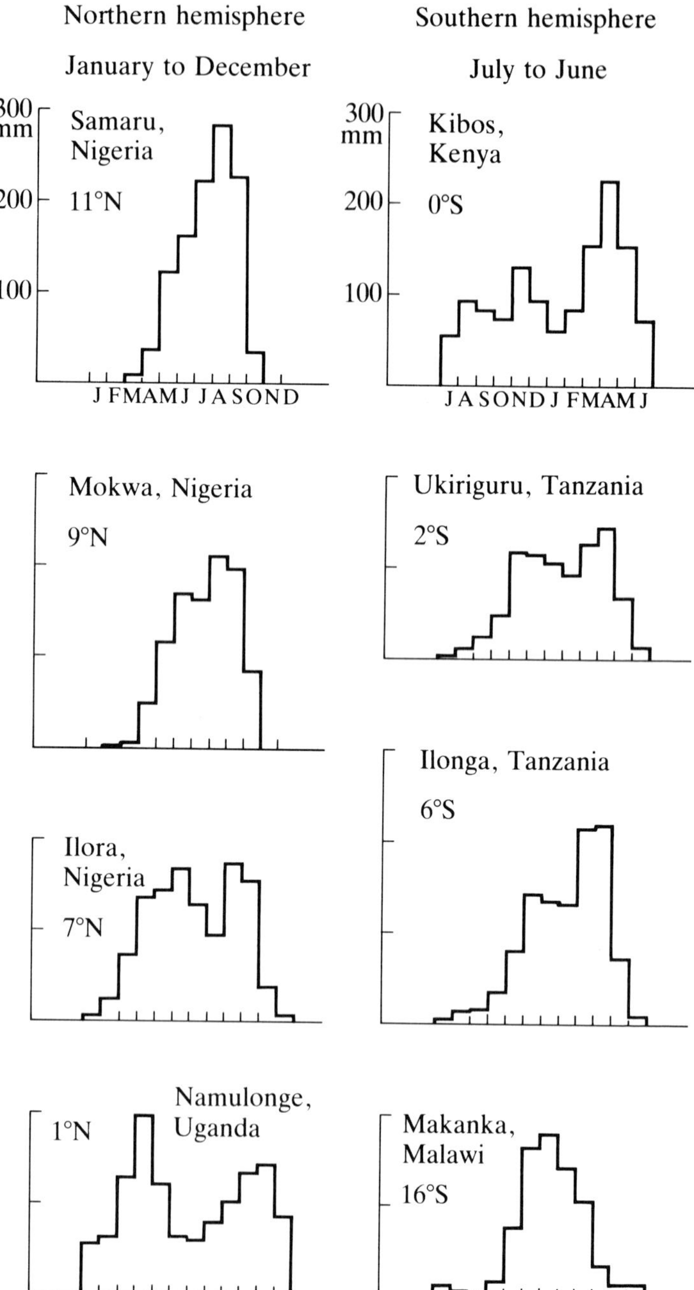

Fig. 10.2   Rainfall distribution and latitude (Cott. Res. Reps.).

(1951, 1956a) made a particular study of rainfall and its bearing on cotton production in Africa, going beyond the traditional yearly and monthly averages to a mathematical expression of the upper and lower limits of rainfall likely to be met with throughout the season. He maintained that 'the value of rainfall to agriculture depends at least as much on its distribution and reliability as on its absolute amount'.

He estimated that the moisture-holding capacity of the soil was sufficient on average to allow the cotton plant to withstand 3 weeks without rain, and chose a 3-week period as the basis of his calculations. In order to obtain a smoother curve he used units of 1 week (later changed to 10 days) and calculated 3-week (20-day) moving totals; for each period he calculated the standard error, and the 1 : 1 confidence limits, such that the rainfall in two years out of four is expected to lie between the upper and lower limits, one year in four above the upper limit, and one in four below the lower limit. Typical rainfall charts from Nigeria are shown in Fig. 10.3, where

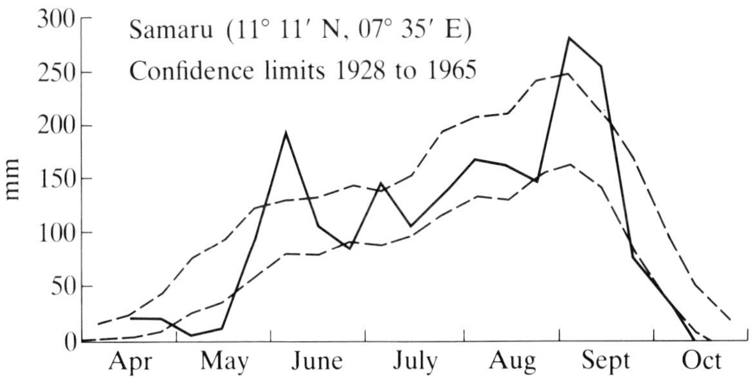

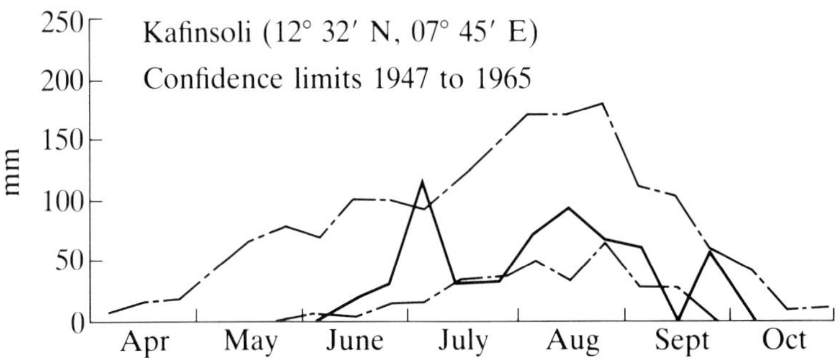

Fig. 10.3 Confidence limits (1:1) of rainfall (broken lines), and actual rainfall in 1973 (solid lines) (Cott. Res. Rep., Nigeria, 1973–74).

the upper and lower limits are shown by broken lines; these provide a reference by which the actual rainfall for the season can be judged. The variability of the rainfall is measured by the distance between the two broken lines: Kafinsoli, less than 300 miles north of Samaru, has a lower rainfall but a wider spread. Its expected rainfall in 20 days only reaches 50 mm in three years out of four, while at Samaru the expectation in August/September is 150 mm.

Manning reckoned that 10 years' data were the minimum required to establish confidence limits, while a period exceeding 20 years was desirable. While the longer period will establish the limits more precisely, it can do nothing to reduce the variability. Computer programmes are available for calculating the confidence limits, and are especially useful when the rainfall data are stored in a form which can be read by the computer (Walker and Rijks 1967, Thorp 1975).

By superimposing the water requirements of the crop on the rainfall chart (Fig. 10.4), various sowing dates can be tested for their suitability in providing adequate water for the growth of the crop. It will also show whether or not the rainfall will be adequate to produce a profitable crop in three years out of four. The apparent water deficit in Uganda (Fig. 10.4) is made up by soil water stored during the first rains. Evapotranspiration and the crop factor are much more stable from year to year than is rainfall, and for this exercise small variations in crop water requirements can be ignored.

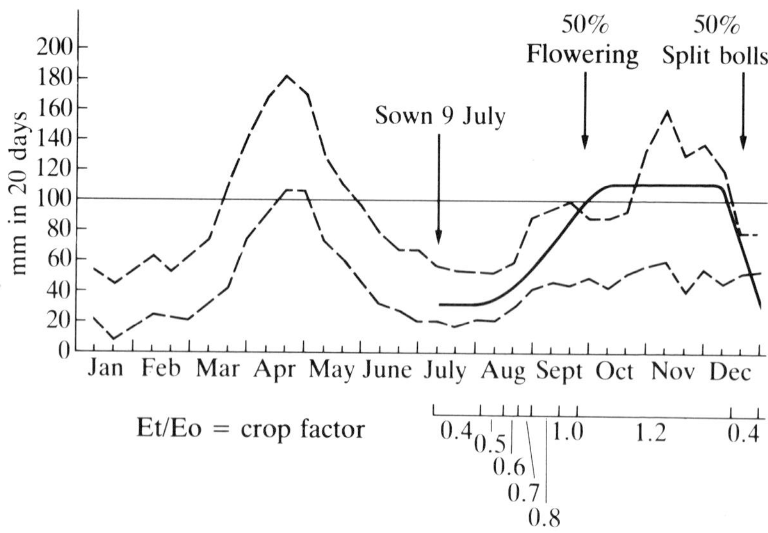

Fig. 10.4 Confidence limits (1:1) of expected rainfall and crop water requirements of cotton at Namulonge, Uganda (adapted from Thorp 1973).

The optimum sowing date depends, of course, on other factors as well, but one of the most important – dry weather at harvest – can also be estimated from the appropriate period on these charts.

Methods similar to those of Manning have been used by a number of agrometeorologists. Manning noted that the rainfall records for any particular period tended to have a skew distribution, and corrected for this by using $\log(x + c)$ in his calculations, where x is the measured rainfall and c a constant; others do not think that this correction is necessary (Glover and Robinson 1953). The number of rainy days in the month has also been used to derive confidence limits, as shown in Fig. 10.5 from Indonesia; the calculations for this are much less tedious.

A slightly different version of the problem of when to sow is met in countries where the rains may be expected to break suddenly at any time within a period of a few weeks; this seasonal event can be mapped to show the probability of the arrival of the main rains on any given date (Lineham 1965).

Rainfall intensity

The type of rainfall, i.e. intense or gentle, is also important to the agriculturist; the 'tropical storm' is no myth. The research station at Henderson in Zimbabwe has an annual average of about nine falls of over 25 mm (1 in.) in a day, and three falls over 50 mm; at

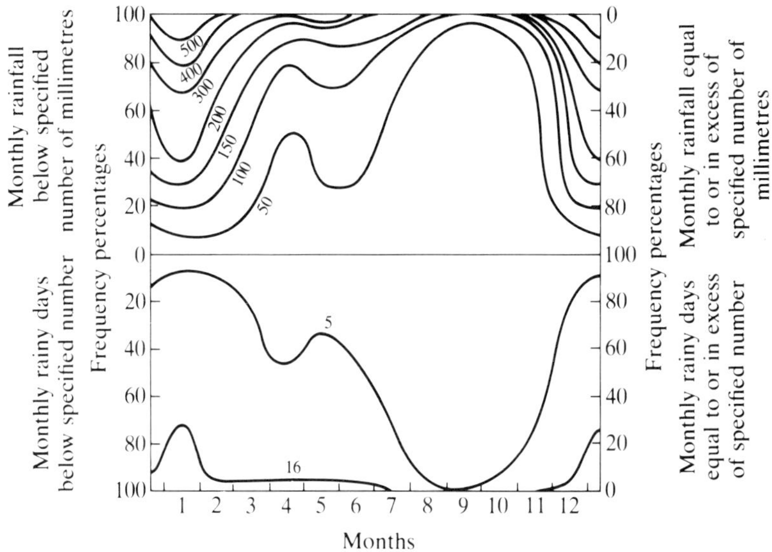

Fig. 10.5 Expectation of rainfall and rainy days; Frequency of specified levels in any month at Jeneponto, South Sulawesi, over 28 years (Indonesia, 1979).

Shinyanga in Tanzania, on one occasion over a quarter of the season's total of 635 mm fell in two heavy storms; such examples have been equalled in other tropical countries.

The battering effect of such storms on arable land is sufficient to break down the structure of the surface soil and increase run-off and soil erosion. An important effect of mulching of bare soil is to protect the surface from the force of the rain, and hence to increase the rainfall acceptance by the soil; Hutchinson *et al.* (1958a) have shown that a planted grass cover (*Cynodon*) or 8 cm of grass mulch improved rainfall acceptance by three to five times during heavy storms compared with bare soil in Uganda. Weeds or cover crops have the same effect. Farbrother considers that this intensity of rainfall marks perhaps the greatest difference between tropical and temperate agriculture, transcending in importance other physical differences of soil, temperature or amount of rainfall.

A storm can be characterized in terms of amount, rate and duration (Hutchinson *et al.* 1958a). Gauges for recording the rate of fall of a storm are usually elaborate and costly, but a simple home-made instrument, particularly suited for tropical storms, has been devised and described by Farbrother (1951). Heavy storms can reach a very great intensity; the peak rate may last for only a minute or two, but during that time it may exceed the rate shown in Fig. 10.6, 10 in. (254 mm) per hour recorded in a storm at Namulonge, Uganda.

Soil water

Available soil water can be defined as the quantity of water stored in the root zone between field capacity and wilting point (Fig. 10.7).

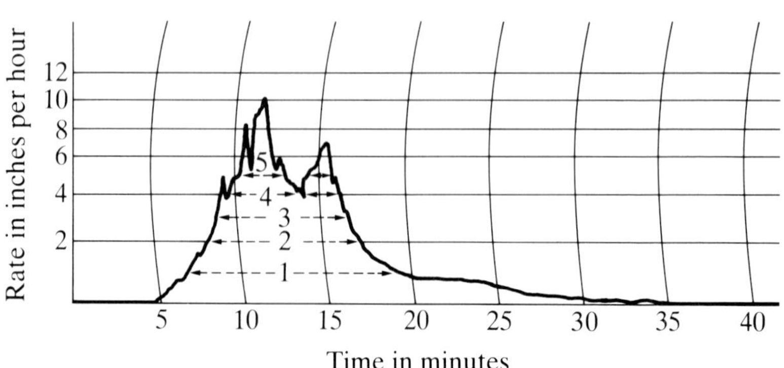

Fig. 10.6 Farbrother rainfall-intensity record chart for a typical storm (Hutchinson *et al.* 1958a).

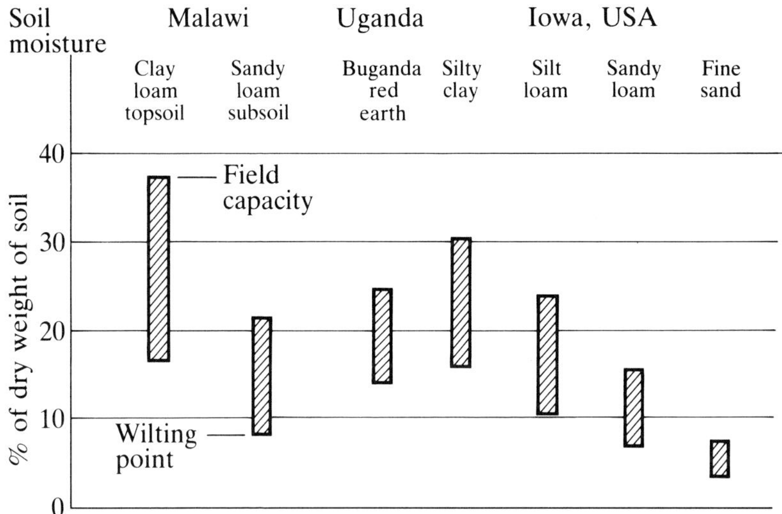

Fig. 10.7 Soil water between field capacity and wilting point in different types of soil: Malawi (Munro and Wood 1964), Uganda (Hearn 1970), Iowa (Russell 1939).

It provides a buffer by which the plant can keep growing during periods between additions of water, either rainfall or irrigation, and is therefore complementary to the rainfall pattern in determining suitable growing conditions; different crops can use different fractions of this water without experiencing undue water stress. The availability of water to the plant depends on the tension (pF) at which it is held in the soil particles and the suction pressure in the plant roots, neither of which is easy to measure in the field.

Field capacity (pF about 2.7) is measured by wetting a column of soil to excess and allowing it to drain under gravity until the soil moisture reaches a stable level; one problem is to decide when this point is reached, as slow drainage may continue for a long time. The height of the column may also affect the result. Wilting point (pF about 4.2) is determined by measuring the soil moisture when an indicator crop reaches permanent wilting point; this depends to some extent on the crop used as an indicator, and on how permanent wilting is defined.

Gravimetric soil moisture determinations can be made quite accurately, but their use on a field scale is limited by the time required and also the disturbance of the soil which frequent sampling entails. Quicker and cheaper methods have therefore been sought, and the most popular of these measure electrical conductivity (resistance units), neutron scattering (neutron probe) and moisture tension (tensiometers), all of which are functions of soil moisture, and hence can be used to measure changes in available moisture and for the control of irrigation.

Resistance units consist of two electrodes held apart at a standard distance, either by embedding them in gypsum or by separating them with a standard thickness of nylon fabric. When the units are buried in the soil, their moisture content maintains equilibrium with that of the surrounding soil, and the resistance between them is a function of the moisture content of the soil in which they are located: the drier the soil the higher the resistance to an electric current. Leads from the electrodes are brought to the soil surface and tagged, and a portable resistance meter is clipped to the leads when readings are being made. The meter incorporates a small dry battery to provide the necessary current.

Within reasonable limits, the measured resistance at field capacity and at wilting point is independent of the soil type and the total moisture in the soil, so that the meter can be calibrated in percentage of available moisture – 100 per cent for field capacity and 0 per cent for wilting point. These resistance units have proved of great practical value both in research and in the control of irrigation, although they have been criticized on various grounds. They are affected by the mineral content of the soil water, and the calibration of intermediate percentages is affected by hysteresis, the wetting and drying cycles being slightly different. Nylon units, in which the electrodes are separated by a nylon mesh, are less affected by soil and fertilizer minerals, and are more sensitive at the wet end of the scale than gypsum blocks (Farbrother and Harrison 1957). The placing of the resistance units must ensure good contact with the soil, and the soil filling the holes in which the units are inserted must be compacted to the same extent as before, as improved aeration can attract an excess of plant roots; filling the holes with a soil slurry is probably the best method of achieving this. The great advantage of resistance units is that they are relatively cheap; they can be installed in large numbers at different depths, and readings can be taken at frequent intervals without any disturbance of the soil.

When moist soil is bombarded with neutrons, they are deflected by collision with water molecules but pass through the solid particles of soil. This deflection causes a number of the neutrons to return to the source, where they can be measured, and the percentage reflected in this way provides a measure of the moisture in the surrounding soil. In practice, holes are drilled in the soil and lined with a tube of material which allows the neutrons to pass freely, and a neutron probe is lowered to the required depth in the tube. The probe is then activated, and a receptor incorporated in the probe receives the reflected neutrons showing a reading on the dial in the top of the probe.

A large number of sampling tubes can be inserted cheaply, and there is no problem of refilling the hole, but the probe is expensive and requires a substantial power source. The reliability of the equipment so far leaves something to be desired, and this can be important where servicing facilities are not readily available. The neutron probe measures total soil moisture, and a knowledge of the field capacity and wilting point are necessary to convert this to available water.

A tensiometer consists of a porous pot linked by a tube to a pressure gauge, and the system is filled with a standard saline solution. The pot is buried in the soil and in close contact with it; when the soil is dry the salt concentration of the soil solution increases, and water is drawn out of the porous pot by osmosis; as the soil is wetted water is drawn in. Each unit requires its own gauge, and the units are relatively expensive; the placing of the units requires the same care as do resistance units. There is also the danger that a prolonged drought may allow air to enter the system, which then has to be dismantled and recharged.

Control of irrigation

Soil characteristics

The principles of water use apply equally to raingrown and irrigated cotton. A well-designed irrigation system will enable the farmer to maintain the water available to the crop between field capacity and a level below which the crop would experience a check in growth. To do this efficiently he has to know the characteristics of his soil, and what are its limits of available water (Fig. 10.7). Average figures for the total available water at field capacity (soil water tension 0.2 atmospheres, pF = 2.6) in different types of soil are as follows:

Water per metre of soil depth	
Fine-textured soils	200 mm
Medium-textured soils	140 mm
Coarse-textured soils	60 mm

The fraction of these totals available to cotton under average conditions, with evapotranspiration running at 5 to 6 mm per day, is 65 per cent. This figure may have to be corrected for higher or lower levels of evapotranspiration, and adjusted for rooting depth and application efficiency (Doorenbos and Pruitt 1977), to give the depth of irrigation water required. With furrow irrigation the minimum practicable application is about 50 mm of water; smaller amounts can be applied by overhead or trickle irrigation, but there is a limit to the frequency of changes in the location of the supply pipes. A pre-sowing irrigation is normally applied to bring the soil in the crop rooting zone up to field capacity.

When to irrigate

Visible symptoms of water stress can be used for timing the application of irrigation water, of which wilting of the leaves is the most obvious. Roe (1950) states that the surest and safest indicator for Pima cotton is wilting by noon with recovery by evening. For Upland cotton Marr and Hemphill (1928) give two other generally accepted indicators: the darker bluish tinge of the foliage in the dry

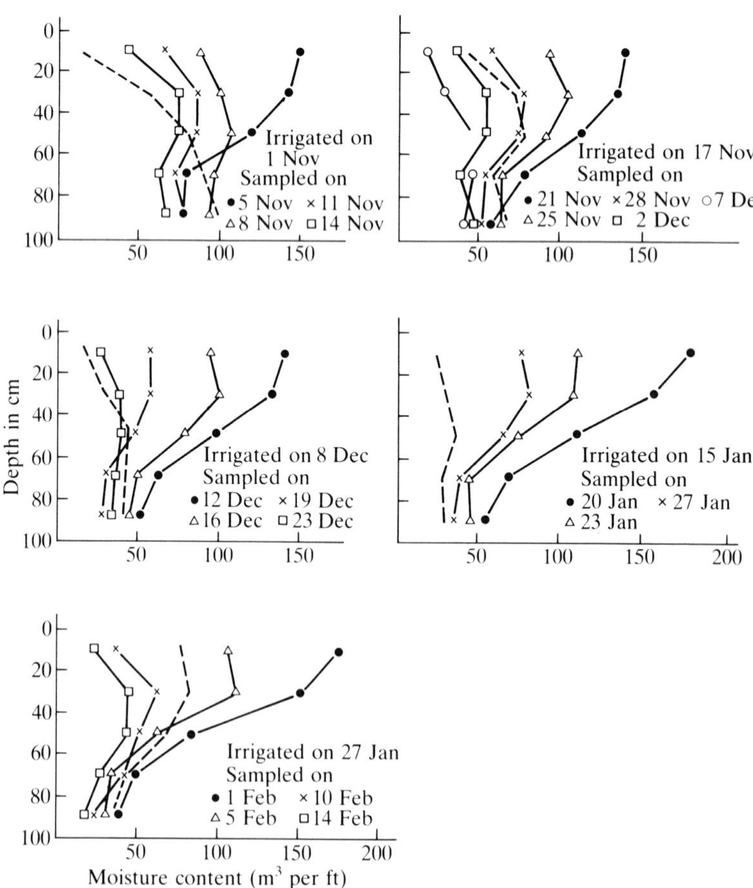

Fig. 10.8   Moisture content of soil in the Sudan Gezira (m³ per ft) down to 1 m for 5 irrigation cycles (October to February); broken lines represent positions immediately before irrigation (Farbrother 1969).

spots of the field, and the rapid expansion of red colour from the main mature stalk into the light green upper part of the stem just below the terminal bud. Some people maintain that when the number of flowers visible in the crop canopy reaches a certain level it indicates the need for irrigation.

Direct measurements of soil moisture for irrigation control are taken in the upper zones of the soil profile where depletion is greatest (Fig. 10.8) at 30 or 40 mm below the soil surface. With gypsum or nylon blocks the percentage of available water can be read directly on a suitably calibrated resistance meter, a level of about 30 per cent indicating the need for irrigation. The soil moisture percentage at this level has to be calculated from the soil characteristics when using a neutron probe, while the reading on the tensiometer will depend on the calibration units – approximately pF 4, 10 atmospheres or 150 lb per in.^{2}.

When meteorological data are used for irrigation control, a balance sheet is maintained to show the depletion of soil water (Fig. 10.9). Starting with the soil at field capacity, the balance sheet is entered daily with rainfall and irrigation on one side and evapotranspiration on the other; the daily balance will show the available water or the soil water deficit, and any surplus over field capacity is usually written off as deep drainage. The irrigation water applied

SOIL WATER BALANCE SHEET

Scheme: *Ralin Ku*   Soil type: *Silty loam*   Total available soil water *25* v%   0–30 cm

Field: *14*   *25* v%   30–60 cm

Farmer: *Lokoto*   *25* v%   60–90 cm

Months: *April – Sept.*   v%   90–120 cm

Pan location: *100 m upwind cropped field*   Crop: *potatoes*. Rooting depth *60* cm

Irrigate when balance is: *115* mm

Date	Days after planting	Epan mm	Wind	Humidity	kpan	kcrop	ETcrop mm	Rain mm	Irr. mm	Balance mm	Remarks
27/4	0	6.3	light	med	.8	.9	4.5	—	80	150	pre-irr
28/4	1	7.2	mod	low	.65	.7	3.5	—		146.5	
29/4	2	6.9	mod	med	.75	.5	2.5	4		148	
											Weeding
											fullcover
										128.5	first flower
9/7	73	8.7	light	low	.7	1.1	6.5			122	
10/7	74	9.8	mod	low	.65	1.1	7.0			115	
									45	160	Irrigation
										142	
30/8	95	11.7	mod	med	.75	.8	7.0			135	
31/8	96	10.6	light	med	.8	.8	7.0	16		144	

Fig. 10.9 Example of soil water balance sheet (Doorenbos and Pruitt 1977).

must be measured with reasonable accuracy, while evapotranspiration is estimated by applying the appropriate crop factor to the evaporation (E_o or E_t) obtained by one of the methods discussed earlier – open pan evaporation or a figure calculated from meteorological data. Irrigation is applied when a predetermined deficit has been reached; this is usually when about 65 per cent of the total available water has been withdrawn from the rooting zone, and the irrigation water applied should make up this deficit, bringing the soil moisture back to field capacity.

Chapter 11

Insect pests

Cotton without insecticides

For nearly a hundred years after the American Civil War (1861–65) the expansion of cotton production was a matter of trial and error. To reduce dependence on the United States, cotton was introduced into many areas where it had not been grown before, but where the climate appeared to be suitable. If the insect pests were at a low enough level to permit a profitable crop, the cotton grew and production expanded; if the insect pests were too damaging, the crop failed and production ceased. Insect pests were second only to climate in determining where cotton could be grown successfully.

At that time the only insecticides available were inorganic chemicals such as sulphur, calcium arsenate, various mineral oils and fumigants, and substances of plant origin like derris, nicotine and pyrethrum; they were not effective enough to pay for their employment on most field crops (Maxwell Darling 1958). Before the Second World War yields of about 100 kg of lint per ha were considered economic in peasant agriculture in India and Africa, and 270 kg per ha was the average yield (1937–39) in the USA; the Sudan and Peru averaged over 300 kg per ha while Egypt had the highest yield at about 550 kg per ha (Table 11.1).

The insects did not have it all their own way, however, and various measures were introduced to reduce pest damage. Probably the best known of all cotton pests is the boll weevil (*Anthonomus grandis*), which spread from Mexico to the cotton belt of the USA; the first record of its entry was in Texas about 1892. It was brought under partial control by a range of measures including the change to early cropping varieties, resulting in a reduction in the number of over-wintering weevils; it caused a shift of the centre of the cotton-growing industry from the deep south to the west of the country (Moss 1914). In Nigeria earlier sowing was recommended to control red bollworm (*Diparopsis watersi*), although the emphasis on food production prevented its adoption except on research stations

Table 11.1   *Changes in yield of cotton lint in certain countries over 40 years*

Country	Average 1937–38 to 1939–40	Average 1977–78 to 1979–80
	kg per ha	
United States	270	599
British West Indies	130	268
Brazil	175	266
Peru	435	649
Argentina	170	292
Egypt	556	836
Sudan	307	340
India	100	164
Pakistan		304

Sources: CEC 1948, ICAC 1983.

(Geering and Baillie 1954). A legally enforced close season was established in most countries in Africa to prevent the carry over of pests like red and pink bollworm (*Pectinophora gossypiella*) which have few or no alternative host plants. Collection by hand of the egg masses of the cotton leafworm (*Spodoptera littoralis*) in Egypt was enforced by law (Brown 1953). Trap crops and poison baits were suggested and tried, but were never accepted as routine controls. Rainey (1948) used paper bags to protect his material for research on boll development, bagging the dated flowers two days after anthesis to avoid insect attack. Finally, Parnell *et al.* (1949) in South Africa showed that jassid resistance was related to the hairiness of the cotton leaf, and his ideas and resistant material spread all over Africa during and after the Second World War.

Chemical control of insects

The discovery of DDT and BHC, synthetic organic insecticides with strong contact action and long persistence, revolutionized insect control in cotton as well as in other fields. DDT was tested against the more important cotton insects in the USA in 1944 (Ivy 1944), and it was not long before it was being used on a commercial scale. It has been followed by a large and increasing number of organic compounds; Green *et al.* (1979) state that there are 800 active ingredients registered in the USA, and 80,000 different trade names. Besides the spectacular increase in yield achieved in the established cotton-growing areas, the new insecticides have enabled cotton production to expand into areas hitherto considered uneconomic because of pest incidence.

The dangers inherent in the intensive use of powerful insecticides were soon recognized: the destruction of beneficial parasites and predators and the development of strains of the pest species resistant to the insecticide. The initial response to these developments, however, was to intensify the use of insecticides, adding new chemicals to control species reaching pest proportions because of the destruction of natural controls, and increasing the dosage and frequency of application to cope with resistance developing in the pest species.

The use of insecticides is dependent on working capital, always a problem with small-holders. In order to stimulate cotton production, with an associated improvement in the standard of living, governments provided subsidies and credits to small-scale farmers in the tropics so that they could buy the recommended insecticides.

Methods of pest management, which include both chemical and biological methods of control, are not generally specific to any one pest, and are discussed in the next chapter. The life history and behaviour of individual species, however, can have a profound significance in devising cultural methods of control, and these are discussed in this chapter in relation to each pest.

One of the first essentials in pest control is insect recognition, because a field of cotton usually supports a very diverse insect fauna, some of which do little or no damage, some are positively beneficial as parasites and predators, and only a few are serious pests (Table 11.2); some understanding of these insects and their way of life is necessary for a better understanding of the principles of insect pest control.

Much of the information on individual pests which follows is taken from *The Insect Pests of Cotton in Tropical Africa* by E. O. Pearson (1958), which is still a standard work of reference on the subject but is now out of print. His field key to the principal disorders of cotton is reproduced with permission in Appendix A. McKinley (1968) provides a key to the lepidopterous larvae which attack cotton in central Africa. A table showing the importance of each pest in various countries is given at the end of this chapter (Table 11.3).

Coleoptera

Anthonomus grandis **Boh.**   **Cotton boll weevil**

Description

The section Rhynchophora, or snout beetles, includes more than 25,000 species; the snout takes various forms, that of the weevils carrying two antennae about half way along the snout (Plate 20). The adult boll weevils vary in size, averaging about 6 mm in length; when they first emerge from the pupa they are light brown, but become grey or almost black within a few days. The snout or

Table 11.2   *Cotton pests*

Order	Major pests	Minor pests
Coleoptera	*Anthonomus*	*Podagrica*
		Syagrus
		Alcidodes
		Sphenoptera
Isoptera		*Hodotermes*
Lepidoptera		
Bollworms	*Heliothis*	*Sacadodes*
	Earias	*Cryptophlebia*
	Pectinophora	
	Diparopsis	
Leaf eaters		*Spodoptera (Prodenia)*
		Spodoptera (Laphygma)
		Alabama
		Sylepta
		Trichoplusia
Leaf miner		*Bucculatrix*
Cutworms		*Agrotis*
		Spodoptera (Laphygma)
Hemiptera	*Dysdercus*	*Helopeltis*
	Empoasca	*Oxycarenus*
	Bemisia	*Calidea*
	Lygus	*Nezara*
	Aphis	*Paurocephala*
Thysanoptera		Thrips
Orthoptera		Locusts
		Grasshoppers
Acarina	*Tetranychus*	*Hemitarsonemus*

proboscis curves downwards, with the mouth parts at the end. The boll weevil is difficult to distinguish morphologically from other species, but if a damaged square is kept for a week or two in a jar or envelope, any weevil which develops is certain to be the boll weevil.

The adult over-winters in what shelter it can find from the cold, hiding under rubbish, in cracks in the ground, in barns, hedges or thickets and in stored cotton seed. Survival may vary from a fraction of 1 per cent to as high as 40 per cent. On emerging in the spring, it may feed on young leaves and terminal buds of cotton until the first squares appear. Both male and female weevils feed on the young squares, and the female lays her eggs in the young bud. Later eggs may be laid in the walls of bolls up to three-quarters of full size, and the female avoids if possible squares or bolls in which an egg has been laid already.

The eggs generally hatch in about three days, producing a larva which is white with a brown head, curved and legless; infested

Pl. 1 Bacterial blight – stem lesions. (D. L. Ebbels).

Pl. 2 Bacterial blight – vein lesions. (J. M. Munro).

Pl. 3 Red leaf. (J. M. Munro).

Pl. 4 Okra leaf. (J. M. Munro).

Pl. 5 Jassid damage (*Empoasca*). (J.M. Munro).

Adult stainer (*Dysdercus fasciatus*). **Pl. 6** (ICI, Plant Protection Division).

Pl. 7 Stainer nymph (*Dysdercus* sp.). (ICI, Plant Protection Division).

Pl. 8 Flared square (*Heliothis*). (ICI, Plant Protection Division).

Pl. 9 *Heliothis* larva. (ICI, Plant Protection Division).

Pl. 10 *Diaparopsis* larva. (ICI, Plant Protection Division).

Pl. 11 *Earias* larva. (ICI, Plant Protection Division).

Pl. 12 *Pectinophora* larva. (ICI, Plant Protection Division).

Pl. 13 Tip-borer damage (*Earias*). (ICI, Plant Protection Division).

Pl. 14 Aphis. (ICI, Plant Protection Division).

Pl. 15 Whitefly (*Bemisia*). (ICI, Plant Protection Division).

Pl. 16 Red spider mite damage (*Tetranychus* sp.) (ICI, Plant Protection Division).

Pl. 17   Leaf-eating larvae (*Alabama*). (ICI, Plant Protection Division).

Pl. 18   Leaf miner damage (*Bucculatrix*). (ICI, Plant Protection Division).

Pl. 19 Boron deficiency; leaf petiole lesions. (J. M. Munro).

Pl. 20 Boll weevil adult (*Anthonomus*). (ICI, Plant Protection Division).

squares flare open and either wilt or shed. The larva feeds inside the square or boll, going through several moults, and pupates seven to ten days after hatching. The adult weevil gnaws its way into the open five or six days later, and starts feeding on tender young squares. The average time from egg to adult is two to three weeks, but each stage is materially influenced by temperature and humidity; the life cycle may be completed in 14 days in hot moist weather, but last as long as seven weeks when the weather is cold and dry. In the USA the adults hibernate during the winter; feeding is restricted to the tetraploid cottons, and in the tropics the weevil may survive on stand-over cotton or on native perennial varieties.

Distribution

The boll weevil was first recorded in Mexico in 1843, and is probably native to Central America. It was found in Cuba in 1871, and entered Texas about 1892; since then it has spread to all parts of the USA cotton belt. The Arizona boll weevil, *A. grandis* var. *thurberiae* Pierce, is regarded as a distinct variety. From Mexico the boll weevil has spread southwards to Venezuela and Colombia, but another species, *A. vestitus* Boh., is found in Ecuador and Peru; this has a similar life history and habits, the adult being rather smaller than that of *A. grandis*. The boll weevil has not been found in other parts of South America, nor in any cotton-growing country outside the Americas.

Symptoms

Unless the buds are examined individually by opening the bracteoles, the first signs of damage are flaring, wilting or shedding of the young squares. The same symptoms are produced by bollworms, so the bud must be examined to confirm the presence of boll weevils. Feeding punctures may be found in any part of the square, and are left open, but egg laying nearly always takes place about half way between the tip and the base of the bud, and the female seals the hole with a yellowish wax to protect the egg. Once the egg has hatched, the larva can be found by cutting open the bud or boll.

Control

The following control methods, supplemented by the use of calcium arsenate dust, were recommended in 1942: (a) destruction of all cotton refuse after harvest by burning or burying; (b) fall ploughing to remove sites for hibernation; (c) rotation of cotton with other crops; (d) early planting and early maturing varieties to hasten crop production; (e) in the 'Florida Method' all the buds or squares are removed in the first week in June to destroy the first generation of eggs; this does not seem to reduce the crop (Fernald and Shepard, 1942).

After a period of increasing dependence on synthetic insecticides, current programmes of integrated pest management and optimum pest management (see Ch. 12) emphasize once again the importance of cultural practices, and are now dealing with communities rather than individual farmers. Complete eradication of the boll weevil is being attempted in parts of North Carolina and Virginia.

Cotton stem borers

Stem borers attack the mature stems of cotton, usually near ground level. Eggs are laid in the bark of the stem, and the larvae tunnel into the wood; besides weakening the stem, they allow access to wood rotting fungi. A heavy attack interferes with the translocation of water and nutrients from the roots, and the plant wilts or dies; the weakened stem may break under the weight of the crop.

Sphenoptera gossypii Cotes is found in India and Pakistan, and occasionally in the Sudan, while *S. neglecta* (Klug.) is widely distributed in Africa. Other genera are found in South America – *Eutinobothrus (Gasterocercodes) gossypii* Pierce in Peru and Ecuador, and *E. braziliensis* Hamb. in Brazil, Argentina and Paraguay.

Uprooting and burning of cotton residues after harvest usually provides effective control; the uprooting must be complete, as the larvae may pupate in the stem below the weakened area, and be left in the stump if the stem breaks.

Stem-girdling beetles

The adult gnaws away the bark and outer wood of the stem at varying heights above ground level, leaving a ring of frayed fibres in which the egg is laid; if the ring barking is complete, the top of the plant dies and may break off at the girdle. There may be no trace of larval feeding, and identification of the pest is then difficult.

Alcidodes brevirostris (Boh.), the stem-girdling weevil, is found in East Africa from Tanzania to the Cape Province, while *A. gossypii* (Hust.) occurs in West Africa. Similar damage is caused by another genus, slender beetles with long antennae, the body yellow with black markings: *Tragiscoschema bertolonii* Thoms. in Malawi and Mozambique, and *T. nigroscriptum* Fairm. in eastern Tanzania.

Leaf-eating beetles

Various beetles feed on the leaves and bracteoles of cotton, sometimes nibbling the surface of green bolls and allowing the entry of disease organisms. The most important genus in Africa is the *Syagrus* beetle, with three species forming an overlapping series: *S. calcaratus* (F) extends from West Africa to Uganda, *S. morio* Har occupies East Africa and *S. rugifrons* Baly occurs in southern Africa.

These beetles can cause serious losses at the seedling stage of the cotton crop, while the larvae are root feeders which can cause wilting and even death of the plant.

Flea beetles

These are leaf-eating beetles which jump like fleas when disturbed. The eggs are laid in the soil, and the larvae probably feed on roots. A number of species attack cotton, and two species of *Podagrica* are important pests of cotton in the Sudan; they both occur in the main cotton-growing areas, but *P. puncticollis* Weise extends to the south and *P. pallida* (Jacoby) to the west of the Sudan. *P. weisei* Jac. sometimes attacks cotton seedlings in Kenya, *Longitarsus gossypii* Bryant in the Sudan and *Zomba gossypii* Bryant in Malawi.

Isoptera

Mound-building termites

Mound-building termites do not often attack healthy living plants, although there have been local reports of serious damage in Tanzania and Kenya, but the construction of the mounds has a considerable influence on soil fertility. The mounds are constructed with soil brought up from depth and cemented together with saliva; if this soil comes from a zone where leached material has accumulated, it will have a high colloid content and contain plant nutrients (Hesse 1955). This is most likely in regions with a dry climate and a light soil, whereas in the humid equatorial regions, the mounds are deficient in nutrients and plant growth is depressed (Griffith 1938). Rainfall tends to run off the mounds, and the tunnels below them facilitate drainage, so they dry out quicker than the surrounding soil. All this affects the vegetation, forming striking circular patterns when seen from the air.

Harvester termites

These termites form extensive underground nests, and emerge in full daylight to forage in the open for green material; they clip this off growing plants, and carry the pieces underground to the nests. *Hodotermes mossambicus* (Hagen) has been recorded as attacking cotton in the drier parts of Tanzania and in Malawi, where damage to young cotton can be severe in the Shire Valley.

Lepidoptera

Bollworms

Heliothis **spp. (syn.** *Helicoverpa***)**   **American bollworm, tobacco budworm**

Description

The larvae pass through six instars, occasionally seven, reaching a length of about 40 mm before pupation. The first two instars are without prominent markings, but the head and other parts are dark brown to black, giving the larvae a spotted appearance. In later instars these black parts fade to brown, and a colour pattern of longitudinal bands develops (Plate 9). The colour is extremely variable, and may be predominantly green, straw-yellow, brown or even black. The mature larvae drop or crawl to the ground and pupate in the soil at a depth of 25–175 mm.

The adult is a stout-bodied moth, broad across the thorax and thence tapering; the body is about 18 mm long, with a wing span of 40 mm; the general colour varies from dull yellow or olive-grey to brown, sometimes with a pinkish suffusion. They are most active between sundown and dark. On cotton, egg laying is closely associated with the period of bud and flower production; the eggs are found mostly on the squares and upper surfaces of the leaves, but may be scattered about the plant. The eggs are 0.4 to 0.5 mm in diameter, at first yellowish-white and glistening, but changing to dark brown before hatching.

The eggs hatch in about 3 days in hot weather, but may take 9 days in cold weather (17 °C mean temp.) and extremes of 14 and 17 days have been recorded. The temperature limits for development in experiments in the USSR were 14 and 38 °C. Temperature affects the development of the larvae as it does that of the egg, the length of life ranging from about 15 to 60 days.

The pupal period is about 16 days; in cooler weather the period is extended, and Valentine (1954) found that the average period in eastern Tanzania varied from 12 days in January (minimum temp. 21 °C) to 25 days in July (minimum temp. 16 °C). Colder weather also induces the formation of diapause pupae, whose emergence depends not on the date of formation of the pupae, but on rising temperatures and the onset of rain. Larvae pupating in April and May in South Africa showed 60–80 per cent of individuals with diapause (Pearson 1958), while at Ukiriguru in western Tanzania Reed (1965) found that the percentage only reached a maximum of 2.6 per cent of the pupae formed in May. No diapause occurs near the equator in Uganda, for instance, and Valentine (1954, 1955) suggests that the minimum temperature must fall below 60 °F (15 °C) before diapause pupae are formed.

The length of life of the adult moth is greatly affected by the availability of food, in the form of nectar or its equivalent; in the absence of food the ovaries do not develop and the moth dies within 3–6 days, and this explains why moth activity is largely confined to the flowering phase of crops. The female moth is the longer lived, averaging about 15 days compared with about 9 days for the male.

Distribution

Heliothis armigera (Hb) occurs in southern Europe, probably the whole of Africa, the Near and Middle East, India, Asia, from Japan to eastern Australia and New Zealand, and a number of Pacific islands including Fiji but excluding Hawaii. *H. zea* (Boddie) occurs in the New World from Canada to Uruguay and has been introduced to the Hawaiian islands. The two species are closely related, and are similar in behaviour and biology; some differences in behaviour have been noted on crops other than cotton. *Heliothis virescens* (F.), the tobacco budworm, is sometimes more important than *H. zea* in the USA (Clower 1980); and *H. punctigera* Wall. than *H. armigera* in parts of Australia (Evenson and Basinski 1973).

Symptoms

The larvae normally confine themselves to the reproductive parts of the plant, feeding on flower buds, flowers and bolls. The contents of the bud are usually eaten completely, leaving a hollow shell, and this often causes the bracts to flare open; flared squares are a reliable indication of bollworm damage, not only of *Heliothis* but also of other bollworms (Plate 8). When feeding on young bolls, the larvae often move about the plant, feeding on a number of bolls without finishing any of them; this habit increases the damage, since the boll will be lost by shedding or attacked by fungi and bacteria entering through the hole made by the larva. Shedding losses may pass unnoticed unless the nodes on the fruiting branches are checked; shed buds and young bolls found on the ground below the plants should be examined for bollworm damage. The moths tend to select the larger and more freely flowering portions of the crop for egg laying, and even large individual plants.

The fully-grown larva habitually feeds with only the front portion of its body inside the hole it has made in the boll, and the frass is deposited outside the hole.

Control

Heliothis can breed on a very wide range of plant species, including many cultivated crops; crops like maize, sorghum, tobacco, groundnuts, cowpeas (*Vigna unguiculata*), dolichos beans (*Lablab niger*), pigeon pea (*Cajanus cajan*) and chick peas (*Cicer arietinum*) are indeed often more attractive to this bollworm than cotton itself.

These alternative host plants allow the bollworm to maintain or increase its numbers prior to attacking cotton or may divert the attack from cotton, depending on the relative acreage, timing and attraction of the cotton crop and the host plants. These factors have been used in retrospect to explain why *Heliothis* is a more serious pest in some regions than in others, and why some seasons are worse than others for bollworm attack; but present understanding of these factors is insufficient to predict the level of *Heliothis* damage to cotton in any one area or year.

The use of other crops to divert egg laying from cotton is an attractive proposition, but it has not been universally successful and evidently needs careful timing if it is to be of value and not a danger to the cotton crop. Where the trap crop (see Ch. 12) has to be sacrificed once its object has been achieved, this method is unlikely to be adopted by peasant farmers; but in the case of maize and sorghum this may not be necessary. In these crops the losses from bollworm attack are relatively slight, and survival of larvae is lower than it is on cotton.

It is in any case good practice to avoid sowing crops at a time when a generation of moths bred in that crop is liable to infest later sown cotton. It will usually be possible to avoid this by modifying the agricultural system, substituting a longer-term variety or crop for the one that initiates the infestation, or a change in sowing dates. Ratooned or stand-over cotton provides an excellent opportunity for the build-up of an early generation of moths.

Strains that can flower prolifically over a long period can compensate for bollworm losses early in the season by producing a late crop after the first severe attack is over; such varieties have been selected deliberately for this character in South Africa (Kerkhoven 1963). At the other extreme, the recent developments in short season cotton in the United States avoid among other things late season attacks by boll weevil and bollworm (Watson 1980).

Varietal differences in resistance to *Heliothis* are known to exist in cotton, but until recently they have not been considered large enough to be worth exploiting. Within the last 10 years or so, interest has revived, and genetic resistance is now being studied intensively in the USA and Mexico. Besides the resistance found in individual strains and hybrids, the characters high gossypol and glabrous leaf confer resistance, and possibly nectariless as well.

Attempts at biological control by release of the egg parasite, *Trichogramma luteum*, in South Africa were unsuccessful, as a high rate of parasitism was never attained (Parsons and Ulyett 1936). Destruction of moths by poison baits and in light traps has been tried, but so far they have proved ineffective as a control measure, though useful in recording population changes. A polyhedral virus disease of *Heliothis* is now being used on a commercial scale in the

USA for bollworm control. The use of pheromones either for trapping the moths or for upsetting the mating process is being investigated (see Ch. 12).

Earias spp.   **Spiny bollworm, spotted bollworm**

Description

There are five larval instars, reaching 15–18 mm long when fully grown. The larva is stout and spindle-shaped, bearing a number of long hairs or setae on each segment and a number of fleshy tubercles which give the insect its name (Plate 11). These tubercles are present on the last two thoracic and on all the abdominal segments; each segment has two pairs of tubercles, one pair dorsal and the other lateral. They may be long and slender or short and rounded, according to the species, and each bears a hair at the apex. The ground colour is light brown, tinged with grey or green, with dark brown or black spots at the base of the setae.

The pupa is chocolate brown, about 13 mm long, without projections other than three small ones on the terminal segments. The pupa is enclosed in a cocoon of tough felt-like silk, usually attached to the food plant or to a fragment of plant debris on the ground. There is no diapause, although growth of the larvae and emergence from the pupa may be delayed by low temperatures.

The adult moth is usually seen at rest, with its wings snugly folded and covered with a dense soft coating of whitish scales; it is about 12 mm long with a wing span of 20–22 mm. The abdomen and hind wings are a uniform silvery or creamy white, while the colouring of the thorax and forewing vary according to the species. The eggs are laid singly on most parts of the plant, with a preference for young shoots, and, as they become available, the peduncles and bracteoles of flower buds and bolls. They are spherical, slightly under 0.5 mm in diameter, light blue-green, and decorated with longitudinal ridges which project upwards to form a crown.

Symptoms

The larvae of *Earias* are distinguished from other bollworms by their marked propensity for stem boring. They attack only soft, growing tissues, especially the terminal bud of the mainstem, where they channel downwards from the growing point. If tunnelling is extensive, the top leaves wilt and the apex of the mainstem collapses (Plate 13), but if only the apical bud is damaged the main axis is replaced by twin stems developing from axillary monopodial buds.

In feeding on flower buds or green bolls, the larva bores right into the part attacked, and the tunnel opening is blocked by excrement. The larva often enters the boll from below, tunnelling upwards at a slight angle to the peduncle. It moves about considerably, leaving the interior of the bud or boll only partially eaten out.

Distribution

The genus is confined to the Old World and Australasia. *Earias insulana* (Boisd.) has an extremely wide range, covering most of Africa including Madagascar, Mauritius and the Canary Islands. It extends northwards to southern Europe and eastwards through the Near and Middle East to India and south-east Asia, being recorded in Japan, Formosa, the Philippines and Australia.

Earias biplaga (Wlk.) is found throughout Africa south of the Sahara, being less common than *E. insulana* in the drier parts of their common range. The habits of the two species do not differ in essentials.

In India, the situation with respect to *E. insulana, E. fabia* (Stoll) and *E. vitella* (F.) is similar. *Earias fabia* does not extend west of the arid zone of the Indus and is less important than *E. insulana* in much of the Punjab, but elsewhere in the subcontinent it predominates. It has been recorded from Formosa to New Guinea, but the precise limits to the south-east are uncertain.

In Australia *E. huegeli* Rogenh., predominates on cotton, and it occurs as far east as Tahiti and the Marquesas Islands; it overlaps with *E. fabia* in Fiji.

Other species of *Earias* have been identified in parts of the Old World, but not apparently as pests of cotton; *E. cupreoviridis* (Wlk.) is, however, reported as a major pest of cotton in China by Chao and Li (1965).

Control

There is no diapause in the life history of *Earias*, and the rate of development depends directly on temperature. Its survival and carry over as a pest of cotton from one season to the next depend therefore on the supply of food throughout the year; and this depends on the presence of suitable host plants. The recorded host plants in the Punjab are virtually confined to the Malvales, and while *E. insulana* breeds on a number of hosts belonging to different genera (*Abutilon, Althaea, Gossypium, Hibiscus, Malva, Malvastrum* and *Sida*), *E. fabia* occurs only on *Gossypium* and *Hibiscus esculentis*. Elsewhere the records on wild host plants rarely record the species of *Earias*, but Reed (1974a) has studied the host plant preferences of the East African species: the main alternative host of *E. biplaga* was *Waltheria indica*, and that of *E. insulana* was *Abutilon*. The importance of *Earias* as a pest will depend on the extent to which the cotton crop and the alternative hosts are complementary, or competitive, in respect of the insect. It is a serious pest only where there is no effective close season between successive cotton crops.

In subtropical regions, where cotton is grown as an annual in the single rainy season, the important host plants are available over the same period. The scanty evidence suggests that *Abutilon* and some species of *Hibiscus* are preferred over cotton as food plants, and where cotton constitutes only a small fraction of the total vegetation, their influence may be important. Carry over occurs on alternative host plants which survive as biennials in fallows, pastures, roadsides and the banks of rivers and canals; on stand-over cotton where a close season is not observed; and on irrigated crops of cotton and *Hibiscus* grown out of season, and the weeds of the other irrigated crops.

In the moist equatorial regions, conditions favour the occurrence of perennial shrubby Malvales. They provide a supply of food all the year round, but there is no compulsion to transfer from these wild hosts to cotton when it is available, and heavy infestations from this source are not usual.

Pectinophora gossypiella Saund. (formerly *Platyedra*)

Pink bollworm

Description

The newly hatched larva is yellowish, with a prominent dark brown head. There are four instars, in the third of which the typical colour pattern appears; this consists of two salmon-pink transverse dorsal bands on each segment, the anterior band being broader than the posterior (Plate 12). The ground colour of the later instars is almost ivory white, and the skin is smooth and shiny. Larvae that have fed on rotting tissues, such as fallen buds and flowers, remain colourless and almost transparent. The fully grown larvae vary from 10 to 12 mm in length, occasionally longer.

The pupa, 7–10 mm long by 2.3–3.0 mm broad, is rather bright brown in colour and covered by a fine pubescence; it is protected by a loose, slight cocoon, which is not fully closed. Pupation may be in the soil, in green bolls, or in the seed or lint of an open boll.

Under certain circumstances, the fully fed larva enters a resting stage or diapause. For this it constructs a tough, closely woven cocoon without a passage to the exterior, in which it may remain for a period varying from some weeks to many months before it finally pupates. A common site for the diapause larva is within a hollowed-out seed, or in a pair of seeds joined together; in the Punjab 95 per cent of the diapause larvae are found in 'double seeds', which are used as an indicator of diapause, but this close association is not found elsewhere, and possibly depends on the size of the seed.

The adult moth is nocturnal, and remains closely concealed by day. It is small and inconspicuously marked, 8–9 mm long with the wings folded, and a wing span of 15–20 mm. The colouring is variable, but generally greyish brown over all the parts visible when the wings are folded; the abdomen is rather paler, and the hind wings silvery grey with a deep fringe of hairlike scales on the posterior edge.

The eggs are small, 0.5 mm long, oval and flattened, with a wrinkled appearance; they are at first pearly iridescent, becoming tinged with reddish orange before hatching. They are laid singly or in small groups on sheltered parts of the plant, with flower buds and immature bolls preferred.

Distribution

The pink bollworm originated in the Old World, probably in India, and was widespread there and in Ceylon, Burma and China before 1910. In that year it spread to Egypt, in 1911 to the West Indies, in 1916 to Mexico and Brazil, and thence to the USA and the rest of Central and South America. It was probably introduced into the coastal regions of Tanganyika and to Madagascar before it reached Egypt, but the main spread took place southward from Egypt through the Sudan to Uganda and the Belgian Congo. Only isolated specimens have been found in central and southern Africa; in South America it was recorded in Ecuador in 1950 (Barlow 1952), but disappeared until a fresh outbreak took place in 1978. The importation of infested seed has contributed in the past to the spread of this pest, and may do so again if strict phytosanitary control is not enforced.

Symptoms

The bollworm does most of its damage to green bolls, but is also found feeding in flower buds and open flowers. In these it feeds on the anthers and style, and sometimes the ovary, when shedding invariably results; the bollworm is often to be seen in apparently undamaged open flowers. In the bolls, the larva feeds mainly on developing seeds, and usually completes its development in a single locule, but may tunnel into a second; in partly grown bolls, the attack on one locule may affect others in the same boll, causing poor development of seeds and lint. Entry to the boll is usually effected below the calyx, and the entry hole is too small to be easily seen. Unlike the larvae of other major bollworms, those of pink bollworm are found feeding in the lint and seed of opened bolls.

When the larva matures inside a green boll it either cuts a neat circular hole about 2 mm in diameter through the boll wall and falls to the ground, pupating in the surface trash, or it pupates inside the boll after preparing an exit hole, over which the cuticle in left intact as a transparent membrane. Pupation on the ground is usual in hot dry climates, while in cooler and more humid conditions the larva tends to remain inside the boll. When bolls are attacked late in their life and open before the larvae have completed feeding, they continue feeding inside the seed. They may pupate among the lint, or they may be picked with the seed cotton and carried into the markets and ginneries.

Control

Although *P. gossypiella* has been recorded on a number of wild and cultivated species belonging to the Malvales, particularly on *Hibiscus esculentis* and *H. cannabinus*, it appears to be so closely associated with the true cottons as to be scarcely able to maintain itself on any other host. Control measures therefore depend on restricting the effective carry over from one cotton season to the next. A close season for cotton between crops is obviously the first essential, accompanied by the destruction of crop residues and volunteer seedlings and prevention of stand-over cotton and ratooning.

Sowing should be spread over as short a period as possible; the initial infestation is likely to be small, and the population builds up fairly slowly, but late-sown cotton will receive a heavy infestation from the population bred up on the early sowings, and tends to produce a higher proportion of diapause larvae.

Diapause larvae, where they occur, require additional precautions. A knowledge of the factors that control the resting state is of great practical importance, but some of the evidence is conflicting, and it is not possible to say precisely why the diapause is more common in some regions than in others. It does not occur, except very rarely, in the equatorial regions between roughly 7 ° north and south; out-side these limits its wholesale occurrence is confined to the short day or winter period, with a day length of $11\frac{1}{2}$ hours or less. Nutrition, moisture and temperature all have an influence on the occurrence and duration of the diapause.

In general, the first generations bred in an annual crop consist wholly of short-term larvae, but in later generations an increasing proportion of the larval population enters diapause. The promotion of a short growing season, by the use of early maturing varieties and of farming practices which hasten the development of the plant, will therefore reduce the formation of resting larvae (Watson 1980). It may be possible to adjust the time of sowing so that the majority of the diapause larvae pupate and give rise to moths before the new crop is suitable for breeding (Squire 1940).

Sanitation on and off the field is important to destroy resting larvae. Crop residues should be destroyed as soon as possible after harvest, and ploughing and irrigation in the winter is recommended in the USA to destroy larvae in bolls left on the ground; where there is no cold season, an irrigated crop may be taken between cotton crops with the same effect.

Seed cotton should be ginned as soon as practicable, and ginneries swept clean of fallen seed, seed cotton and debris. Ginneries and stores should be fumigated or sprayed at least once a year as soon as ginning is completed. Many larvae are found in the waste from the openers and cleaners; this waste should be burned or treated to kill the larvae before they pupate. Seed for oil extraction should be crushed before the moths emerge, and planting seed treated by heating or fumigation. Heating the seed to 55–65 °C for 5–10 minutes will kill the larvae without affecting germination; in the Sudan, this is achieved by spreading the seed in a thin layer in the sun. Methyl bromide, carbon disulphide and hydrocyanic acid gas are effective fumigants.

Results so far achieved by the encouragement of parasites, predators and diseases have been inconclusive, and no recommendations for control on these lines have been made. Resistance has been observed in some of the wild species of *Gossypium*, notably in *G. thurberi* (*trilobum*): this species has been used in the production of triple hybrids in the USA and the Ivory Coast. In Arizona, Wilson and George (1981) are concentrating on earliness, smooth leaf, nectariless and okra leaf in a resistance breeding programme.

Diparopsis spp.   Sudan or red bollworm

Description

The larvae pass through five instars, reaching a length of 25–30 mm before pupation. The newly hatched larva is pale cream or greyish white, with the head and some other parts black; the characteristic rose-red markings appear in the second and subsequent instars, three dorsal red lines forming a forward-pointing arrowhead on each segment of the body except the first and last (Plate 10). The fully grown larva has a distended, slimy appearance.

The pupa is enclosed in an irregular earthen cell, consisting of soil particles cemented together, with a silvery silk lining. The adult has a wing span of 25–35 mm; the fore wings are usually more densely coloured than the hind wings, and are divided by transverse lines into four distinct segments; the colours of the fore wings are variable. The eggs are 0.5 to 0.7 mm in diameter, rather larger than those of *Heliothis*; they are spherical except for a slightly flattened base, with a mesh marking and bearing minute spines. The eggs are laid singly, and have a characteristic sky-blue colour, changing to grey before hatching.

Distribution

The genus *Diparopsis* is confined to Africa, except for a single area in South Yemen (formerly Western Aden Protectorate). *Diparopsis castanea* Hmps. occurs south of the Equator, from South Africa to

Zambia; *D. watersi* (Roths.) is the species found north of the Equator, from Senegal to Somalia and northwards to the Sahara desert, the Sudan and Ethiopia.

Diparopsis tephragramma B.–B. is found in Angola, and is probably the species which attacks cotton there. The main cotton-growing areas of the equatorial region, Zaïre, Uganda, Kenya and all but the extreme south of Tanzania, are free of this pest, and it has not reached Egypt.

The habits and life history of the two main species, *D. castanea* and *D. watersi* are similar as regards their main features; there is practically no information on *D. tephragramma*.

In South America, *Sacadodes pyralis* Dyar is closely related to *Diparopsis*, both morphologically and biologically (Pearson 1954).

Symptoms

The larvae attack the reproductive parts of the plant, feeding on flower buds, flowers and bolls; before buds are available the larvae may occasionally bore into the terminal bud of a shoot, and a shortage of fruiting bodies after a heavy attack on young cotton may induce the larvae to become stem borers, entering the stem at one of the nodes.

The young larvae will attack the nearest flower bud or flower, and eat out the entire contents. The bracts will usually open as flared squares, as occurs with *Heliothis*; if shedding occurs the bud often remains suspended by a silk thread left by the larva when entering the bud. If green bolls are available, the partly grown larva will usually enter one and complete its larval life within it. It enters the boll completely, and most of the frass excreted remains inside the boll and plugs the entrance hole. It generally enters a part of the boll protected by a bract, which remains stuck to the boll as the frass dries. As movement of the larvae usually takes place after dark, it is extremely rare actually to see one on the plant without cutting open a damaged boll.

Control

Since the publication of Pearson's (1958) book, detailed studies of *D. castanea* have been made in central Africa by Tunstall *et al.* (1959), as the basis of a successful programme of pest control which has been widely adopted in southern Africa (Tunstall and Matthews 1966). The incidence of *D. watersi* in relation to watering and the close season in the Abyan delta east of Aden has been reported by Proctor (1962).

Three characteristics of the red bollworm are important in devising control measures. Firstly, it is strictly confined to cotton; although the moth and its eggs have occasionally been found on other species, it is unlikely that significant numbers can survive on them.

Secondly, this species possesses a facultative diapause in the pupal stage; pupae formed in the early part of the infestation all give rise to moths within two to three weeks, but a large proportion of the later-formed pupae do not emerge until after periods ranging from a few weeks to many months. High soil temperatures (36–37 °C and above) impede the emergence of diapause pupae, but when temperatures fall at the start of the rains all the pupae formed in successive months renew their development together, and emerge as moths to give one, sometimes two, peaks of infestation during the cotton growing season. The second peak is probably the result of waterlogging of the earthen cell surrounding the pupae which delays the emergence of some of them by about three months. The third characteristic is that the larvae feed inside the cotton bolls and pupate in the soil, being thus protected from parasites, predators and insecticides for most of the larval and pupal period.

The carry over of *Diparopsis* from one season to the next can be limited either by minimizing the production of diapause pupae, or by preventing the survival of those produced. Only short-term pupae are produced early in the season, and by limiting the cotton season to the times when such conditions prevail, the production of diapause pupae should be prevented (Pearson and Mitchell 1945). This reasoning was applied to the Lower River area of Malawi, and in 1952 early sowing and pure stand cotton replaced late sowing interplanted with maize; this reduced the damage caused by red bollworm appreciably, but did not eliminate the pest (Munro 1958).

At the same time, it is important to prevent the breeding of the moth on ratooned or stand-over cotton, and on perennial cotton (Kidney cotton and Hindi Weed) grown as garden plants in many parts of Africa. A legal close season can be imposed to limit ratooning and enforce early uprooting after harvest. Where some degree of mechanization is the practice, the crop can be ploughed out, or uprooted with a stalk puller (Pothecary 1968).

Where there is any latitude in the time of sowing, especially with irrigation, this can be used to arrange that the diapause pupae emerge when no food is available for their offspring. In the Sudan, where diapause pupae emerge in mid-July to mid-September, riverain cultivation of cotton along the Nile was infested with *Diparopsis*; when the Gezira scheme was started, this riverain cotton was eliminated, and the sowing date under irrigation was put back to August. The moths emerge before there is any food available, and the pest has been eliminated from the area; it does, however, persist in the Gash Delta and is often of importance, although sowing dates are similar to those in the Gezira (Tunstall 1958).

The survival rate of *Diparopsis* appears to be very low; it is rarely more than 20 per cent between the egg and the fifth instar larva and the bulk of the loss occurs in the early stage of its life. Egg parasitism is very exceptional, and only rarely do larval parasites exercise any important degree of control. Various species of ant are important predators of larvae and pupae in different parts of Africa, and survival from egg to larva may be reduced by dry conditions. In the pupal stage losses probably range from 20 to 40 per cent, being greater among diapause pupae and the later generations of pupae in the crop. Not enough is known of the effect of weather and natural enemies to provide a full explanation of the high mortality which is known to occur.

The stage of development of the crop at the time when an infestation occurs has a considerable effect on the damage which it receives, and the pattern of crop formation can be altered by the pest attack itself. An early attack will cause losses of buds and flowers by shedding, and the crop will necessarily develop from later formed fruiting points; if environmental conditions allow these to be produced in sufficient numbers, and they are not damaged by later bollworm generations, a normal sized but late crop will be produced. Once the bolls are more than 10 days to 2 weeks old, they remain on the plant even if damaged, and provide a large and secure source of food; as soon as older bolls become available, the larvae tend to attack them in preference to buds and young bolls. A pest attack occurring when bolls have already started to form will therefore be accompanied by less shedding of buds and young bolls, but will result in a high proportion of damaged bolls.

Cultural control methods such as ploughing to expose the pupae on or near the surface have been tried, but in fact only those actually raised to the surface were killed by heat, while high but sub-lethal temperatures could delay emergence from diapause and result in a more damaging attack on the subsequent crop. Crushing with heavy rollers did not give results sufficient to justify the direct costs and damage to soil structure; disc harrowing for weed control in the Gash Delta was found to destroy about half the pupae. Cultivation also increases the accessibility of the pupae to predators.

When irrigation water is available, prolonged waterlogging of the soil between crops destroys a high proportion of the pupae. Mulching with plant residues or encouraging weed growth in the dry season to maintain a soil cover and reduce soil temperatures was tried in Malawi, but it turned out that the increase in the mortality of pupae arose from predators, and not from any change in soil temperature.

Sacadodes pyralis **Dyar**   **South American red bollworm, false pink bollworm, Colombian pink bollworm**

The counterpart of the African red bollworm (*Diparopsis*) in South America is *Sacadodes*; 'the differences between *Sacadodes* and any given species of *Diparopsis* are in general little, if any, greater than those separating the four species comprising the latter genus' (Pearson 1954). The forms of attack on cultivated cotton resemble each other very closely, and diapause pupae have been reported from Venezuela.

Sacadodes is found from the Equator northwards from Ecuador through Colombia and Venezuela to Central America as far as El Salvador and the West Indies. Its wild host plants are restricted to cotton and its near relatives; the main alternatives to cultivated cotton are the native perennial cottons, *G. barbadense* var. *brasiliense* and *G. hirsutum* race *marie-galante*, but it has also been found breeding on *Cienfugosia affinis* Hoch. in Venezuela.

Crytophlebia leucotreta Meyr.   **False codling Moth**

The history of the false codling moth (*Cryptophlebia*, formerly *Argyroploce*) in Uganda is an interesting example of the changing status of a pest. In the 1930s it was regarded as the most important of the bollworms (Tothill 1940), but it was barely mentioned by the entomologists at Namulonge in the 1950s. When insecticide spraying became standard practice in the 1960s, however, it was noted that *Cryptophlebia* was not being controlled, and by 1969 it was by far the most damaging pest of sprayed cotton, with over 90 per cent of the late bolls damaged (Reed 1974b). This suggests that some form of biological control was operating in unsprayed cotton, but the controlling factor has not been identified. The picture is confused by changes in the timing of the maize crop on Namulonge, and by an increase in the cultivation of maize around the station.

The false codling moth lays its eggs on the surface of unripe green bolls; it avoids very young bolls and buds as well as fully ripe bolls. The larva tunnels about in the boll wall, and eventually enters the boll cavity and starts feeding on the seeds; it finally leaves the boll and pupates in debris on the surface of the soil. Boll-rotting organisms colonize the tunnels made by the bollworm, and many of the bolls attacked are totally destroyed by bollrot. It seems that the problem of boll rotting at Namulonge is largely due to this bollworm.

The false codling moth has no resting stage, but has a catholic taste in alternative hosts; for a long time it has been known as a pest of citrus fruit in South Africa. Spending its whole larval period inside the cotton boll, it is protected from parasites and predators, and it is not easy to reach with insecticides; it probably requires a higher dosage rate of insecticide than most other bollworms (Reed 1976).

Cutworms

Cutworms typically spend the day curled up in the ground near the roots of cotton seedlings and emerge at night to feed on the succulent stems, cutting them off just above ground level. They have a wide range of food plants, both wild and cultivated, of which cotton is only one; some species behave only occasionally as cutworms, such as the lesser armyworm and the cotton leafworm (*Spodoptera* spp.).

The black or greasy cutworm (*Agrotis ypsilon* Hfn.) is typical; it is of world-wide distribution in the temperate and subtropical zones, but as a pest of cotton it is confined to Egypt, the Sudan and South Africa. The larva is smooth-skinned, at first pale yellowish green with black markings, but in the later stages slate-grey with darker longitudinal lines, and reaching 35–45 mm in length. It feeds at night, and lies curled up just below the soil surface during the day. It pupates in the top layer of the soil where it forms an earthen cell; the pupa is smooth, pinkish brown, 20 × 5 mm.

The fore wings and body of the moth are patterned in sombre shades of brown, the hind wings being pearly with a brownish margin. The wing span is about 50 mm, and the moths are nocturnal in flight. Eggs are cream-coloured, dome-shaped, and are laid singly on moist soil or on green leaves.

The common armyworm *Spodoptera* (formerly *Laphygma*) *exempta* (Wlk.) hardly ever atacks cotton, but the lesser armyworm, *S. exigua* (Hb.), has become a serious pest in the northern Sudan, and attacks on cotton have been recorded all over the world. Its behaviour is similar to that of the common armyworm: large numbers of larvae hatch out together and march across country in a swarm as their food supplies become exhausted. In America the fall armyworm, *S. frugiperda* (A. & S.), often causes great destruction to cotton and corn.

Leaf-eating caterpillars

Numerous species of leaf-eating caterpillars are found on cotton, but most of them have non-specialized feeding habits and only occasionally give rise to serious damage. A few are major pests, different species assuming this role in different parts of the world. Their colour often varies according to the type of leaf on which they are feeding, but other habits may be characteristic: thus there are loopers and semi-loopers, leaf-rolling caterpillars and leaf miners. Except for the leaf miners, they are all exposed on the surface of the leaf, and are fairly easily controlled with insecticides.

Spodoptera littoralis (Boisd.)   **Cotton leafworm**
(formerly _Prodenia litura_)

This moth is widely distributed in Africa, Asia and Australia, but its main area of importance is around the Mediterranean – Egypt, Turkey, Israel, Syria and Spain. In Egypt it is a major pest in the northern delta, where it breeds on a succession of crops ending with the winter crop of Egyptian Clover (*Trifolium alexandrensis*) where the larvae and pupae are protected by the lush growth. This is cleared for cotton planting, except for small patches kept for seed; as the land is cleared, there is an invasion of the young cotton in April and May by the larvae moving like armyworm. The main pest attack, however, comes from moths which emerge from the pupae in the clover seed plots in late May and early June, just before the cotton starts to flower. Their eggs are laid on the underside of the cotton leaves in clusters containing several hundred eggs, which will hatch in about four days.

The traditional method of control is to pick off the pieces of leaf carrying the egg masses and burn them; to be effective this collection must be repeated every four days, and was a routine carried out every year by squads of children.

Further south in the Nile delta the weather becomes drier and warmer, conditions less favourable to the pest, and a severe outbreak occurs only when weather conditions are suitable; this happens about one year in 10 in Upper Egypt as far as Minia and Assiut (Brown 1953).

Alabama argillacea Hbn.   Cotton leafworm

This leaf eater is a native of tropical America, and now extends from Bolivia to the southern states of the USA. The eggs are laid singly on the leaves of cotton, and the larvae eat out the leaf tissue between the main veins; a severe attack can defoliate a field of cotton within two weeks if not controlled (Plate 17). The caterpillars are at first yellowish green, later developing a black stripe along the middle of the back with a fine central yellow line, and black dots on each segment. The fully grown larva webs a leaf or two together, and pupates inside the web (Fernald and Shepard 1942).

Sylepta derogata (F.)   Cotton leaf roller

The leaf roller is perhaps the commonest leaf eater attacking cotton in Africa and Asia, but it rarely causes sufficient damage to warrant special control measures. The eggs are laid on the leaf, and the young larvae congregate in a roll of the leaf secured by threads which they spin. When partly grown they disperse and each forms an individual roll; they pupate in the leaf roll, or in plant debris on the ground. In West Africa the leaf roller is reported as surviving the dry season as a resting larva.

Loopers and semi-loopers

The various species of loopers and semi-loopers are all leaf-eating caterpillars, usually with a wide range of host plants and only occasionally attacking cotton. Their common names – loopers, inchworms, span worms, measuring worms – are derived from the way they move because of the absence of prolegs from the central segments of the body. The normal caterpillar has six legs on the first three segments, which correspond with the thorax of the adult; the 10 or so abdominal segments are similar in appearance to the thoracic segments, but are supported by prolegs which have a totally different structure. Prolegs are normally found on abdominal segments 3, 4, 5, 6 and 10, but in the looper family (Geometridae) the prolegs are missing from segments 3, 4 and 5. The caterpillar moves by bringing the hinder end of the body up close to the front end forming a loop, which straightens out as the front end is moved forward.

Different species of loopers are of local importance on cotton in different countries, the most widely distributed being *Plusia* spp. in the Old World, and *Trichoplusia* and *Pseudoplusia* in the New. Important genera of semi-loopers are *Cosmophila* (*Anomis*), *Xanthodes* (*Acontia*) and *Tarache*. *Cosmophila flava* (F.) is the species found in Africa, Asia and Australia, while the New World species are *C. erosa* (Hb.) and *C. texana* Ril. *Xanthodes graellsii* (Feisth.) is a pest of cotton in Africa, India and Pakistan. *Tarache nitidula* (F.) occurs in the same areas, and at least three other species of *Tarache* have been recorded as cotton pests in India and Pakistan.

Leaf miners

Bucculatrix thurberiella Busck   **Cotton leaf miner**

The larvae of *Bucculatrix* enter the leaves of cotton directly from the eggs, and the first two instars mine between the upper and lower surfaces (Plate 18). The second moult occurs under a horse-shoe shaped silky covering, and it is the later larvae which cause major damage by eating holes in the leaf; they can reduce the leaves to skeletons, stunting the plants so severely that the buds and bolls are shed. The larvae grow to about 6 mm in length, and pupate on the stems and petioles of the plant, forming conspicuous cocoons of white silk. The adult is also white, about 7 mm long with a pronounced crest of white hairs on the head; they can be seen flying around when the crop is disturbed. The eggs are small, bullet-shaped with vertical ribs, and are laid singly on the leaves, bracts and bolls.

This pest only occurs on cotton in the New World, where it is found in countries ranging from Peru to the United States. The worst affected areas are in Mexico, Central America, Ecuador and Peru; losses occur in some seasons in the USA, particularly Arizona, and in Colombia. The damage reaches serious proportions late in the season, and at harvest time whole fields appear grey in colour with the leaves reduced to a network of veins. Early maturity of the crop helps to avoid losses, and destruction of stalks immediately after harvest will reduce the carry over from one season to the next, as this species only survives on *Gossypium*. In Ecuador, the synthetic pyrethroids have proved very effective in controlling *Bucculatrix*.

Hemiptera

Dysdercus spp.   **Cotton stainers**

Description

There are five nymphal instars, in the first of which the insect is yellow, turning to orange-red, without markings, and about 2 mm long (Plate 7). The nymphs are similar in shape to the adults, but without wings, although rudimentary wings are visible from the third instar. After the first instar, the adult colour pattern appears on the thorax and abdomen. The first instar nymphs do not feed, but all the later stages possess the typical rostrum folded beneath the body. The length increases to 12 mm in the final instar.

The adult is about 15 × 4.5 mm broad, the size varying with the species. Most of the dorsal surface of the insect is reddish orange to light fawn in colour, with a white prothoracic collar and in some species a black posterior band on the thorax; each segment of the abdomen (hemelytron) usually has a black mark in the broadest part, varying from a round spot to a transverse band (Plate 6). The underside of the insect is ivory white, marked with red, orange, yellow, or black according to the species. The rostrum is a segmented tube containing the stylets through which the insect feeds, and is folded beneath the body reaching to the second abdominal segment or beyond.

The eggs are ovoid, 1.5 × 0.9 mm, creamy white in colour, changing to orange as the embryo develops. They are laid in shallow holes in the soil or debris, and covered with the same material by the female.

Distribution

The genus *Dysdercus* is distributed throughout the tropics, each continental area having its own group of species, the largest number being in the New World. The species differ remarkably little in their structure and biology. The genus is closely associated with plants of the order Malvales, and many species are pests of cotton. It is a serious pest of cotton in many tropical countries, but only of minor importance in Africa north of the Sahara, Pakistan, and the northern parts of India and the United States.

Symptoms

Stainers feed on developing seed in the green bolls and on ripe seed when the bolls open, using the rostrum to penetrate the boll wall and the ripe seed; very young nymphs (second instar) may not be able to reach the seed in fully grown green bolls, and cannot penetrate very hard seeds.

No symptoms are visible until the boll ripens, and the presence of the insects is the only outward sign that damage is taking place. If a sample of green bolls is cut open, however, recent attack is indicated by dark watery spots on the inside of the boll wall; earlier attacks will have caused rotting of the immature seeds, often accompanied by a yellow staining of the lint.

The yellow staining is caused by a fungus, *Nematospora gossypii*, whose spores are found in the gut and salivary glands of the stainer. They do not pass through any cycle of development in the insect vector, and appear to be introduced into the green boll with the salivary fluid when feeding. The damage caused by *Nematospora* depends on the stage of development of the boll when it is attacked. Bolls up to two weeks old either shed or abort, the embryo being killed; bolls infected when three to four weeks old may be slightly distorted in shape, the lint is heavily stained to a dirty brown or yellowish colour and when the boll opens the carpels are twisted, the lint does not fluff out, and characteristically the fibres adhere to the edges of the carpels and are stretched out to form a web; bolls over five weeks old complete their development normally, but the lint is diffusely stained to a depth and extent depending on the time available to the fungus between infection and boll opening, and increasing from straw yellow to dark brown with time.

Not all stainers carry *Nematospora* spores, and the direct effect of stainer feeding is much less severe. When the bolls are less than a month old, only the seeds directly punctured are affected, the embryo shrivels up and the lint hairs remain unthickened and mat together. In older bolls, seeds in which the embryo is fully formed seem able to withstand non-infected punctures without any affect on the lint, and the boll opens normally showing no sign of damage. Sometimes bacterial infections in apparently intact green bolls are associated with puncturing by *Dysdercus* and other Hemiptera. The evidence suggests that bacteria from the boll surface are carried through the boll wall on the insect's stylets, or enter through the puncture, but such infections are not of great importance.

The predator *Phonoctonus* displays an interesting case of mimicry, resembling the stainer in general shape and showing very similar colour patterns to those of the stainer species on which it feeds. The predator can be distinguished by its larger size, and by having a distinct neck between the head and the thorax. Sweeney (1960) found that the predators always lagged behind the stainers when the latter moved from their wild host (*Sterculia* spp.) to the cotton crop; *Phonoctonus* also feeds on another insect, *Odontopus*, which is found on the same wild hosts as the stainers, and remains there for some time after the stainers have migrated.

Empoasca spp. (syn. *Amrasca*, *Austroasca*)   Jassids

Description

The nymphs are pale yellowish green in colour, somewhat frog-shaped and flattened, from 0.5 mm in length in the first instar to 2 mm in the fifth and final instar. They are very active, always seeking the underside of the leaf in daylight, and move across the leaf with a characteristic sideways run, like a crab.

The adult is small, elongate and wedge-shaped, about 2.5 to 3.0 mm long, pale green with semi-transparent shimmering wings; they have a sideways walk like the nymphs, but are quick to hop and fly when disturbed. The female lays her eggs embedded in the midrib or in a large vein on the underside of the leaf.

The eggs hatch about 6–10 days after they are laid, the nymph takes rather longer to develop and the whole life cycle occupies 3–4 weeks in warm weather; in cold, dry weather the adults may live for up to two months. *Empoasca devastans* is estimated to have 11 generations a year in India, each female giving rise to an average of up to 30 nymphs under favourable conditions.

Distribution

Empoasca facialis (Jacobi) is the common species in southern, Equatorial and West Africa, and *E. lybica* (de Berg) in North Africa and the Sudan; the two species overlap in East Africa. *Empoasca devastans* Dist. is the common jassid of India, and *E. terrae reginae* (Paoli) of Australia. Other species of jassid occur on cotton, but are of minor importance.

Symptoms

Jassids feed by sucking the sap from the leaves of the plant, but the effect is not localized at the point of feeding; they apparently introduce a toxin into the leaf which can impair its photosynthetic efficiency roughly in proportion to the numbers of nymphs present. The first symptoms are a paling of the green colour at the edges of the leaf in zones more or less bounded by the reticulate veins; the colour change develops from pale green to yellow, red and finally to brown as the tissue dies (Plate 5). The colour changes in young leaves are followed by a downward curling of the leaf edges, as growth continues in the middle of the leaf after it has ceased peripherally. A sustained attack will cause the leaves to dry up and shed; if the plant is young when attacked, growth may be wholly arrested.

Control

Natural enemies seem to be of little importance in controlling jassids, but the intensity of the attack depends to a large extent on climate. Jassid attack is a feature of regions with a monsoon climate, characterized by a single rainy season followed by several months of more or less complete drought, in which temperatures are at first low, but rising to very high just before the break of the rains. At the other extreme, in the moist Equatorial regions with a two-peaked rainfall and medium-high temperatures without marked extremes, jassids do not occur on cotton in any great numbers. Between these two extremes, an intermediate situation exists, and jassid attack is not severe unless growing conditions for the crop are poor.

The discovery that hairy cotton plants resisted jassid attack was first consciously exploited in South Africa in the early 1920s, and resistant strains are now widely used in regions where jassid attack is severe. The degree of resistance has been adapted to suit the level of attack, so that glabrous cottons are grown in the Equatorial regions which are practically free of jassids and moderately hairy varieties are sufficient to control jassids in intermediate situations. This ideal picture is complicated in the Sudan, for instance, where hairy varieties are more susceptible than glabrous to whitefly, a pest of equal or greater importance. The development of leaf hair is discussed in Chapter 4, and breeding for jassid resistance in Chapter 14.

Bemisia tabaci (Gennadius)   **Whitefly**

Description

The minute eggs are attached by stalks to the underside of the leaves. Only the first instar larva is mobile; later stages settle down like scale insects on the underside of the leaves, and are flat, pale yellow, with a narrow waxy fringe. The adult is only about 1 mm long, and is covered with a white waxy bloom; it is easily seen flying around if the leaves are disturbed.

Distribution

Whitefly is best known as a pest of cotton in the Sudan, where it is the vector of leaf curl virus, and produces *asal* or honeydew which contaminates the lint; this contamination, or stickiness, interferes seriously with the processes of ginning and spinning.

Whitefly has been recorded on cotton in practically every country where the crop is grown in the Old World and in Latin America, but until recently it was regarded as a major pest only in the Sudan and the Punjab. Following the use of broad spectrum insecticides, however, it has emerged as an important pest of cotton under tropical and subtropical conditions in many countries. A whitefly of the same family and of similar habits, *Trialeurodes abutilonea* (Haldeman), is reported as attacking cotton in Tennessee, and similar species are pests of greenhouse crops and citrus in the USA.

In the Sudan it flourishes in hot, humid, weather, especially when there is an abundance of weeds, many of which are alternative host plants. Peak populations are found early in the cotton season in November and December, spreading northwards with the warm and humid southerly winds. The spread is checked by colder, dry northerly winds, and as a result there is less damage in the northern Gezira than in the south and in the Tokar Delta. Intensified crop rotations have provided a continuous source of food, and the application of pesticides has exacerbated the situation.

Symptoms

Leaf curl disease is virtually restricted to the *barbadense* cottons in Africa, and to some malvaceous hosts; it is characterized by swelling of the veins of the leaves and bracteoles, resulting in an upward or downward curling of the edges of the leaf. This may be followed in severe cases by foliar outgrowths from the veins, twisting of inter-nodes, reduction in leaf size and shedding of fruiting bodies (Nour and Nour 1964). Considerable progress has been made in breeding leaf curl resistant varieties with the lint quality required in the Sudan, such as BAR XL 1 and VS 1 (Hughes 1969).

Since the introduction of leaf curl resistant varieties, the direct effects of whitefly have assumed more importance, both in *barbadense* and *hirsutum* cotton, in particular the sticky lint caused by honeydew. The level of infestation may vary from the presence of a few adults on the underside of the leaf to a complete cover of white, waxy larvae, again on the underside, with droplets of honeydew dripping on to the leaves and bolls below (Plate 15).

The saliva surrounding the feeding stylets has a slightly toxic effect on the adjacent leaf cells, producing chlorotic areas where a number of sessile larvae have been feeding continuously. Severe infestations weaken the leaves, and cause premature shedding of leaves, buds and bolls.

Control

The emergence of whitefly as a serious pest of cotton in the Sudan coincided with the control of jassid with DDT; the reason appears to have been the removal of jassid as a competitor rather than the destruction of natural predators, which do not appear to exert much influence until late in the season (Hassan 1969). The same result is obtained if jassids are controlled by leaf hairs. In the absence of insecticides, plots of smooth-leafed cotton show severe damage by jassid and few whitefly; adjacent plots of hairy-leafed varieties show heavy infestation with whitefly and few jassids. The scope for cultural or biological control of whitefly is limited, and Hassan (1969) stated that biological methods have never been used for controlling cotton pests in the Gezira, but uprooting is recommended instead of cutting down the cotton plants after harvest to prevent regrowth.

Lygus spp. (syn. *Taylorilygus*)   Capsid or mirid plant bugs

Description

The eggs are minute, buried in the soft tissues of the host plant. The nymphs are fragile in appearance, ovoid with long, slender legs and antennae, pale apple-green with darker eyes. There are five instars, 1–4 mm long; the later instars show brownish mottling on the legs and abdomen. The adult has a small head with a broad and strongly convex pronotum; the colour varies with the species, those attacking cotton being brown, variegated with shades of green or yellow, with strongly marked black or dark brown spots.

Distribution

Lygus attacks cotton growing in fairly cool, moist conditions, particularly in the high-altitude, equatorial regions of Uganda and Zaïre and in Angola. The same species, *Lygus vosseleri* Popp., has been found in specialized areas in many countries of Africa south of the Sahara. The tarnished plant bug, *L. lineolaris* (Pal. de Beauv.) and related species, are important pests of cotton in the USA.

Symptoms

Both nymphs and adults feed only on very young tissues of cotton, namely unopened vegetative buds, flower buds and very young bolls. The feeding punctures appear as very minute, black specks, but the damage is not noticeable until the leaves start to unfold; as the lamina expands, it ruptures where it has been punctured, presenting an increasingly tattered appearance. Flower buds and young bolls attacked by lygus will shed, and the sympodia may be truncated, resulting in a severe reduction in the number of bolls produced. A severe infestation will change the whole appearance of the plant, which grows tall and straggling with short lateral branches and often a number of monopodia growing strongly from the base of the mainstem, while the foliage is tattered and reduced.

Control

Lygus can breed on a wide range of host plants, and the adults feed on an even wider range. In Africa severe attacks on cotton are encouraged by the tendency to plant the food crops before cotton; large populations build up on cereal crops such as sorghum, maize and finger millet (*Eleusine*). When the grain matures or is harvested, there is a sudden exodus of the pest and a massive transfer to the cotton crop. Manipulation of sowing dates offers some possibility of reducing the attack on cotton, as does the exploitation of feeding preferences. Where sorghum is ratooned to provide a second harvest, as in the north-west of Uganda, this practice is thought to account for the rarity of lygus damage to cotton (Taylor 1945); Stride (1968, 1969) has shown that a common weed in Uganda, *Cissus adenocaulis* Steud. ex A. Rich, was more attractive to lygus than cotton and could be used as a trap crop; in California strips of lucerne (alfalfa) between cotton fields attract virtually all the lygus away from the cotton (see Ch. 12).

Aphis gossypii Glov.   **Cotton aphid.**

Description

These small, ovate, soft-bodied, sluggishly moving and gregarious insects, with slender legs and antennae, are characterized by a pair of tubular processes projecting backward and upwards from near the end of the abdomen.

The first arrivals on young cotton plants are winged females, 1.2 to 1.8 mm long, blackish green in body colour, with two pairs of transparent wings. Reproduction is carried on exclusively by viviparous, parthenogenic females; the migrant females give birth to nymphs similar to the adult, which develop into wingless (apterous) females, slightly larger and more globular than the winged form. Multiplication is carried on by the apterous females until circumstances require migration of the colony, when winged forms reappear. Pearson (1958) stated that no eggs or sexual forms of this species had been described, but according to ICI (1979, 1981), eggs are sometimes produced in zones with cold winters.

In the USA up to 51 generations a year have been recorded, and the average life of an individual aphid is 28 days; the rate of development increases between 17 and 28 °C, but the greatest rate of reproduction (2.7 offspring per day) occurs at 19–20 °C.

Distribution

Aphids occur in every cotton-growing country in the world, and tend to become increasingly important following the use of insecticides against other pests.

Symptoms

Aphids feed mainly upon young, growing tissues and are therefore associated with the early stages of the crop. Feeding on young shoots and on the underside of the young leaves, they cause the edges of the leaves to curl downwards and the surface to crumple (Plate 14). The excess of sugars in the sap on which they feed is excreted in the form of honeydew; this sticky secretion covers the foliage, interfering with respiration through the stomata and providing a subtrate for the growth of a black sooty mould, *Capnodium*. If the attack occurs or persists later in the season, the honeydew may drip on to the open bolls, and is one cause of 'stickiness' of lint in the spinning process; sooty mould may discolour the seed cotton.

Control

A high degree of control is normally exerted by predators and a few parasites. The most important predators are the adults and larvae of ladybirds (Coccinellidae), the larvae of lacewings (Hemerobiidae and Chrysopidae) and of hover flies (Syrphidae); the larvae of ladybirds and lacewings have elongate, rather spiny bodies. An entomophagous fungus, *Empusa frensii*, can control an aphid attack in wet weather. A black ant, *Myrmicaria natalensis*, has been observed in southern Africa to transfer wingless female aphids from one plant to another, and feeds on the honeydew.

A number of wild host plants have been recorded, but information on their status is scanty; it is presumed that the aphid passes the non-cotton season on these alternative hosts.

Unless the attack is very severe, it is better to leave control to natural predators, as the early application of insecticides may result in heavier attacks by more serious pests later in the season.

Calidea dregii Germ.   Blue iridescent bug

The adults are quite large, up to 17 mm long, and brilliantly coloured, the upper surface being an iridescent blue or green with a pattern of dark spots or stripes. They feed like cotton stainers, puncturing the green boll to feed on the seeds, and can probably transmit fungi causing yellow staining of the lint. Although several species of *Calidea* are found on cotton in Africa, they rarely do much damage, and *C. dregii* is a serious pest only in Tanzania and Zambia. It does not normally breed on cotton, but the adults can invade a cotton crop in very large numbers when green bolls are present, feed for a short time, and disappear as quickly as they came; they leave an apparently healthy crop, as the punctures are not visible on the surface of the bolls, but when the bolls ripen they are distorted, with damaged seeds and stained, immature lint. It is only by cutting open the green bolls that the damage can be appreciated before the bolls start to open. This happens in parts of the Lake Regions of Tanzania and Kenya, and in northern Zambia, and is one of the main factors preventing cotton growing in the Western Region of Tanzania.

Nezara viridula (L.)   **Green shield bug**

This species is very widely distributed throughout the world, and is usually only a minor pest of cotton. Both nymphs and adults have a distinctive shape resembling a shield, and give off an unpleasant odour when disturbed. Three colour varieties are recognized, the commonest being apple-green. They feed on the green bolls of cotton causing damage precisely similar to that of cotton stainers, and have been shown to transmit internal boll disease in the West Indies.

Helopeltis schoutedeni **Reuter**   **Mosquito bug**

Helopeltis is associated with a moist, warm climate, and various species are pests of tea, cacao and cinchona in India, Indonesia and Africa. As a pest of cotton it occurs in regions formerly occupied by Equatorial forest; thus in Africa it is found in West Africa and parts of Kenya, Uganda, eastern Tanzania, Mozambique and Malawi, being absent from the dry savannah areas.

The adult is a slender, fragile insect with a body about 8 mm long; the males are predominantly yellow in colour, but while some females are yellow, the majority are blood-red. The insect carries a distinctive thorn about 2 mm long projecting from its back.

It feeds by sucking the juices from young, green tissues near the growing point, resulting in leaf tattering like lygus damage, but accompanied by scars and callouses on the stems and branches. In a severe attack the whole growing point of the plant is distorted and growth is stunted; cotton can recover from a light attack. Outbreaks are sporadic and unpredictable, and numbers can increase very suddenly.

Paurocephala gossypii **Russell**   **Psyllid**

This pest appears to be important on cotton only in Zaïre and in parts of Malawi, causing a disorder called 'psyllose' by the Belgians. The insect punctures the leaves and branches to suck the sap; the upper surface of the branches turns a dark reddish colour, while the fully grown leaf turns dark green or reddish and is shed. All parts of the plant are reduced in size, and flower buds and young bolls are shed. The adult is about 2 mm long, pale yellow with transparent wings; the nymphs are relatively immobile, feeding on the leaf veins.

Oxycarenus hyalinipennis **(Costa)**   **Cotton seed bugs, dusky cotton stainer**

This is another widely distributed species which is found in large numbers in the open bolls of cotton. Feeding and breeding are practically confined to ripe seed, so boll development and lint quality are not affected; but the weight of the seed can be substantially reduced and percentage germination lowered. The adults are quite small, about 4 mm long, dull black or brown in colour; both nymphs and adults produce a powerful smell when crushed.

Thysanoptera

Thrips, Caliothrips, Frankliniella spp.   **Thrips**

The nymphs resemble the adults, but are wingless; they pass through three instars, and a fourth, quiescent, stage in the soil without feeding. The adult is 1–1.25 mm long, slender, with narrow wings fringed with long, slender hairs giving a comb-like structure, and mouth parts adapted for rasping and sucking. They feed on the underside of the cotton leaves, which later turn brown on the upper side and silvery on the underside before being shed. They migrate from weeds and other host plants to cotton when conditions are suitable.

Thrips tabaci Lind. is found in the USSR, India, Pakistan, Egypt and the Far East. *Frankliniella* is the main genus attacking cotton in the United States and Latin America. *Caliothrips fumipennis* Bagn. & Cam. and *C. sudanensis* Bagn. & Cam. are the two main species of cotton thrips in the Sudan, and this genus also occurs in Argentina and Brazil.

Orthoptera

Grasshoppers

Grasshoppers and locusts belong to the same family, the Acrididae, the locusts being distinguished by their migratory behaviour (see below). They feed on the green leaves and tissues of a wide range of wild and cultivated plants; they have no particular preference for cotton, which usually suffers only minor damage. There is usually only one generation a year, the eggs taking roughly six months to hatch, and the nymphal and adult stages covering a similar period.

The eggs are laid in a cluster in the ground, surrounded by a frothy fluid which hardens to form an egg pod about 25 mm long. The nymphs resemble the adults in general form, and tend to be gregarious and sluggish; they can cause localized patches of severe defoliation, especially when the hatching of the eggs coincides with the seedling stages of the raingrown crop. Many species of grasshopper have been recorded on cotton, but only a few are likely to be of any importance.

In Africa the most important and widespread are two species of *Zonocerus* which do not attack the Gramineae to any extent. The body, legs and elytra are boldly patterned in bright colours: black, red, yellow, orange and green. The elegant grasshopper, *Z. elegans* (Thnb.), is common in southern and eastern Africa, while *Z. variegatus* (L.) is distributed north of the equator from Kenya to Sierra Leone. In the United States species of *Melanoplus* are found nearly everywhere, and *Schistocerca americana* (Drury) is reported as damaging cotton in Oklahoma.

Locusts

The migratory forms of grasshopper are usually called locusts, and have permanent breeding grounds from which they fly outwards in swarms and settle on temporary breeding grounds, often in areas growing cultivated crops. The wingless nymphs, when they hatch out, crawl along the ground in great numbers looking for food, preferably grass and grain crops, but stripping the green leaves from any plant they find on the way.

In Africa and the Middle East the Desert Locust Control Organization has been successful in controlling outbreaks at or near the permanent breeding grounds, and locusts have caused little or no damage to cotton in recent years. The desert locust, *Schistocerca gregaria* (Forsk.), breeds in the Arabian peninsula and may cross the Red Sea to the Sudan, where the Gash and Tokar Deltas used to be invaded regularly. The red locust, *Nomadacris septemfasciata* (Serv.), breeds in the Lake Rukwa area of Kenya, and used to spread into all the cotton-growing regions of central, eastern and southern Africa in the 1930s. The African migratory locust, *Locusta migratoria migratorioides* (R. & F.), breeds in the flood plains of the middle Niger, and may spread into the cotton-growing regions of western Africa.

In the United States the Rocky Mountain locust is a form of the lesser migratory grasshopper, *Melanoplus mexicanus* Sauss., which invades many of the states west of the Mississippi River to some extent every year; the swarms have at times caused serious crop losses.

Crickets

The crickets, family Gryllidae, sometimes attack young cotton, clipping off the seedlings below the cotyledons in the same way as cutworms. *Brachytrupes membranaceus* (Dru.) is widespread in the rainfed cotton areas of Africa. Among the mole crickets, the *changa* from Puerto Rico, *Scapteriscus vicinus* Scudd., attacks cotton in the southern United States, and *Gryllotalpa africana* P. de B. has been found damaging cotton in Kenya and West Africa. In Pakistan,

Acheta and *Gryllus* spp. will destroy the kernels in cotton seed as well as damaging young seedlings.

Acarina

Tetranychus spp.   **Red spider mite**

Description

The red spider mite passes through three larval or nymphal stages, and feeds actively in all three. In the first stage it has six legs, but the other two stages are eight-legged and similar in shape to the adult. With eight legs, no antennae and no wings, the spider mites are not true insects, but belong to the group Arachnidae. The adults are minute, only 0.3 to 0.5 mm long, and usually bright red in colour, although the colour of the females is more variable. They feed mainly by sucking the juices from the lower surface of the leaves, which they cover with a fine web. The eggs are spherical, semi-translucent or creamy, nearly 0.1 mm in diameter, and laid singly on the leaf or web; the females can reproduce partheno-genetically, which gives rise to males only.

Distribution

Spider mites attack a wide range of host plants, and are widely distributed throughout the world, often appearing as greenhouse pests. They have only become a serious pest of cotton following the use of organic insecticides against other pests. A number of species have been recognized, but as they all have similar habits it is not necessary for the non-specialist to identify them precisely.

Symptoms

The underside of the cotton leaf becomes silvered, as a result of the aggregation of white marks caused by the feeding punctures. Yellow mottling appears on the upper surface, usually concentrated at the base of the leaf near the junction of the main veins (Plate 16); as the attack develops, the leaves of the plant lose their green colour, turning yellow or red, and eventually wither and are shed. The mites on the underside of the leaf can be identified with the help of a hand lens.

Control

As infestations are likely to develop following the use of organic insecticides, the longer their use can be delayed the better. The attack often spreads from isolated groups of infected plants, so that spot spraying with an acaricide may control the infestation if it can be identified before it has spread too far. The defoliation caused by

Table 11.3   *The distribution and importance of the major cotton pests*

no francophone Africa?

	Tetranychid mites	Tarsonemid mites	Aphis gossypii, etc.	Bemisia spp.	Nezara viridis	Lygus spp.	Empoasca spp.	Dysdercus spp.	Helopeltis spp.	Psallus seriatis	Bugs (others)	Thrips	Anthonomus grandis	Anthonomus vestitus	Syagrus spp.	Podagrica spp.	Sphenoptera spp.	Myllocerus maculosis	Estigmene acrea	Pectinophora gossypiella	Pectinophora scutigera
USA	○		●			○	○			○		●	●						*	●	
Mexico			○	○	*	○	○					●	●						*	●	
Colombia	○		○	○	*		*	○		*		*	●							○	
Peru	○		●			○	○	○				●		●							
Nicaragua					*	*		*		*			●						*		
Brazil	●	○	●	*		○	*	○	○			●								●	
Argentina	*		●		*			*		○		●								●	
Greece	○		*	*	*					*										*	
Spain	○		●	*			○			*										○	
USSR	○		○							○	○									○	
Iran	●		○	○		*	*					○								●	
Syria	○		●	○		○	○					○								○	
Turkey	●		●		○	●	○					○								○	
Israel	○		*				*													○	
Egypt	○		○	*	*		*					○								●	
Sudan			●	●	*		●	○	○			○			○					○	
Uganda	*	*	○			●	*	○												●	
Kenya	*	*	○			○	*	●												○	
Tanzania	○	*	○		*	*	*	●	●											○	
Burundi	*	●	●			*	*	○												●	
Malawi	○		○				○	○												○	
Mozambique			○			○	○	●				○								●	
Angola			○				○	●				○								●	
Zimbabwe	●		●		*	○	○	●	○			○								*	
South Africa	●		○									○			*						
Nigeria	*	○	*			*	○	●	*						*					○	
India	○	*	○	●			●	●		○	●							*	●	●	
Pakistan	○		○	●			●	*			○								●	●	
Burma	○		○				○				○									○	
Thailand			○				○					○								○	
China	●		●						○			○								●	
Australia	○		○				○	*				○								○	*

Source: after Ceiba Geigy 1972.

* pest present but of little or no economic importance
○ pest economically important, control measures required
● pest of major economic importance, control measures regularly required often several sprays per season

Bucculatrix thurberiella	Sylepta derogata	Sacadodes pyralis	Alabama argillacea	Cosmophila spp.	Spodoptera (Laphygma) spp.	Spodoptera littoralis	Spodoptera ornithogalli	Plusia spp.	Trichoplusia ni	Diparopsis castanea	Diparopsis watersi	Earias insulina	Earias biplaga	Earias fabia	Earias spp. (other)	Heliothis armigera	Heliothis virescens	Heliothis zea	Xanthodes graellsii	Argyroploce spp.	Cryptophlebia leucotreta	Eutinobothrus brasiliensis	Cutworms	Grasshoppers, locusts

a severe attack makes picking easier, but if it starts before the bolls open, the yield of the crop will probably be reduced. Some differences have been reported in the effectiveness of acaricides against different species and forms of *Tetranychus*.

Among other genera of mites found on cotton, the most common is probably the tea mite, *Hemitarsonemus latus* Banks. Shade and high humidity favour this mite, and the only serious manifestations on cotton have been in the forested areas of Brazil, Zaïre and southern Nigeria. The mites feed on the underside of the leaf, which acquires a dark green, glazed appearance, and the edges of the leaf roll downwards; later the lamina develops irregular, short cracks, and fragments of the leaf fall away. The mites can be seen with a hand lens: they are yellowish and translucent, 0.15 to 0.22 mm long, and the male is extremely active. The symptoms have been recorded as 'acariose' or 'leaf roll' in some reports.

Chapter 12

Insect Control

Insecticides

The introduction of new and effective insecticides since 1945 has
resulted in a tremendous increase in the consumption of agricultural
chemicals; Green *et al.* (1979) quote the following figures for the
world output of pesticides:

	'000 tons
1945	100
1955	400
1965	1000
1975	1800

From 1974 to 1984 world expenditure on pesticides has risen from 5
billion to 13.8 billion US dollars (British Agrochemicals 1985); of
the total in 1984, 1.6 billion was spent on cotton pest control, 75 per
cent on insecticides and 18 per cent on herbicides (Wood Mackenzie
& Co. pers. comm.). The increase in consumption of pesticides has
occurred not only in the developed countries; insecticides are being
used in countries which never used them before, and cotton growing
has been established in areas which would be devastated by insects
if they were not controlled with chemicals. This chapter therefore
starts with a discussion of the application of insecticides in the field
before going on to current ideas on integrated pest management.

Insecticides can be classified into six main types, as shown in
Table 12.1. It is not proposed here to quote detailed recommendations for insecticides and dosage rates in different countries, as
these are so diverse and change so frequently that they would be out
of date by the time a textbook of this kind is published. New
products are continually being put on the market, and older products
may be banned for health or environmental reasons. Trade names
and formulations for the same product vary from country to country,
and some products may not be available in countries where imports

Table 12.1   *Types of insecticide*

Type of insecticide	Examples
Inorganic	Paris green, lead arsenate
Natural	Nicotine, derris, pyrethrum
Organochlorine	DDT, dieldrin, endosulphan, toxaphene
Organophosphorus	Monocrotophos, parathion, malathion, dimethoate
Carbamate	Carbaryl, aldicarb, carbofuran
Pyrethroid	Permethrin, cypermethrin, fenvalerate, deltamethrin

are strictly controlled. The toxicity of an insecticide to human beings and other mammals is considered to be of overriding importance in some countries, while it is ignored in others.

Fixed or variable spray regime

Insecticides can be applied according to a fixed schedule, or by reference to the pests present in the crop. The fixed schedule is typified by the recommendations in Uganda at the peak of its production, when Bowden (1970) stated that 'the spray schedule recommended for control of the bollworm/bug complex is four sprays of DDT at 1 lb active ingredient per acre applied at fourteen day intervals and commencing thirty-five days from seedling emergence.' The schedule has several advantages. The farmer knows exactly how much insecticide he will need and when he should apply it: the system is therefore very suitable for the introduction of insecticides into small-holder farming in underdeveloped countries with little or no experience of pest control by chemicals. The insecticide can be packaged to provide the correct dosage per acre or hectare; it allows a routine procedure to be organized, and simplifies contracts for spraying by ground equipment or aircraft. On the other hand, it depends for its effectiveness on a fairly regular and predictable pest attack every year; Reed (1976) was doubtful of this when he wrote 'throughout Uganda, however, variations in pest attack, in species and severity, in different seasons, are such that a fixed regime cannot be ideal. An optimum fixed regime will be inadequate under heavy pest attacks and wasteful if pest attacks are light.'

To cope with these variations in pest attack, Matthews and Tunstall (1968) developed the technique of regular scouting for insect pests in the crop; the numbers of each insect pest found on a sample of cotton plants were recorded, and the data used to determine when the insecticide should be applied and which insecticide should be used. They showed that a flexible system like this was more efficient than a fixed spraying schedule, giving better yields and usually requiring less insecticide.

The main difficulty in introducing a programme of scouting is the training of a sufficient number of efficient and reliable scouts. In small-scale farming the scouts may operate under the extension service, which in turn will advise the farmers which insecticides to apply and when to apply them; better farmers will want to learn the technique for themselves, while large farms often employ their own scouts, and decide for themselves what the spraying schedule will be. In some countries the extension service operates spraying teams, organizing the farmers in each area to assist them (Zambia, Mozambique and Indonesia). The cost of insecticides and the range of products may be greater or less than with a fixed schedule, and is certainly less predictable, and a different form of credit may be needed. To simplify organization, spraying is commonly timed to take place once a week during the main season, on each occasion choosing the appropriate chemical according to the scouting records. The weekly spray is omitted if the records show little pest activity, or an extra mid-week spray may be added in case of a sudden heavy attack of some pest; some latitude must be allowed for interruption of spraying by rain (Fig. 12.1). A fixed day on which to start spraying every week allows for better organization of other farm

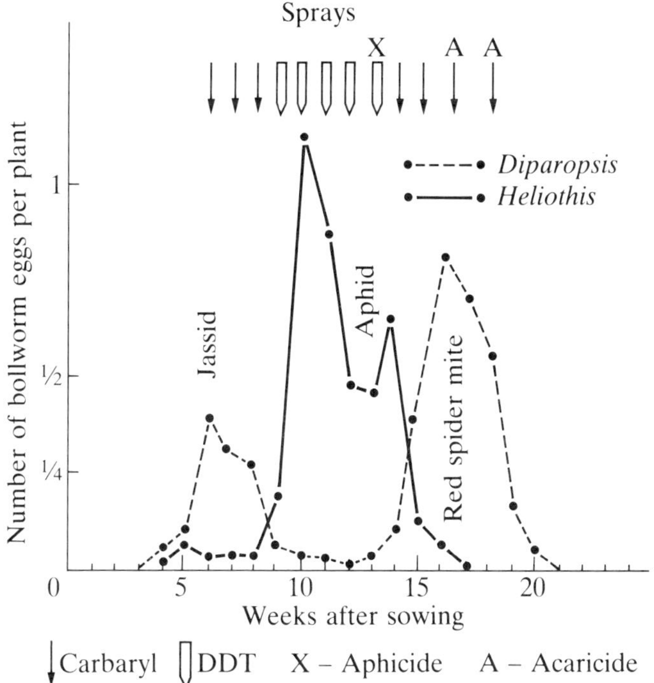

Fig. 12.1 Example of a spraying programme in Malawi (Malawi 1976).

work; where a travelling spray team has to cover a district, or a contract has been made for aerial spraying, the fixed day becomes almost obligatory. The interval of a week between sprays is not entirely arbitrary, as many pests are particularly vulnerable in the young stages and are more easily controlled in the first week after hatching.

Scouting technique

The technique recommended by Tunstall *et al.* (1962) for scouting a field of insects, and adopted throughout central Africa and elsewhere, consists of two parts: in one part at least 24 plants spaced more or less equally along the two diagonals of a field are examined twice a week for the more important pests – in Zimbabwe these were red bollworm (eggs), American bollworm (eggs), jassid and perhaps red spider mite; in the second part the remaining insects are assessed once a week by a general examination of groups of plants on a zigzag course across the field (Fig. 12.2).

Some insects are readily seen, and a count or estimate made of their numbers; others are far from easy to find, and characteristic signs of damage are used to reveal their presence and status. Early warning of a pest attack is important, because in general the older the larvae and the later the nymphal stage the less susceptible they are to pesticide chemicals and the more damage has been done. For example, it is possible to identify and count the number of eggs of American or red bollworm rather than the larvae or flared squares, thus gaining an earlier warning of attack; but the correlation between egg numbers and larval damage needs to be checked, as egg counts are not always a reliable guide to an impending infestation.

Different methods of recording are used for different pests; the eggs of red and American bollworm, larvae of spiny and pink bollworm, buds punctured by boll weevil, leaf tattering by lygus, and so on. As plants get older, the count may be made on a limited sample of leaves or buds: for example, the number of jassids on the top four fully expanded leaves. Small insects like aphis, whitefly and red spider mite may be too small or too numerous to count individually, and are recorded as low, medium or high infestation.

Recording is easier if specially prepared forms are used; these forms also ensure that the required number of plants are examined and are a reminder that all the relevant pests are recorded. The form used in Malawi is shown in Fig. 12.3 as an example, but the number of plants examined and the pests recorded will vary according to local conditions. A simple system of recording the eggs or larvae of a limited number of pests is the peg board (Fig. 12.4), developed in Malawi by Beeden (Malawi 1976). This consists of a strip of blockboard, approximately $20 \times 6 \times 1.3$ cm, with 3 lines of

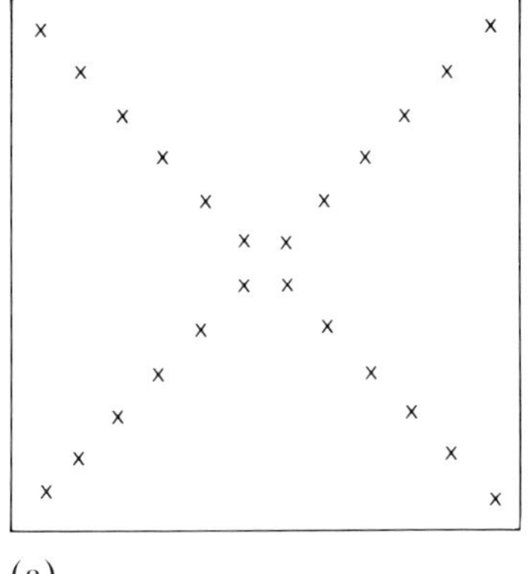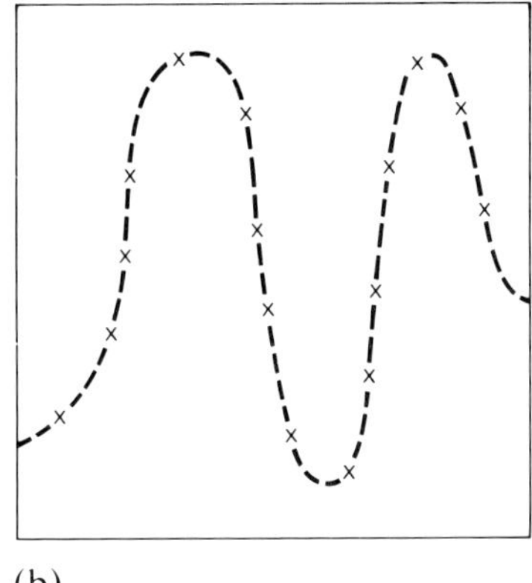

(a)   (b)

Fig. 12.2 Scouting for insect pests (adapted from Malawi 1976): (a) main pests recorded twice a week on 24 or more plants along two diagonals; (b) at least 12 plants chosen at random once a week on a zigzag course across the field, and examined for other pests.

25 holes each in which a peg or matchstick can be inserted. The first line (white) records the number of plants scouted, the second (blue) and the third (red) recording the numbers of two different pests. Lines can be drawn to show the pest level at which spraying is recommended (a level of 5 is shown in Fig. 12.4). The white line is divided into 2 halves, corresponding to 12 plants in each of 2 diagonals scouted in the field.

Critical levels

The cotton plant can tolerate a certain amount of damage; it normally produces more leaves than it needs, and can compensate for the loss of a proportion of the buds and bolls produced (Dale 1962). Current ideas on pest control admit a certain amount of damage and only resort to insecticides when a critical level is reached. This reduces the number of sprays, which in its turn reduces costs directly, allows the development of beneficial insects, and reduces the likelihood of the pest developing resistance to the insecticides.

Critical levels above which spraying should start have to be determined for each pest, and may be set at zero; in the case of a pest known to be a regular menace spraying should start at the first sighting. When the critical level is above zero, say 5 per 100 plants, there should be some discretion in its use; a slow build-up might be permitted to exceed the critical level in the expectation that it would be controlled by natural enemies, and higher levels might be permitted towards the end of the season when the crop is less vulnerable.

(a) Name of Scout: *C. B. MCHIRIMBA*   Name of Farmer: *W. MAUGENA*

Area: *KASUPE WEST*   Village: *CHANTHUNYA*

EGG COUNT FORM

(b) GENERAL SCOUT FORM

Instructions:

1. Choose 12 plants at random while making a zig-zag tour of the garden.

2. Examine each plant to see if one of the four indicated pests is present. If so place a tick in the appropriate place on the form. At the end of the scout, count the number of ticks to help remember on how many plants the pest was found, i.e., if red spider mite was found on four plants, the box will have four ticks.

3. Refer to the following table to judge whether or not to recommend spraying against the pest.

0 - 3 ticks	Nil or Light Attack	No need for special action except in case of a severe aphid or red spider attack restricted to a part of the garden.
4 - 7 ticks	Medium attack	Recommend spraying the required insecticide or insecticides.
8 - 12 ticks	Heavy attack	Recommend spraying the required insecticide or insecticides.

4. Combine the recommendation based on the General Scout with the recommendation based on the Egg Count to make a final recommendation to the farmer.

Fig. 12.3   Example of insect scouting record forms: (a) Bollworm eggs (b) General pests (Gower and Matthews 1971).

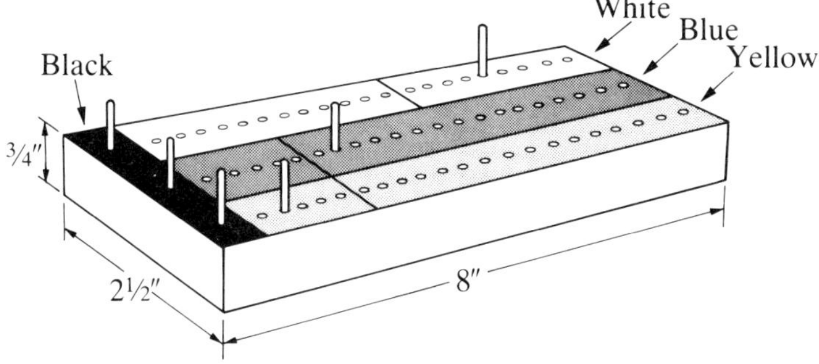

Fig. 12.4   Peg board for recording insect counts (Malawi 1976).

Localized or spot spraying can be employed against pests like red spider mite, which tend to start in small colonies and spread outwards to give a general infestation; if these colonies are controlled at an early stage, the attack can be contained and a general infestation prevented.

Application of insecticides

Insecticides are supplied in the form of dusts, granules, liquids or wettable powders. Granules are suitable for systemic insecticides which can be applied to the soil and taken up by the plant roots, but most insecticides have to make contact with the insect itself, either in flight or crawling on the surface of the leaves and flowers. To do this the active ingredient (a.i.) must be evenly and finely dispersed over the surface of the plant, not an easy matter with a dosage rate for most modern pesticides of 1 kg or less of a.i. per hectare. One litre or one kilogram applied evenly to a hectare of cotton with a leaf area index of 10 will provide only 1 particle 100 microns in diameter per cm^2 of leaf surface; to provide adequate coverage, the particles of a.i. must be much smaller than this (Green *et al.* 1979). This is usually achieved by diluting the pesticide with an inert carrier so that the number of droplets is increased, but each one contains only a fraction of a.i.; in a dust, the carrier is a finely ground mineral, but sprays are usually diluted with water, the safest and most readily available diluent.

Concentrates for dilution with water are of two main types: emulsifiable concentrates (EC) in which the product is insoluble in water, but dissolved in oil together with an emulsifying agent; and wettable powders (WP) in which the product is partially or wholly soluble in water, but mixed with an insoluble filler to allow it to be finely ground to a rapidly soluble state and to prevent stickiness or caking in storage. Only a few products are fully soluble in water, and can be sold as concentrated solutions, but various technical problems make this unusual.

The amount of dilution depends on how the pesticide is to be applied. Application rates can be divided into high volume (HV), low volume (LV) and ultra-low volume (ULV), but there is no generally accepted definition of these classes. Low-volume sprays usually range between 50 and 200 litres per hectare, but the Malawi (1976) schedule ranges from 56 to 280 litres according to height of plant. Morton (1973) calls an aerial application of 10 to 20 litres 'very low volume', and includes anything up to 10 litres per hectare in the ULV class; Johnstone (1971) and Mowlam *et al.* (1975) give 5 litres as the upper limit for ULV spraying.

A large volume of liquid can be used to give complete coverage of the foliage – spraying to run-off – but HV sprays like this are seldom used on cotton because they need a plentiful supply of water, large tanks and powerful pumps. Low-volume sprays are commonly used for hand-operated and tractor-mounted sprayers on the ground and for aerial spraying, but ULV spraying is becoming increasingly important.

In the cotton-growing areas of the tropics and subtropics, with rainfall averaging less than 1000 mm per annum, it is often difficult to find sufficient water for LV spraying. It usually has to be carried some distance uphill from a stream or well, and is stored in the cotton field in oil drums (200 litres) or similar containers which are refilled during the intervals when spraying is not being done. Ultra-low-volume spraying is therefore an attractive proposition, requiring less water and covering the ground more quickly, but the finer droplets cannot be directed on to the plant surfaces; it takes skill to make use of wind currents to deposit the spray evenly.

Increasing the volume of the spray or dust by diluting it makes it more visible, so that the spray pattern can be checked by eye and a blocked nozzle can easily be spotted. Whatever the volume, the distribution of a spray can be monitored more precisely by spraying a solution of a brightly coloured dye on the crop, using the same equipment and technique as for the routine application of insecticides. A suitable dye is lissamine scarlet, 15 cc (1 tablespoonful) mixed in 12 litres of water (G. A. Matthews pers. comm.); if the droplets are very small and difficult to see on the green leaves, white cards or pingpong balls can be attached to various parts of the plant to show them up. Fluorescent tracers can be used in the same way, or added to the pesticide solution.

Spraying equipment

Some form of equipment is necessary to put the chemical insecticide in contact with the insect pest in the field. The chemical should be applied at the right time, in the right place, with the best combination of effectiveness and cost. The timing has been discussed in the previous sections; placement will depend on the pest and its habits; cost-effectiveness involves a wide choice of chemicals, formulations, dilution, droplet size and the availability of labour, working capital, water and services. Equipment may vary from a bucket of dust applied by hand, or a bucket of liquid sprinkled over the crop with a brush, to aircraft fitted with rotary atomizers or electrostatically charged sprays which are attracted to the plant.

The habits of the insect pest will determine where the insecticide should be placed. Some insects, such as aphis, whitefly and red spider mite, tend to feed on the underside of the leaf, and an upward-directed spray may be necessary for their control; some larvae, such as *Heliothis*, are mobile and the spray coverage need not be so complete as it is for those which are static or only move for a brief period, like red and pink bollworms; *Heliothis* tends to feed on the upper branches of the cotton plant, while red bollworm feeds at all levels, and is controlled more efficiently with a tailboom (see below); cutworms and some stem borers attack the base of the stem, and the spray may be more effective when applied to the soil rather than to the plant.

Complete coverage of the plant is fairly easy with high-volume sprays, but with lower volume the droplet size becomes increasingly important; ULV sprays are dispersed in very fine droplets to distribute a few litres of spray over a hectare of cotton. Scarcity of water is often the main reason for the use of ULV sprays, scarcity of labour for the use of tractor-mounted sprayers or aircraft. On the other hand, lack of capital and service facilities may limit the use of such machines, and the grower has to do the best he can with cheaper and simpler equipment.

Nozzles and atomizers

Dispersion of the insecticide is usually achieved in one of three ways: through nozzles directly on the plant, through a nozzle into an airstream which carries it into the crop, or fed on to a spinning disc which breaks it into droplets which are thrown off by centrifugal force, and disperse like a mist or fog.

When spraying directly on the plant, the dosage rate and droplet size depend on the combination of pressure, nozzle design and aperture. The jet from the nozzle may be in the form of a fan or a cone, the latter being most common for insecticide application, and the design of the nozzle may affect the droplet size.

Airblast sprayers for cotton do not require a pressurized tank, as a stream of air from a fan directed over the nozzle draws the chemical from it as in the carburettor of a motor car engine; the aperture in the nozzle or jet and the rate of airflow control the dosage rate and the droplet size.

The spinning disc or rotary atomizer is fed either by gravity or pressure through a nozzle, the rate of flow being determined by the size of the nozzle and the pressure behind it. This type of atomizer can not only produce finer droplets than the other two types, but also the size of the droplets is much more uniform.

Developments are taking place in electrodynamic spraying (Matthews 1982). The application of an electrostatic charge to the spray droplets helps to keep them separate from each other; the charged droplets are attracted to the plant, giving a better coverage of the plant surface, reducing the proportion that falls on the ground and minimizing the risk of downwind drift. The application of a high voltage charge to the liquid before dispersion can be used to give very accurate control of droplet size by adjusting the voltage and flow rate.

Nozzles are easily blocked by small particles of dirt in the spray. To prevent this filters are installed in the spray line or in the nozzle assembly, but water used in mixing chemicals should be strained through wire or plastic gauze. Nozzles and filters should be cleaned thoroughly after use with clean water, before the insecticide residues can dry out in the system. The nozzle aperture enlarges gradually with use, and nozzle tips should be checked for wear at least once a year. Plastic is being used more and more instead of brass for nozzles and spray fittings; the best of these have good resistance to erosion, and are less expensive so that they can be renewed more often (Matthews 1982).

Knapsack sprayers

Most small-holder cotton is sprayed by hand, as the capital cost is relatively small and spraying can be carried out by the farmer himself, his family or by hired labour.

The simplest form of sprayer has a lance fitted with one or two nozzles and a reciprocating pump incorporated in the handle; the pump is operated by moving a slider with one hand while the other hand holds the lance; insecticide is carried in a container on the back or at the waist of the operator.

Probably the most common type of knapsack sprayer has a metal or plastic tank carried on the back, with an air reservoir which is pressurized by a handle at the side operating a pump below or inside the tank. While in use the pressure should be kept fairly constant, usually between 30 and 50 psi, and a pressure gauge is a useful addition. These sprayers can feed up to about eight nozzles arranged either on a horizontal, hand-held lance or on a tailboom hinged to the top of the tank behind the operator.

The tailboom (Fig. 12.5) was developed in central Africa by Tunstall *et al.* (1961, 1965) to give spray coverage to all parts of the

Fig. 12.5   Knapsack sprayer fitted with tailboom.

plant including the underside of the leaves, which is particularly important in the control of red bollworm (*Diparopsis*). The output and direction of the spray can be varied in relation to the growth of the plant, while the concentration of insecticide in the spray remains constant throughout the season. The tailboom consists of a vertical boom attached to a pivot at the top of a frame mounted on the back of a knapsack sprayer; it is held by a spring to the bottom of the frame, to allow the boom to yield if it meets an obstruction. The nozzles are carried in pairs on crossbars which can be bolted to the vertical boom at any height required, and the number of pairs used can vary from one to four, depending on the height of the cotton plants (Fig. 12.6).

The tailboom adds considerably to the weight of a knapsack sprayer, but is widely used in southern Africa where red bollworm is a serious pest. Tailbooms can be mounted on ox-drawn equipment and fitted to tractor sprayers, where the weight is less important.

A typical knapsack sprayer holds 12 to 15 litres of spray, and will

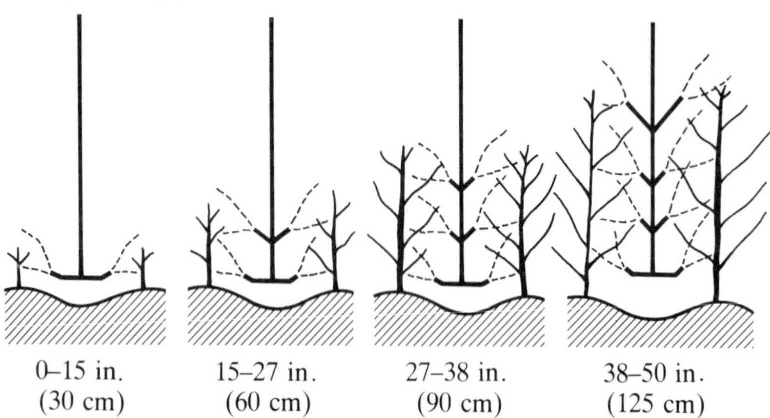

Fig. 12.6 Nozzle positions on tailboom for different heights of cotton plant (Malawi 1976).

need to be filled about 20 times (300 litres) to spray a hectare of fully grown cotton.

A less common type of knapsack sprayer has an airtight metal tank which is partly filled with spray liquid. The tank is then closed and air pumped in to a pressure sufficient to discharge all the liquid before the pressure falls too low for the nozzles to function satisfactorily. These tanks are heavy, and the change in operating pressure as the tank empties is a disadvantage.

Motorized knapsack sprayers have a small petrol engine mounted on the tank, driving a fan to produce a blast of air through a flexible tube; insecticide is fed into this airblast from a nozzle, and the mouth of the tube is directed by hand at or into the crop. The airblast disturbs the leaves and drives the insecticide into the foliage.

Although not strictly knapsack sprayers, light and portable sprayers have been developed for ULV spraying (Matthews 1981). They comprise a spinning disc atomizer driven by a small electric motor (Fig. 12.7); the motor is powered by batteries carried in the handle of the sprayer, and insecticide is fed by gravity through a nozzle on to the disc. A small bottle of insecticide screws into a socket behind the spinning disc, and is inverted when in use by rotating the handle, allowing the insecticide to drop by gravity on to the disc. The spray is carried downwind, and can cover from six to two rows of cotton, as the size of the plant increases. Although designed to apply oil-based ULV formulations, so that the droplets do not dry out before making contact with the plant, the cost of the solvents make these much more expensive than water-based formulations. Mowlam *et al.* (1975) found that high-quality wettable powders and emulsifiable concentrates could be used successfully at an application rate of 6 to 12 litres per hectare. In field tests the

Fig. 12.7 Battery-powered sprayer used to apply insecticides at ultra-low volume (S. J. Brown).

water-based formulations applied with hand-held ULV sprayers were cheaper than other methods of application, and gave equally good insect control and yields.

Tractor-mounted sprayers

A tank and spray boom can be mounted on a tractor, usually at the back so that the tractor driver is not contaminated with the spray. The tank is filled with insecticide mixed with water, and a small centrifugal pump driven from the power take-off (PTO) delivers the insecticide from the tank to the nozzles on the boom at a constant pressure. Some of the liquid may be recirculated to the tank to agitate the mixture and prevent any solids from settling out to the bottom. The rate of application is controlled by the tractor speed, the pressure from the pump and the size and number of nozzles on the boom. In some models the same pump can be used to fill the tank with water from a pond or canal.

The clearance under a standard tractor may be insufficient to allow it to pass over a mature crop without damaging the tops of the plants, and there is a limit to the height at which the boom can be set. Special high-clearance sprayers have been designed (e.g. the High Boy) in which the engine and spray boom are raised well above the crop; as these units are only used for spraying, they are economic only when sufficiently large areas have to be sprayed to justify the extra capital cost.

The passage of a sprayer of whatever kind – knapsack, tractor-mounted or high-clearance – will cause some damage to a well-grown crop where the branches from adjacent rows have met and interlocked. It is generally reckoned that the losses caused by this damage are small compared with the benefits obtained, but they should be minimized as far as possible. For repeated sprayings, the tractor can be driven along the same rows in the same direction each time, thus confining the damage to these rows. Some farmers leave empty rows at suitable intervals at sowing time, so that the sprayers can be driven along these rows. As in skip-row planting, the adjacent rows will compensate for the gaps to a large extent, by producing more crop. The damage can be avoided entirely by aerial spraying.

Aerial spraying

Aerial spraying does no damage to the crop, and can be carried out irrespective of ground conditions, as when the soil is too wet for a tractor to operate. It is fast – an aircraft can easily cover 40 ha of crop in an hour – leaving the insect pest no time to migrate back into the sprayed area from that not yet sprayed. Water and chemicals do not have to be distributed about the field, as all mixing and loading are carried out at the airstrip. In a well-grown crop, the spray penetration may be less good than that achieved with a tailboom or airblast ground sprayer, but this is only important with certain insects such as red and pink bollworm; in general, aerial spraying is only slightly less effective than ground spraying. Penetration is sometimes better with a helicopter than with fixed wing aircraft, because the down-draught of the propeller forces the spray downwards into the crop, but it covers the ground more slowly. In windy weather the problem of spray drift can be more serious than it is with ground equipment.

The ideal conditions for aerial spraying are large fields with a long uninterrupted run between turns, level ground and no obstacles such as trees and telephone or power lines. The aircraft, mostly fixed wing (Fig. 12.8) rather than helicopters, are normally owned and managed by a firm specializing in crop spraying; the farmer makes a contract with the firm to have his cotton sprayed at weekly intervals for perhaps eight weeks, using the insecticide appropriate to the pests of the week as determined by scouting. Ground signals are provided by the farmer to guide the pilot in order to ensure complete coverage of the area. Technicalities such as swath width (commonly 18 m), volume (commonly 20 litres per ha) and droplet size (commonly 200 microns) are kept under constant review (Miller 1969). Most aircraft are fitted with a spray boom and nozzles, but increasing use is being made of rotary cage nozzles and spinning

Fig. 12.8   Aerial spraying of insecticides (Union Carbide).

discs, which give a finer and more uniform droplet size (Matthews 1982).

Care of equipment

All spraying equipment must be thoroughly washed out at the end of each day's spraying, as spray residues are difficult to remove if they are allowed to dry, and will cause blockages in the spray lines. Nozzles and filters should be dismantled and washed in clean water before they dry out, and the tank, pump and pipes flushed out with water. The nozzle tips must be treated gently, and not cleaned out with wire or pins; operators must be warned against clearing an obstruction by putting piping or the nozzle to the mouth and blowing. After the sprayer has been reassembled, clean water should be pumped through the system, checking the flow from each nozzle. The outer surfaces should be kept clean, and any unpainted metal and metal joints kept lightly oiled.

Safety precautions

The need to inculcate a proper respect for the poisonous qualities of insecticides can hardly be overstressed; practically all insecticides are more or less poisonous, and where there is a choice between two equally effective products, the less poisonous should be chosen. Safety precautions are usually given on the labels of the containers, but many users are illiterate, or may not bother to read them. Operators must be taught to handle insecticides with care, avoiding any direct contact with bare skin and the inhalation of fumes or spray; the hands and face should be washed with soap and water after handling insecticides, especially before eating and smoking.

The use of protective clothing presents a problem in hot climates. Ideally anyone handling insecticides should wear overalls, gumboots, gloves and a mask; overalls and gloves should be washed regularly – immediately if insecticide is spilt over them. The individual farmer should be made aware of these dangers, but cannot be forced to take full precautions. An employer is, however, responsible for the safety of his employees; he must ensure that all the necessary safety equipment is provided and available, and endeavour to see that it is utilized as far as possible, especially by those handling the concentrate before dilution. A supply of clean water and soap should always be available close at hand.

Insecticides should be stored in a dry place, away from sunlight, footstuffs, drinking water, animals and children. If possible they should be kept in a locked cupboard or store. Empty containers should be collected, so that they cannot be picked up and used by unsuspecting persons; cardboard and polythene can be burned, while bottles, tins and drums should be rinsed out, crushed and buried. It is absolutely impossible to clean out a container sufficiently well to make it safe for use as a cooking utensil, or for storing foodstuffs and drinking water (Malawi 1976).

Care must be taken that supplies of drinking water are not contaminated with water used for rinsing out containers, cleaning the sprayer or washing protective clothing. The water should be disposed of in a soak-away pit well away from streams, ponds, dams or wells. Fish are highly susceptible to very small amounts of insecticide. If livestock are grazed on cotton residues after harvest, a sufficient interval should be allowed after the last spray for the insecticide to become inactive; the interval will depend on the active ingredient – seven days for carbaryl and dimethoate, three weeks for endosulfan (Malawi 1976) – and no grazing should be allowed after using DDT or highly toxic chemicals.

If insecticide poisoning is suspected, medical attention should be obtained as soon as possible, and the doctor will want to know what insecticides have been used.

Integrated pest management

The cost of controlling insect pests with chemicals has tended to increase to such an extent that it is rapidly becoming uneconomic, and growers have eventually had to accept the advice of the entomologists who had understood the dangers at an early stage. The environmentalists have also been protesting against the indiscriminate use of agricultural chemicals; the use of DDT and certain other chemicals on field crops has been banned in the USA, and subsequently in several other countries.

Peru was the first country to enforce a more rational approach to insect control, which included limitations on the use of organic insecticides, and has become known as *control integrado*, integrated pest control or integrated pest management (IPM). The credit for *control integrado* in Peru goes to Dr J. E. Wille, a German entomologist appointed to La Molina agricultural research station in 1929. He encouraged the use of insecticides for the control of cotton pests, but suggested about 1954 that the integration of biological and chemical methods would provide a cheaper and more effective system of pest control. The new system was tried out on a field scale from 1957, and was made compulsory for all cotton farmers by legislation passed in 1961–62.

The system included recognized insect control measures such as a close season, restricted planting dates and the destruction of alternative host plants, which are discussed in more detail in later sections. Bollworm egg parasites of *Trichogramma* sp. were bred locally for release in the cotton fields. Individual farmers were free to spray or dust their cotton with calcium and lead arsenate, and with sulphur, but official permission was required before any organic insecticide could be applied; this permission was only given after the crop had been inspected by qualified observers, highly skilled entomologists who could make an independent assessment of the pest situation and confirm that other methods of control had proved inadequate.

The number of applications of organic insecticides has varied from season to season, usually averaging between one and two, and rarely exceeding four per season. This compares with 6–10 applications in Equador, 12–14 in Colombia, and as many as 30 or 40 in Central America; it is estimated that the cost is less than a quarter of what it would have been with uncontrolled use of insecticides, while yields have been maintained or even improved at the same time (Ordish 1969, Bachini 1980).

Conditions in Peru were especially suitable for such an approach, as the cotton is grown under irrigation in coastal river deltas separated by dry desert areas; these closed communities have no problem of reinfestation from alternative host plants in the surrounding areas. But the effectiveness of IPM programmes does not depend on these special conditions, as was at first thought. The major research programme mounted against the boll weevil in the USA has developed similar recommendations, and has led eventually to legislation establishing controlled communities and weevil-free zones in the cotton belt. Since the 1970s, IPM programmes have been adopted all over the USA, some by voluntary cooperation among farmers, others promoted by government agencies (Keller 1981).

The current approach to insect pest control is to reduce the dependence on chemical insecticides to the minimum, and to put much more emphasis on other measures to reduce infestation (Adkisson *et al.* 1982). While complete elimination of the pest may be possible in some cases, more usually the application of insecticides is delayed as long as possible, tolerating a limited level of pest damage; a completely pest-free crop is not the objective, but one in which the losses due to pest attack will not greatly exceed the losses which occur naturally in a healthy crop.

Integrated pest management takes many forms, which usually depend on the life history and habits of the insect concerned. For this reason a thorough study of the insect pest is essential before suitable measures can be planned to divert it from the crop, some of which have been discussed in the previous chapter under individual pests. Measures which are effective against one pest may encourage another, and it is then necessary to decide which is the most important, or which can be more easily controlled by other means; an example is leaf hairiness, which deters jassids but encourages whitefly.

Cultural methods

The most important of the cultural methods of control is probably the destruction of crop residues immediately after harvest, burying or burning the cotton stalks and damaged bolls left on the plant or on the ground (see Ch. 17). This provides effective control over pests and diseases which can survive only on cotton or closely related species, of which the most important are the boll weevil, the leaf miner *Bucculatrix* and bacterial blight of cotton. Control is rather less effective against pests which feed freely on other malvaceous species besides cotton, but it makes their survival more difficult. There are quite a number of such pests, including the red, pink and spiny bollworms, cotton stainers and several species of leaf-eating caterpillars and beetles. Crop destruction is usually supplemented by a legal close season, which lays down a date by which all cotton residues should be destroyed, and the date before which no cotton must be sown. This breaks the reproductive cycle of the pest, and may also reduce the value of the diapause in the carry over from one season to the next; diapause occurs most often in larvae or pupae produced late in the season, so early destruction reduces the numbers of diapause forms; the close season may be so arranged that emergence from diapause occurs when no satisfactory food supply is available (see red and pink bollworm, Ch. 11).

Loose seed cotton lying in fields, at loading points, along road sides, in stores and ginneries, should be collected and burned or buried; stores should be cleaned out and fumigated at least once a year as soon as they are emptied.

The time of sowing may be arranged so that the peak pest population occurs at a time when the crop is not vulnerable. Several pests tend to build up late in the season, and early maturing varieties or early sowings may set their crop before the insect reaches pest proportions. In Malawi a series of sowings which used to run from November to April was replaced in 1946 by a regime which concentrated the sowing of the crop in December; this resulted in an earlier harvest, greatly reducing the formation of diapause pupae on which the red bollworm depended for its survival between seasons (Pearson and Mitchell 1945). There is a wide choice of sowing dates with irrigated cotton in the tropics, but within each area the range of sowing dates should be kept as narrow as possible to prevent pests moving from early to late sown crops. The timing of other crops which are alternative hosts may affect the attack on cotton – in Uganda the main lygus attack occurs when sorghum is harvested, and the adults migrate to the cotton crop (Pearson 1958); on the other hand the alternative host crop may be so timed as to attract the pest away from cotton.

Alternative host plants

Weed control, in addition to reducing weed competition, reduces the pest population by eliminating a source of infestation by pests which can breed on the weeds. Spiny bollworm in Uganda was shown by Reed (1974a) to depend largely on the presence of *Abutilon* as a weed in or near the cotton crop; stainers often breed on weeds of *Hibiscus* spp. before the bolls of cotton open and allow breeding to continue in the crop. Infestation may also occur from patches of weeds surrounding the crop, and can be reduced by removing these weeds or spraying them with insecticides.

Sweeney (1960) in Malawi found that four tree species, of which *Sterculia africana* was the most common in the lower Shire valley, were the only hosts from which stainers fly directly to cotton in any large numbers. Control methods applied in test areas included the destruction of these trees within a 10-mile radius and treating the soil around the trees in another area with insecticides, to prevent the initial invasion of the crop from the alternative hosts. Stainer numbers in the subsequent seasons were in fact too low to provide an adequate test of these methods.

Trap cropping

Trap cropping depends on diverting the pest from the cotton crop to another host plant, and is therefore most suited to the control of pests with a wide range of hosts. The diversionary crop must be more attractive to the pest than cotton, and must be at its attractive stage during the period when cotton is vulnerable to attack; the timing of the trap crop is often critical, as a wrongly timed crop may increase instead of decrease the attack on cotton. Preferably the damage to the trap crop should be small, so that it yields a useful return. If the proportion of trap crop is small, it will be more economical to attract the pest to it and then apply insecticide to the trap crop, rather than to spray the whole cotton crop (Stride 1969). The pest population on the trap crop can sometimes be useful in building up a population of parasites and predators, to such an extent that the next generation of the pest is much diminished; this is claimed by Harland (1953) to happen with *Heliothis* on maize in Peru. This must be taken into account when considering the desirability of applying insecticides to the pest on the trap crop.

The interaction of cotton and maize in attracting *Heliothis* is probably the most widely researched example of trap cropping. Maize is attractive for 15–20 days during the tasselling period, while cotton is susceptible for a much longer period of about three months; in practice, however, the important attack on cotton usually takes place in the early part of the flowering period, and a short succession of maize sowings is sufficient. It is theoretically possible to plant a series of maize crops to attract *Heliothis* from cotton, cutting down each maize crop when the infestation has reached its peak after tasselling, but the bollworm does little damage to the maize cobs, and it is unlikely that a peasant farmer can be persuaded to destroy a useful crop. Maize may be inter-planted in the cotton crop to increase its effectiveness and one acre of maize to five of cotton was found to be sufficient to reduce *Heliothis* to the status of a minor pest in South Africa (Parsons and Ulyett 1934). Harland (1953) describes a method used successfully in Peru, where a row of maize is planted every 10th row through the cotton field, the maize acting both as a trap crop and a reservoir for bollworm parasites. Similar forms of inter-planting are common in many countries, though the reason for doing so is not clear. Belts of late sown maize mixed with cowpeas were recommended long ago by Quaintance and Brues (1905) in the USA; in Queensland, Australia, Wells (1942) found that 6 per cent of the cotton area sown with 3 successive sowings of maize acted successfully unless the moth population was very high. On the other hand, maize in Southern Rhodesia (Zimbabwe) has not shown the same attractiveness as elsewhere (McKinstry and Prentice 1937) and has failed as a diversionary crop in the Sudan (Bedford 1937).

The tarnished plant bug (*Lygus* spp.) in California and Arizona illustrates how migration to cotton from an alternative host crop can be discouraged, as well as the use of the alternative host directly as a trap crop. The alternative host crops are alfalfa (*Medicago sativa* L.) and safflower (*Carthamus tinctorius* L.), both of which are preferred to cotton but suffer little damage from lygus; egg laying on cotton occurs when the alternative hosts become less favourable for feeding by maturity or harvest. The alfalfa is cut for hay once a month. A scheme has been devised for sowing and harvesting the alfalfa fields in alternate strips, in which half the strips are harvested at mid-month; the adult lygus will lay their eggs on the half-grown hay of the other strips rather than migrate to cotton, and the nymphs do not have time to mature before the hay in these strips is cut two weeks later (Moore 1973). For safflower, a single application of insecticide is recommended when the lygus nymphs have nearly completed their development. As a trap crop for lygus, Stern *et al.* (1969) sowed strips of alfalfa 6 m wide and 150 m apart in cotton, the strips being cut for hay every month. The population of adult lygus on the cotton remained very low, the numbers being only about 1 per cent of those on the alfalfa.

Breeding methods

The use of early maturing varieties to reduce the damage caused by late season pests has already been mentioned. At the other end of the scale, varieties which continue to flower prolifically late in the season may be able to compensate for losses of fruiting bodies caused by early season pests. It has been shown that the deliberate removal of early buds does not necessarily reduce the final yield, in fact in certain circumstances it can enhance it (Eaton 1955, Goodman 1956, Brown 1965, Evenson 1969), and differences between varieties in their reaction to bud removal have been noted. Although yield may not be reduced, the crop will ripen later and the plants will grow taller. This applies to many varieties produced in those parts of Africa where little or no chemical pesticides are used (Reed 1965); the harvest will naturally be later in these circumstances.

The breeding of pest-resistant varieties is an attractive proposition, but is limited by a scarcity of parents showing appreciable resistance. The most striking success in cotton has been the jassid resistance of varieties with hairy leaves (Parnell *et al.* 1949). This resistance has been confirmed in India and elsewhere, although it is claimed there that other factors besides hairiness are either involved in the resistance or can act independently in smooth-leaf cotton (Batra and Gupta 1970).

Resistance to insect attack may take three forms: host preference, where cotton is less attractive to the insect than alternative host plants; tolerance, where some damage may occur, but not sufficient to affect the crop seriously; and antibiosis, where some constituent of the plant disturbs the normal development of the feeding insect. The following characters are known to confer some resistance, and information is accumulating about their resistance mechanisms and the insect pests against which they are effective (Maxwell and Jennings 1980, Maxwell 1980):

Absence of nectaries	Frego bract
Deciduous bract	Gossypol content
Okra leaf	Glabrous leaf
Red plant colour	Hairy leaf
Male sterility	Factor X

Probably the most positive resistance comes from gossypol, which is secreted along with other terpenoids in the small blackish glands found on various parts of the plant, and especially on the cotyledons. It is the converse of the glandless cotton bred to produce seed which is free of gossypol, so that the meal may be used as a protein food for human consumption and for feeding pigs and poultry (Cross 1962). It was found that gossypol-free cotton was attacked by several insect species which had not previously been recorded as cotton pests (Gillham 1965, Lyon 1970); the normal level of gossypol, or one of the related terpenoids, obviously conferred some resistance to insect pests, and as a corollary, high gossypol should confer more resistance. This was supported by the resistance observed in certain species of cotton – Tanguis (*G. barbadense*) in Peru (Lukefahr *et al.* 1966) and some of the American diploid species (J. A. Lee 1966) which have a high density of glands on the plant body and boll walls. The breeding of high gossypol varieties which can compete with commercial varieties in yield and quality has turned out to be more difficult than expected, but progress is being made, and regional tests of bollworm resistance have been run in the United States for the last seven years (Sappenfield and Dilday, 1980).

The main problem in breeding for resistance is in testing the resistance achieved in terms of yield. Besides the usual difficulties with plot size, control plots and guard rows common to all entomological field trials (see Ch. 16), any one factor is unlikely to have a large enough effect on yield to be measurable: the significant difference in a normal field trial is usually about 20 per cent of the mean. Smaller differences can be detected by increasing the number of replications, or repeating the trial at different sites. It has been mentioned, however, that resistance may be specific to a particular pest; the problem then is how to control other pests in the trial without affecting the particular one being studied.

Resistance is usually measured by the reduction in numbers of the insect population. This reduction may well be statistically significant, and is less affected than yield by the presence of other pests, but the practical grower wants to know what effect this reduction has on the final crop, and whether it is sufficient to justify a reduction in the application of pesticide chemicals. One approach being tried is to combine three or four resistance factors in a single variety, in the hope that their combined effect will be of a measurable size (Lukefahr *et al.* 1975).

Parasites and predators

Insect populations are limited to a greater or lesser extent by parasites and predators; the trouble with insect pests is that the natural enemies build up more slowly than the pest itself, which may have already ruined the crop before it is itself controlled. It has been shown that it is usually worth while to accept a certain amount of damage by early season pests such as aphids and thrips to avoid killing off the parasites and predators of more serious pests such as bollworms, which arrive later in the season. The best advice is not to spray if it can be avoided, and it is usual to set critical levels below which spraying is not recommended.

Trap crops may be used as breeding grounds for predators, giving them time to breed up in sufficient numbers to control the pest before it can move into the threatened crop.

Predators can be bred up and released in the crop to control a pest attack. This has been tried unsuccessfully against several pests, the main cause of failure being the rapid dispersion of the predators once they are released, leaving insufficient numbers in the crop to control the pest. It has, however, been sufficiently successful to justify the commercial rearing of the chalcid fly, *Trichogramma minutum* Riley, which is sold in the USA for use against various bollworms (Fernald and Shepard 1942). *Phonoctonus*, the mimic predator of cotton stainers (*Dysdercus*), was raised in large numbers in Malawi for use in this way (Sweeney 1960), but was never properly tested because stainer numbers fell so sharply in 1964 and subsequent seasons that control was rendered unnecessary.

Diseases of pests

Disease may be deliberately introduced in a pest population to control it without the use of chemicals. One of the earliest to be tried was *Bacillus thuringiensis* against cotton bollworm larvae, but at first it did not provide sufficient control to be used commercially; it is now claimed that more virulent strains of the bacterium have been selected, which are being produced and sold commercially in the United States. The bacteria are applied to the crop through ULV sprayers, mixed with cottonseed oil as a bait (Patti and Carner 1974).

Another disease which has been exploited commercially is a polyhedral virus disease of *Heliothis*. Stocks of the bollworm are reared in the laboratory and infected with the purified virus; the diseased bollworms are then dried and crushed to a powder, which is mixed with water and sprayed on the cotton crop. Smith and Falcon (1973) found in California that 'commercial formulations of the virus have performed as well as the most effective chemical insecticides in suppressing the bollworm'; the addition of cottonseed oil to the formulation acts as a bait for the bollworm, and Andrews (1975) reports that this gave better results than water alone. The cannibalistic tendencies of *Heliothis* make it necessary to rear them in individual compartments in the laboratory. The farmer himself can collect diseased bollworms from his crop to make up similar sprays for his own use (Coaker 1958).

Light traps

Besides being used to monitor changes in the insect population, light traps have been used to control pest numbers in a crop, but the number collected is usually too small to affect the population appreciably. Light traps are less effective on bright, moonlit nights (Pearson 1958). Infra-red (black) light and ultraviolet light are more effective in attracting certain insects.

Light traps have also been used to collect males for sterilization, which are then released to give a sterile mating with the female. One problem has been to find a form of sterilization which will provide a mating which satisfies the female sufficiently to prevent her seeking another mate. It has been calculated that the number of sterile males must be at least 60 per cent of the male population to reduce reproduction significantly, and this level may be difficult to achieve. More effective baits for insect trapping have recently been produced in the form of synthetic sex attractants.

Chemicals affecting insect behaviour

The use of chemicals to influence insect behaviour has made great strides in recent years. Insect repellants much more effective than oil of citronella were discovered during and since the War, and have reduced the incidence of malaria and other tropical diseases transmitted by insects, but they have not found any practical application in agriculture. Attractants, and particularly sex attractants, have shown much more promise, and several have reached the stage of commercial production. Investigations are proceeding on chemicals which stimulate or prevent the feeding of insects (Stride 1964), and it is a chemical, so far unidentified, which prevents more than one egg being laid in a bud by the boll weevil.

Pheromones are chemicals secreted by an animal which stimulate some physiological or behavioural responses in another member of the same species. The sex attractant chemicals are pheromones; many of them have now been isolated and identified, and some have been synthesized. Those of the boll weevil and *Pectinophora*, *Heliothis* and *Diparopsis* bollworms have been found to be mixtures of four or five different components which only act in conjunction. The boll weevil attractants are produced by the male, those of the bollworm moths by the female.

So far their main use has been as bait in trapping the insects to monitor populations, and to signal the imminence of a severe pest attack. This has proved to be no simple matter. First there is the design of the trap itself – the same bait exposed in traps of different design can yield significantly different results. The pheromones are extremely volatile products, and the amounts required are incredibly small; the rate of release must be fast enough to be effective, but controlled so that it lasts for a fairly long period before recharging, and provides a uniform discharge throughout the period. Where there is more than one component, the emission of each must remain in the right proportion to the others throughout. The carrier in which the pheromone is dissolved or diluted may affect its attractiveness, and it may be sensitive to light and air. For slow release, the pheromones are absorbed in cotton wicks, other fibrous material, rubber or synthetic polymers; a three-layer polymeric laminated dispenser has been registered for use against boll weevil (Plimmer *et al.* 1980, Hartstack *et al.* 1980).

The most effective methods of employing pheromones to reduce the pest population on a field scale are still a matter for experiment and discussion. Possible methods include disruption of mating by saturating the atmosphere sufficiently to confuse the attracted sex; trapping sufficient numbers of males so that females remain unfertilized; sterilizing trapped males and releasing them again, so that a large proportion of matings are unproductive.

Other chemicals with some potential in pest control are mating inhibitors and insect growth regulators; among the latter, analogues of the juvenile hormone can check the normal growth and development of most of the main pests of cotton to a greater or lesser extent, so that the adults do not mature sufficiently to breed, but the hormones are most effective when applied in the last larval instar and early pupal stages when it is difficult to achieve effective contact. Another growth regulator, diflubenzuron, has been shown to disrupt the synthesis of chitin in the boll weevil, resulting in the production of infertile eggs (Bull 1981).

Chapter 13

Diseases

In the tropics as a whole, diseases of cotton are much less important than insect damage, but under certain conditions they can be a major constraint in producing a profitable crop. The progress of a disease depends on conditions unfavourable to the crop but favouring the disease. Many disease organisms are widespread, but only assume importance when the balance of conditions is favourable to them; control measures aim at tilting this balance in favour of the crop.

First in importance is vigorous growth of the cotton crop, aided by good cultivation, weed control and adequate and balanced nutrition; if the water supply can be controlled, this can also play an important part. Second in importance is the level of initial infection, which may be seed-borne, trash-borne, soil-borne or come from infected alternative host plants as airborne spores or transmitted by insect vectors. Seed-borne infection can be reduced by selection of disease-free crops as a source of seed, and seed may be treated with chemicals to reduce the amount of infection that it carries. The distinction between trash-borne and soil-borne infection is not always clear, as the organism may depend on small quantities of organic matter in the soil for its survival; both types can be controlled by crop rotation and field sanitation. Rotations should include crops resistant to the disease, and avoid susceptible crops which could build up the numbers of disease organisms. Field sanitation includes the destruction or rendering harmless of infected crop residues and control of weeds. Alternative host plants may be weeds in the crop, wild host plants in surrounding vegetation and other crop plants; if other crops are suspected, they should either be prohibited or distanced as far as possible from the cotton crop. The application of fungicides in the field, except against seedling diseases, has never been found to provide effective and economic control, and is not recommended.

Fungal and bacterial diseases are usually associated with cold wet weather. Low temperature slows down the growth rate of the crop, and moist conditions are usually the most favourable for infection and the development of diseases. Waterlogging can be prevented by adequate drainage or by growing the crop on ridges, and rank growth should be controlled if possible, as it encourages a humid microclimate within the leaf canopy; free circulation of air within the crop is beneficial.

Weather conditions favouring virus diseases may be quite different. The leaf roll virus in Thailand is transmitted by aphids, and dry weather favours an increase in their population.

As weather conditions can be so important in the occurrence of severe outbreaks of disease, and are not under the control of the farmer, disease-resistant varieties offer the only certainty of avoiding damage to the crop. Fortunately, individuals have been found in commercially acceptable varieties which show resistance to most of the most serious diseases of cotton, and resistance breeding has been successful in reducing them to an acceptable level (Nelson 1973).

Seedling diseases

Damage to seedlings tends to occur sporadically, especially when low temperatures delay emergence; wet soil and lack of sunshine keep the soil temperature low during the day and provide favourable conditions for pathogens, and poorly drained areas are often notorious for seedling disease. The pathogen may be seed-borne or soil-borne or both, and infected seedlings may be killed before or after emergence or they may survive and overcome the disease. Loss of stand is likely to be the main cause of crop loss. Even if the crop overcomes the disease, the check to development and fruiting may be considerable.

Replanting may be necessary if losses of stand are severe; not only does this involve additional expense in seed and labour, but later planting usually results in a considerable reduction in yield. Refilling the gaps in the stand is nearly always better than complete clearance and resowing, as vigorous growth of the established plants, few as they may be, compensates for the gaps; it is often remarkable how well they will cover the ground within a month or so. Where a direct comparison has been possible between refilling and complete resowing, the original stand with refills has outyielded the resown crop (Prentice 1972).

The most common seedling diseases of cotton are sore shin (*Rhizoctonia*), damping off (*Colletotrichum*) and bacterial blight (*Xanthomonas*); wet weather blight (*Ascochyta*) can also appear at the seedling stage.

Chemical seed treatment is an insurance against seedling diseases. Dusts are not only effective in killing any pathogens carried on the surface of the seed or in the fuzz, but also afford some protection against soil-borne diseases by diffusing into the surrounding soil after sowing. Acid delinting is very effective against seed-borne diseases, but does not protect the seedling against soil-borne pathogens; it is often supplemented by treating the seed with a fungicide after acid treatment.

Leaf spots, blights and wilts

Almost every cotton crop carries a variety of leaf spots; most of these are unimportant, or only become important when conditions are favourable for their development. The aim must be to delay any attack as long as possible, so that the crop is produced before the disease can get a hold; it is therefore especially important to delay and reduce the initial infection by seed treatment and field sanitation, and to develop as much genetic resistance as possible. Early maturing varieties may allow the crop to be set before disease can develop to serious proportions.

Leaf spots are commonly caused by *Alternaria, Ramularia* and *Cercospora*, as well as a number of other fungi and the organisms causing various leaf blights. Nutrient deficiencies can sometimes increase the incidence of leaf spots by reducing the vigour of the crop.

Leaf spotting caused by blights commonly results in the shedding of the affected leaves, and the disease may affect the stems as well. The most important is bacterial blight (*Xanthomonas*); once a plant is infected, the disease can spread through the tissues even in very dry weather, and develop a mass of bacteria ready to be dispersed by any rain that occurs. *Ascochyta* blight is only serious in prolonged periods of wet weather.

The two main wilts of cotton are caused by *Fusarium* and *Verticillium*. They can be identified by a blackening of the cambium just below the bark; *Fusarium* wilt develops from the top of the plant downwards, with the leaves and stems turning black and dying after they have wilted; *Verticillium* wilt causes a yellow mottling of the leaves, which later turn brown and shed, and the lower leaves usually suffer most.

Bollrot

Very few diseases can attack undamaged green bolls of cotton; probably bacterial blight (*Xanthomonas*) and pink bollrot or anthracnose (*Colletotrichum*) are the only two; Leakey and Perry (1966) report that anthracnose cannot penetrate the boll wall directly, but develops below the surface cuticle and requires an insect puncture or bacterial infection to enable it to enter the interior of the boll. Other bollrot organisms enter through insect punctures as secondary infection (Reed 1976), or when the bolls start to split open at ripening. Control of bollworms and sucking insects which can penetrate the boll wall is therefore a measure for reducing bollrot. Cotton stainers (*Dysdercus*) act directly as carriers of internal boll disease (*Nematospora*). Infection can take place through the sutures when the bolls open in wet weather; damp and cool conditions slow down the splitting of the bolls, and fungi colonize the moist boll contents before they dry out. Attempts have been made to breed cotton with thin boll walls which will open more quickly. An alternative is to improve the aeration within the crop so that boll opening is speeded up, by bottom defoliation, a more open canopy, and okra-leaf types of cotton. While excellent in theory, the control of bollrot by these measures has not been as successful as was expected.

It is usually difficult to identify the primary pathogens causing bollrot because of the presence of so many secondary infections: up to 20 different fungi have been isolated from a sample of rotted bolls from a single plot. If the primary pathogen cannot be determined precisely, it may be estimated from the frequency with which it is found in a sample of bolls, or by studying those bolls in which rotting has not proceeded too far.

Principal diseases

Xanthomonas malvacearum   **Bacterial blight**
(E. F. Smith) Dowson   **Includes angular leaf spot, vein blight,**
blackarm and bacterial bollrot.

Description

Angular leaf spot was first described in 1892 by G. F. Atkinson in Alabama, who showed that it was caused by a bacterium. E. F. Smith (1920) described and named the bacteria, and was the first to demonstrate the connection between the leaf, stem and boll symptoms of the disease.

The bacteria are rod-shaped with one polar flagellum, and produce a yellow slime on the surface of the culture medium. They do not produce spores, but the bacteria themselves are very resistant to desiccation, dry heat and sunlight (Dowson 1957).

Hunter *et al.* (1968) identified 12 races of *X. malvacearum* in cultures obtained from different parts of the world, using a range of host differentials. Using the same differentials, the identity of local races has been confirmed in other countries, but Arnold, Innes and Brown (1976) in Uganda questioned the value of their contribution to a more effective breeding programme under local conditions. Hayward (1964) distinguished two types of bacteria in African cotton, based on their susceptibility to bacteriophages, but here again Cross (1964) showed that the relationship between phage typing and pathenogenicity was not absolute. Arnold and Brown (1968) also investigated the relationship between virulence and phage type, and concluded that in *X. malvacearum* they were not related.

Symptoms

Fresh lesions have a typically watersoaked appearance, dark green or translucent; they later turn brown or black, but while the lesion is spreading, the edges remain watersoaked. The name angular leaf spot describes the outline of the leaf lesions, whose spread is checked by the network of small veins in the leaf; on the cotyledons the spots are small and circular. Some bacteria are transported through the xylem, and cause lesions in the leaf petiole, stem or leaf veins; in young expanding leaves, the attack may be limited to the tissue on either side of the main veins (Plate 2). Stem infections are sunken, may extend to several inches in length and even girdle and kill the branches, the leafless blackened branches giving the appearance from which the name blackarm is derived (Plate 1). The bolls may be attacked at all stages of development; young bolls may be shed, while older bolls show round, sunken, dark green spots, which later turn black (Smith 1953, Dawson 1957).

Leaf spots and other lesions produce a bacterial exudate which is readily dispersed in rain-water, and serves as an inoculum to spread the disease; severe outbreaks often occur after rain-storms. The bacteria normally gain entrance to the host tissue through the stomata, being transported by water to the stomatal cavity where they invade the thin-walled parenchymatous tissue. Mechanical damage to the plant exposes parenchymatous tissue, and allows the bacteria to gain direct access to it; besides bruising caused by the passage of men or machines through the crop, hail-storms may result in a severe outbreak of the disease. Insect punctures in the boll wall may provide a means of entry, but bacterial bollrot commonly starts at the base of the boll, which is infected by bacteria in rain-water collecting in the calyx cup after the petals are shed and the boll is still small (Wickens 1953). Boll lesions caused primarily by bacterial blight allow other boll-rotting fungi to invade the boll and intensify the rotting of the seed and fibre (Smith 1953).

Distribution

The disease occurs in most cotton-growing countries; although unimportant in some areas, in others it can cause serious crop losses and sometimes complete crop failure. The losses result from reductions in stand, defoliation, stem blight, shedding of small bolls, bollrot and reduction in grade because of lint staining. It is thus a disease of considerable complexity, and Wickens (1953) illustrates this with a quotation from Nowell (1930):

In the various West Indian Islands, where Sea Island cotton is grown in the same island from year to year, the differences in incidence are enormous. Moreover, they extend both to the disease as a whole and to its different manifestations on the bolls, leaves and stems. You may get any one of these forms present to a devastating extent without the presence of the other two.

Any assessment of the importance of the disease in various countries must take into account the climatic differences between seasons and their effect on the development and spread of the disease in the field.

Bacterial blight is of minor importance in California, South America, South Africa, Egypt and Indonesia. Very destructive outbreaks of the disease have occurred in the Sudan Gezira (El Nur 1969) and in the West Indies. In Africa south of the Sahara the disease is not so devastating, but resistant selections give consistently better yields than susceptible, and 'control of bacterial blight is of major importance in increasing yields of cotton in these territories' (Wickens 1953). In India and Pakistan the disease is serious only in years when the monsoon rains are exceptionally heavy.

Carry over of infection

Infection in a new crop can arise from contaminated seed and from the debris of a previous crop. Seed infection arises in bolls damaged by bacterial bollrot; the bacteria are usually found on the surface of the seed or in the fuzz hairs, but in severe cases of boll infection the seeds may become contaminated internally through prolonged soaking in the bacterium-laden boll fluid, or by movement of the bacteria through mechanically or insect damaged seed coats (El Nur 1969). Under dry conditions the bacteria enter a dormant stage and can easily survive from one season to another on the seed or in other plant remains. Dry debris such as broken leaves are light enough to be spread by wind from one field to another: the bacteria have been reported to survive on such dry debris for up to seven years (Ark 1958). When the plant debris contains viable seeds, volunteer seedlings may be a source of infection for the new crop (Andrews 1938).

The importance of debris as a means of carry over of the pathogen and as an agency by which it is disseminated has been studied extensively by Massey in the Sudan; a full list of his publications is given by El Nur (1969). Infection takes place at soil temperatures between 18 °C and 32 °C, the optimum being between 24 °C and 27 °C; it increases progressively with increased soil moisture, but at the same time the bacteria do not survive for more than a few weeks in moist debris (Brown 1969). Trash infection is therefore important in dry climates with seasonal rains coinciding with the planting period; moist weather conditions between harvest and sowing reduce the virulence of bacteria in trash, especially where the trash is buried in the soil.

Control

Seed-borne infection can be controlled by seed treatment with fungicides. Wet acid delinting of seed is very efficient in this respect, while dry acid and flame delinting probably have a similar effect. Mechanical delinting on the other hand may spread the infection by distributing the bacteria from infected seeds to those not infected. Fungicides are best applied at the ginnery before the distribution of planting seed; they are usually applied in powder form at rates varying from 1 part of fungicide to between 100 and 250 parts of seed by weight, and for treating naked or tufted seed a sticker has to be added. Seed treatment reduces the level of primary infection (Wickens 1957) but secondary infection may build up, albeit more slowly, to the same level as that in crops grown from untreated seed (Brown 1976). These dusts have little or no effect on dormant bacteria and resting spores, and the critical period for their action is the period between sowing and emergence of the cotyledons (Dransfield 1968).

Seed dressing on a commercial scale costs so little that there is no doubt that it is economic; Wickens (1958) estimated that it cost the equivalent of 3 kg of seed cotton per hectare, and although costs have risen this is substantially true today. The benefits are difficult to estimate precisely, but Manning and Kibukamusoke (1958) estimated that in Uganda yield has been increased by seed dressing by at least 9 per cent; most of this increase has occurred in northern and eastern Uganda, where bacterial blight was usually more prevalent and severe. In trials spread over three seasons in northern Nigeria, Dransfield (1972) recorded increases of 6 to 10 per cent in yield following seed dressing. In the Sudan trials with seed disinfectants started in 1927, and are summarized by El Nur (1969); they show a consistent reduction in primary infection, but no estimate is made of yield increases.

Commercial seed dressings against bacterial blight fall into two main groups, based respectively on mercury and copper. Mercurial compounds are more efficient in controlling primary seed-borne infection: Dransfield (1968) found that the best copper seed dressing was 80 per cent efficient, compared with 96.4 per cent efficiency obtained with the best mercurial dressing. Against this the mercurials are much more toxic to mammals and hence dangerous both to the operators of the seed dressing machines and to the farmers handling the seed; the dressing should include a coloured dye to give warning of its presence. In East Africa the less toxic copper compounds have been preferred (Thomas 1970); in Nigeria organomercurials were used up to about 1970 (Dransfield 1968), but more recently copper dressings have been recommended for use in the batch drum mixers, retaining the mercurials in mixers with a continuous flow of seed (Dransfield and Beeden 1974). New formulations are being tested to find a dressing which will combine the efficiency of the mercurials with the safety of the copper dressings at little extra cost; one of these, bronopol, has been adopted as the standard seed dressing in several countries (Ebbels 1980).

Liquid seed dressings are efficient and avoid the health hazards of poisonous dusts, but Dransfield (1968) showed that some of them have a phytotoxic effect on the seed during storage, so that treated seed must be used within a month or so of the treatment to avoid deterioration. There are no physical problems in treating acid delinted or naked seed in this way, but when applied to fuzzy or mechanically delinted seed, the machines quickly become clogged with damp fibres (Dransfield 1968).

Tarr (1960) recommended treatment of seed with a wet slurry containing the fungicide. He cites the following advantages compared with a dry powder dressing: the immediate killing of the pathogen; the fine film of chemical remaining round the seed offering protection at sowing; in the Sudan treated seed can be dried quickly, so that the slurry treatment provides a rapid method of seed disinfection. This method is probably more suited to naked or tufted than to fuzzy seed, but no tests of phytotoxicity after storage have been reported and it has never been adopted on a large scale.

For the control of trash-borne infection, all refuse should be burned or buried, preferably in moist soil, as soon after harvest as possible. Burning of refuse is most necessary when the soil remains dry between harvest and sowing; in the Sudan the cotton fields are swept clear of trash and the sweepings burned (Pothecary 1968, El Nur 1969). Where cotton stalks are an important source of domestic fuel, they should be piled in large heaps so that the lower layers remain moist long enough to kill the bacteria.

The most effective method of control, however, is the use of resistant varieties, a subject which requires a section to itself.

Breeding for blight resistance

The pioneer work in resistance breeding was carried out in the Sudan by R. L. Knight (1946, 1954). A satisfactory technique for artifical inoculation had first to be developed before progress could be made, and Knight found that he could get uniform infection by soaking diseased leaves in water and using the liquid as a spray; resistance was determined by comparing the resulting leaf lesions with a series of standards (Fig. 13.1), graded from 0 to 12, 10 being

Grade 2

Grade 4

Grade 6

Grade 7

Grade 9

Grade 12

Fig. 13.1   Some of the leaf lesion grades used by Knight (1944) showing resistance to bacterial blight.

fully susceptible in *G. hirsutum*, 12 in *G. barbadense*. Knight identified five major dominant genes controlling resistance, and used the backcross technique to incorporate them in commercial varieties; these genes he designated B_1 to B_5. The genes varied in the amount of resistance they contributed, and the effects were generally additive.

Knight (1963) later identified five more genes, B_6 to B_{10}; Lagière (1960) found two genes in West African Allen cotton, B_{9L} and B_{10L}, which were not homologous with those of Knight; Green and Brinkerhoff (1956) detected three dominant genes for resistance in

Table 13.1 *List of major genes or polygene complexes conferring resistance to bacterial blight*

Gene symbol	Description and source
B_1	Weak, dominant gene obtained from Uganda B31 (*Gossypium hirsutum*)
B_2	Strong, dominant gene from Uganda B31; also recorded in Albar from West Africa, UKBR from Tanzania, varieties in the USA (all *G. hirsutum*)
B_3	Partially dominant gene from Schroeder 1306 (an off-type *G. hirsutum* var. *punctatum*)
B_4	Partially dominant gene from Multani strain NT 12/30 (*G. arboreum*)
B_5	Partially dominant gene from Grenadine White Pollen (a perennial *G. barbadense*)
B_6	Recessive gene from Multani strain NT 12/30 (*G. arboreum*); and possibly from Tanzania, UKBR 61/12 (*G. hirsutum*)
B_7	Gene from Stoneville 20 and other stocks from the USA (*G. hirsutum*). (Dominance of this gene is dependent upon the genetic background)
B_8	Recessive gene from *G. anomalum*, an uncultivated diploid species from Africa
B_{9K}	Strong, dominant gene from Wagad 8, an Indian commercial variety (*G. herbaceum*)
B_{9L}	Strong, dominant gene from Allen 51–296 from West Africa (*G. hirsutum*)
B_{10K}	Weak, partially dominant gene from Kufra Oasis in Libya (*G. hirsutum* var *punctatum*)
B_{10L}	Weak gene from same source as B_{9L}
B_{11}	Weak gene from same source as B_{9K}
B_{In}	Dominant gene from an unknown variety from the USA (*G. hirsutum*)
B_N	Dominant gene from Northern Star, a variety from the USA (*G. hirsutum*)
B_S	Dominant gene from Stormproof 1, a variety from the USA (*G. hirsutum*)
B_{Sm}	Polygene complex found in Stoneville 2 B and Empire, varieties from the USA (*G. hirsutum*)
B_{Dm}	Polygene complex found in Deltapine, a variety from the USA (*G. hirsutum*)

Source: Arnold, Innes and Brown 1976, p. 177.

American Upland stocks, B_N, B_S and B_{In}. The major genes are listed in Table 13.1 taken from Arnold, Innes and Brown (1976), and there appears to be a wealth of minor and modifier genes which interact with the major genes.

The gene B_2 is the basis of all worthwhile resistance in *G. hirsutum*. It is reinforced by B_3 in the West African *G. punctatum* and in breeding work in the Sudan. B_6 is a recessive gene which confers near immunity in combination with B_2 and B_2B_3. The resistance of Stoneville 20 is due to a single major gene B_7 and several minor genes. Apart from the Sudan, however, the main programmes for resistance breeding have centred round naturally occurring resistant selections identified by inoculation and screening such as Stoneville 20 in the USA and Albar in Africa south of the Sahara.

The name Albar is derived from the initial letters of the Allen variety and black arm resistance, and was given to a series of blight-resistant selections made at Namulonge in Uganda by J. B. Hutchinson. Hutchinson had noted the natural resistance of *G. punctatum* in West Africa, and argued that some of this resistance might have been acquired by the commercial Nigerian Allen variety by introgression. He brought samples of commercial seed from the

Table 13.2   *Culture media for Xanthomonas malvacearum*

The medium most commonly used at Namulonge in Uganda was sucrose chalk agar; Bird (1966) used potato carrot dextrose agar at Texas A & M. Other media are given by Wickens (1953) and Last and Dransfield (1959). Liquid broths have the same composition, without the agar.

Procedure
  Mix the ingredients in a conical flask.
  Make up to 1 litre with distilled water.
  Melt in an autoclave @ 10 lb pressure.
  Remove, mix and adjust pH if necessary to ± 7.0 by adding $CaCO_3$.
  Sterilize at 15 lb for 15 minutes.
  Pour into sterile tubes or petri dishes.

Ingredients	*Sucrose chalk agar grams*	*Potato carrot dextrose agar grams*
Peptone	5.0	2.5
K_2HPO_4	0.5	
$MgSO_4$		0.3
$MgSO_4.7H_2O$ (Epsom salt)	0.25	
$CaCO_3$	10.0	0.2
Carrot juice (canned)		15 ml
Yeast extract		0.5
Sucrose	20.0	
Potato dextrose agar		40.0 *
Agar	17.5	10.0

* Commercial (Difco)

a

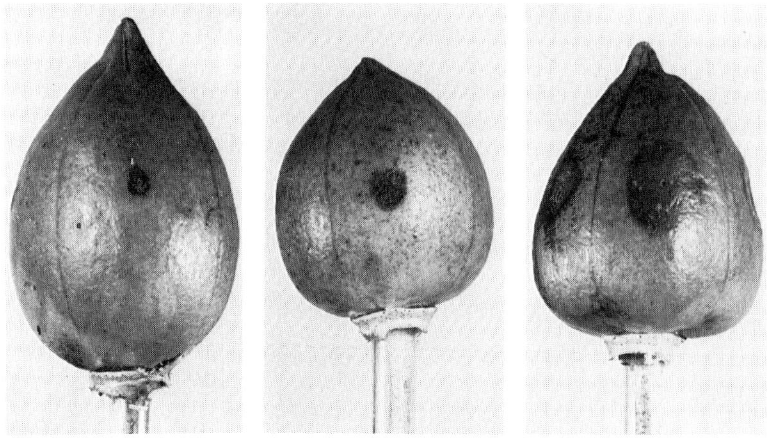

b

Fig. 13.2   Boll inoculation with bacterial blight (Brown 1976, Logan 1958): (a) the needle, which has been dipped into the bacterial suspension in the bottle, is inserted into the boll wall; (b) lesions three weeks after inoculation, showing (left to right) resistant, intermediate and susceptible bolls.

Funtua ginnery in Nigeria to Uganda, and from a sowing of about 6000 seeds, 39 apparently resistant plants were isolated. Single plant progenies of 26 of these were grown in the following season, of which 3 appeared to be homozygous for resistance, 19 heterozygous and 4 homozygous susceptible. Selections were made from the resistant progenies, which were distributed throughout Africa by the Cotton Research Corporation, and used successfully in most of the breeding programmes from South Africa to the Sudan. Albar 637 became the commercial variety in South Africa, Malawi, Zimbabwe and Mozambique; BPA and SATU in Uganda were local selections from Albar (Arnold *et al.* 1968); the Ukiriguru seed issues from UK61 onwards in Tanzania, and the Ilonga multiline IL62, were selected from hybrid material originating in crosses between Albar and locally adapted varieties (Arnold 1970).

Although Knight was satisfied that leaf and stem resistance were correlated, and adopted leaf lesion development as his sole criterion, it is now evident that in different environments resistance to one phase of the disease is not necessarily correlated with resistance to others (Wickens 1953). For example, Stoneville 20, on which most of the early resistance breeding in the USA was based, showed no cotyledon and leaf resistance in Uganda, although it is on these characteristics that it was isolated and used in breeding work in the USA. 'Breeding for bacterial blight resistance is therefore a separate problem for every region, requiring, at least in the early stages, a variety of inoculation techniques to induce the different phases of the disease.' (Wickens 1953).

Too many techniques have been described and used to be detailed here. Weindling (1944) used pure cultures of bacteria on potato dextrose agar in petri dishes at a standard dilution with water; recipes for culture media for *Xanthomonas* are given in Table 13.2. Seed inoculation is usually achieved by soaking the seed in a standard suspension of bacteria. Cotyledons are infected by using abrasives, sandpaper or needles, to allow the inoculum to penetrate the epidermis. Leaf inoculation usually depends on forcing the suspension into the stomata on the underside of the leaf, by high pressure sprays. Stem inoculation is done with a fine sewing machine needle charged from a petri dish culture and thrust into a young internode so that the eye of the needle penetrates approximately to the middle of the stem. Bolls can be inoculated by spraying inoculum into the calyx cup soon after the petals are shed, or by needle inoculation into each of the locules (Fig. 13.2).

Fusarium oxysporum Schlecht *f. vasinfectum*   *Fusarium* wilt (Atk) Synder and Hansen

The fungus

Fusarium wilt is caused by a soil-borne mycelium, which penetrates the plant through the roots and spreads upwards in the vascular tissue, blocking the vessels and interfering with the movement of water up the plant. It can be transmitted through infected seed. It produces two types of spores, conidia which are short-lived and chlamydospores, a resting stage which may live longer. Both types give rise to the mycelium which infects the plant roots.

Many *Fusarium* species are found associated with cotton, and even the pathogenic *F. oxysporum* is not uncommon on the roots of apparently healthy cotton plants. It is only under extensive infection that disease symptoms occur (Ibrahim and Khalifa 1969). This extensive infection is usually associated with the presence of nematodes, which reduce root growth and increase the susceptibility of the cotton by providing numerous openings for the wilt fungus. The most common species of nematode associated with *Fusarium* is the root-knot nematode, *Meloidogyne incognita*; other nematode species sometimes associated with the disease are the meadow nematode (*Pratylenchuris pratensis*) and the sting nematode (*Belonolaimus gracilis*). Both the fungus and the nematode can cause direct losses in cotton yields without the other, but where serious damage to the crop occurs, it usually results from the wilt-nematode complex.

Armstrong and Armstrong (1960) differentiated four races of cotton wilt *Fusaria* based on differential pathogenicity; races 1 and 2 are found in the USA, race 3 in Egypt and race 4 in India; a 5th race has been proposed by Ibrahim (1966), who has shown that isolates from the Sudan were very similar to race 3, but could infect the Ashmouni variety which races 3 and 4 could not. Tests on Tanzanian strains of *Fusarium* indicate that they belong to race 1 (Ebbels and Little 1975).

Symptoms

Symptoms of wilt may appear on cotton plants at any stage of development. In older plants the upper leaves wilt and droop, and this is accompanied by yellowing of the leaf margins and along the veins; the affected areas enlarge, and the leaves are usually shed. A characteristic symptom is discolouration of the phloem, seen as a brown or black ring just below the bark in a cross-section of the stem or branch. It is this growth of the fungus in the vascular tissue which restricts the supply of water to the leaves and causes wilting. Sometimes the discolouration is dispersed through the woody cylinder, and in advanced cases the stem, branches, petioles, peduncles and bolls all blacken and die. When seedlings and young plants are affected, the cotyledons and leaves turn yellow and brown before shedding, and the bare stems blacken and die (Smith 1953).

Distribution

The disease is distributed throughout the world wherever cotton is grown, and is most prevalent in acid alluvial sandy soils. It is important in parts of the USA, Tanzania, Bolivia and India; elsewhere it is not considered to be a serious factor in the commercial crops.

Control

In the Sudan the only serious outbreak of *Fusarium* wilt has occurred in an experimental plot sown continuously with cotton since 1918; cotton grown under the normal rotation, once in every four years, is generally free from the disease (Ibrahim and Khalifa 1969). Crop rotation is therefore a useful practice, but depends on the other crops in the rotation; several crops that are particularly susceptible to root-knot nematodes, such as tobacco and sweet potatoes, tend to intensify the disease complex, and fields planted to cotton after an interval of 25 years have shown severe wilting in the first year (Smith 1953).

The use of balanced fertilizers to maintain vigorous growth of the crop is recommended in the USA, with special emphasis on potash, which has been shown there to reduce crop losses from wilt (Tisdale and Dick 1942); similar experiments in Tanzania have given inconclusive results (Ebbels 1975).

Elliott (1923) found that 6 per cent of the seed from a severely infected crop in the USA was carrying internal infection. Levels in excess of this have been found in seed from selected plants, but Perry (1962b) considered that 'the figure of 0.2 per cent ... probably represents a more normal rate of seed-borne infection' in Tanzania. It is therefore recommended that planting seed should be taken only from wilt-free zones, especially if there is a danger of the disease spreading to hitherto disease-free areas.

Smith (1953) found that *Fusarium* wilt can be controlled indirectly by destroying the nematodes in a wilt-nematode complex. The economic value of chemical nematocides is doubtful: they are at present expensive, require special equipment for their application, and are not always effective (Perry 1962a).

Resistant varieties are the most satisfactory method of control. E. L. Rivers produced the first wilt-resistant variety about 1900, a Sea Island variety that bears his name. He had been joined by W. A. Orton in 1899; they established the plant to row selection method essentially as it is used by modern breeders, picking out individual plants that had survived on severely infected soils. Upland cotton does not possess factors for the same high degree of resistance as does Sea Island cotton, but a series of resistant Upland varieties was produced, culminating in 1942 with the introduction of Coker 100 wilt, which combined wilt resistance with agronomic characters equal to those of non-resistant varieties. In Tanzania Arnold and Arnold (1959) found that wilt appeared sooner or later in all the imported varieties reputed to be wilt-resistant, but good resistance has been achieved by selection in local varieties and hybrids (Ebbels 1975).

Verticillium dahliae Kleb   *Verticillium* wilt

The fungus

Verticillium is a soil-borne disease, invading the plant through the roots and blocking the vascular system in the same way as *Fusarium*, but differs in that it is not transmitted through infected seed. The fungus causing the disease in America was commonly referred to as *V. albo-atrum* Reinke and Berth., but the distinction between the two species is doubtful (Isaac 1967), and the species attacking cotton in the USA is now named *V. dahliae* (Cotton Disease Council 1981).

Symptoms

'The outstanding symptom is the chlorotic areas on the leaf margins and between the principal veins, which make it look mottled' (Presley 1953). In mature plants the symptoms usually appear first on the lower leaves, spreading upwards later in the season; the chlorotic areas spread, and eventually cause the leaf to shed. *Verticillium* wilt tends to attack scattered individual plants in a field, while *Fusarium* is likely to be found in patches, spreading out from a focal point of infection. Discolouration of the diseased vessels in the stem is common to both wilts, *Fusarium* and *Verticillium*; it is seen as brown streaks if the bark is stripped off a diseased branch, or as a characteristic ring of dark dots, the ends of the diseased vessels, if a wilted leaf is snapped off at the base of the petiole or a branch is cut in cross-section.

Distribution

Verticillium dahliae attacks many horticultural and agricultural crops in both temperate and tropical countries; cool wet weather favours the disease. It attacks cotton in most areas of cotton production, with the exception of West Africa and Mexico (Ebbels 1980); Brown (1953) states that it has never been noted in Egypt. While the disease has given cause for concern in the Sudan, Uganda and South Africa, and Ebbels says that it 'ranks as one of the most serious diseases of cotton in most areas of production', severe outbreaks do not seem to occur regularly in tropical countries; when they do occur, resistance is usually found in some of the local varieties, which may have been selected unintentionally for this character (Brown 1976). The disease is of major importance in the lower Mississippi valley and the irrigated areas of the south-west of the USA (Presley 1953).

Control

Like other soil-borne diseases, *Verticillium* wilt is discouraged by rotation of cotton with resistant crops, but this does not provide complete control. Presley (1953) reports that a rotation with alfalfa reduced the amount of wilt in the following cotton crop, but succeeding cotton crops were often attacked more severely than when no alfalfa had been grown. Outbreaks are rather unpredictable; although the disease organism can persist for some years in the soil, a heavily infected field may turn out practically free from the disease when sown again to cotton in the following season.

Any cultural practice that increases the soil temperature tends to reduce the amount of wilt, for example restricted watering under irrigation or increasing the size of the ridges to expose more soil surface. Pullman *et al.* (1979) have demonstrated that soil solarization produces a remarkable reduction in soil-borne populations of *Verticillium*; this involves a fallow year, when moist soil is heated by covering it with clear polythene sheeting for about four weeks during the summer months. Increased vigour as a result of fertilization does not reduce the incidence of wilt, but the crop withstands the disease better at a high level of nutrition (Wickens 1950). In the USA wilt has been found to be particularly severe on soils high in organic matter (Presley 1953).

Breeding for wilt resistance is the most promising method of control, although neither in the field nor in the greenhouse is it easy to ensure a uniform level of infection; there is always a possibility that apparently resistant plants have in fact escaped infection. In the greenhouse, stem inoculation is the usual method (Bugbee and Presley 1967), but Brown (1976) reported 100 per cent infection when seedling roots were dipped in a suspension of the pathogen before replanting; differentiation between varieties, however, was greater when infection rates were lower.

No varieties have been found to be completely immune, and tolerance is perhaps a more appropriate term than resistance. *Gossypium barbadense* varieties have generally been found to be more resistant than *G. hirsutum*. In California, Acala 4–42 produced in 1950 has some tolerance to wilt (Presley 1953), and the newer Acalas, SJ2 and SJ5, have shown superior tolerance (Garber *et al.* 1981). Multi-adversity resistance breeding in Texas is improving resistance to wilt, though the rate of gain has slowed down (Bird *et al.* 1981). In Mississippi the variety Hartsville was found to be highly resistant, though otherwise unsatisfactory, and is being used in breeding programmes to incorporate resistance.

In Uganda B181 was selected for resistance to wilt and bacterial blight in 1934, and was used in breeding the S varieties grown in the north and east of the country in the 1950s (Arnold 1970). These were replaced by the Albar derivative, SATU, which had not been screened for wilt resistance but was later shown to possess equal or better resistance; the same was true of BPA, which replaced BP52 in the rest of Uganda (Brown 1976). In South Africa *Verticillium* wilt is not listed by Kerkhoven (1963) as a major constraint on the yield of cotton, but work on the disease is being carried on at Barberton.

Other fungal diseases

Sore shin is the name used to describe the symptoms produced in cotton seedlings by *Rhizoctonia solani* Kiihn. This is a soil-inhabiting fungus, which attacks cotton seedlings under conditions such as cold, moist weather or when the seedlings are weakened by insect attack such as thrips. Brown lesions develop on the stems near ground level before the formation of foliage leaves, and spots disfigure and distort the cotyledons; in severe attacks the lesions become cankers which encircle the stems or penetrate so deeply that the seedlings fall over and die (Neal 1953). To reduce losses from this disease, sowing on ridges is recommended to ensure good drainage, along with liberal fertilization and delinted seed to ensure quick germination and a vigorous start. Seed treatment with Arasan or Brassicol 75 (300 g per 100 lb seed) is recommended by Sión and Carvajal (1976) to prevent attack.

In Aden *R. solani* is the cause of Abyan root rot in cotton, which results in the wilting of older plants. This form of the disease is found only in South Yemen; wilting occurs most frequently in plants from 6 to 12 weeks old, although the initial penetration of the plant probably occurred within the first few weeks of growth (Daniels 1965).

Seedling blight and damping off may be caused by anthracnose, *Colletotricum* (*Glomerella*) *gossypii* Southworth, which is transmitted by spores adhering to the fuzz. R. Weindling and P. R. Miller (quoted by Neal 1953) found that the size of spore load on the seeds did not affect germination but seemed to influence the amount of post-emergence damping off. The disease is only important in the USA in the cotton-growing states east of the 40 in. rain belt of Oklahoma and Texas; west of this belt dry conditions and high temperatures prevent the survival of the fungus between the seedling blight and the bollrot stage. Cool, moist weather allows the disease to spread and intensify. Affected seedlings show reddish or dark brown lesions on the stems and roots below the soil line; the cotyledons may also be attacked, producing brownish spots that enlarge as the disease develops. Organic mercurial dusts, such as Ceresan, have been shown to increase seedling survival by about 30 per cent compared with untreated seed.

Ascochyta blight, or wet weather blight, is caused by the fungus *Ascochyta gossypii* Woronichin; Smith (1953) considers it to be the most sporadic of all cotton diseases. Plants three to eight weeks old are particularly susceptible and as the name implies it occurs in prolonged periods of wet, cool weather. The disease starts with small circular white spots on cotyledons and leaves, which enlarge, coalesce and become brown and roughened; they often fall out, leaving a ragged appearance. Additional infections girdle the stems and kill the terminal buds and adjacent stem tissues, resulting in loss of stand. Dark brown stem cankers at the branch axils may develop in older plants, but usually cause only minor damage. The same disease can cause bollrot, which starts with rough, circular and brownish lesions, and later invades and destroys the seed and lint. Field sanitation and seed treatment are usually adequate to prevent the disease from being carried over from one season to the next.

Alternaria leaf spot (*Alternaria* spp.) may cause severe shedding of leaves. In the early stages, the spots are pale green with an irregular margin and may be mistaken for bacterial blight. As the area enlarges, however, the colour of the older parts become straw yellow and then rusty brown, forming irregular concentric zones in roughly circular lesions; spores may be formed on the surface of the lesions.

False mildew (*Ramularia gossypii* (Speg.) Ciferri) is often found on the leaves of mature plants; it is rarely serious, but may cause leaf shedding. The spots occur on the underside of the leaf and have a frosty appearance like mildew; they are angular and bordered by the leaf veins as in angular leaf spot. Viewed from above, the lesions are bright green to yellowish green, and only occasionally show the typical white coating.

Towards the end of the season, *Cercospora gossypina* Cooke is a common cause of leaf spotting, but rarely does any serious damage. The spots are roundish or irregular, with purple borders and white centres, not more than $\frac{1}{4}$ in in diameter, and finally fall out of the leaf.

Root rot caused by the soil-inhabiting fungus *Phymatotrichum omnivorum* (Shear) Duggar affects a wide range of wild and culti-vated plants, and is the pathogen of Texas root rot of cotton. It is limited to alkaline soils in the south-western USA and northern Mexico, where losses from this disease fluctuate from year to year, but can be alarmingly destructive (Blank 1953). Symptoms are most apparent during the period of flowering and fruiting; occasional plants in the crop are affected, and the infection spreads to the surrounding plants, resulting in patches of brown dead plants, shading off abruptly through a narrow zone of wilted plants, into normal green plants. The fungus attacks the roots, destroying the cortical tissues, and a white mycelium invades the central cylinder; the mycelium turns tan or brown as the disease develops. Other stages of the same disease produce spore mats on the surface of the soil, and a sclerotial or resting stage. The use of organic manures and green manures seems to offer the greatest promise of practical control.

Virus diseases and similar disorders

Tarr (1964) lists 11 virus diseases of cotton, in some of which a virus is suspected but has not been confirmed; Smith (1972) lists eight, of which cotton virescent disease is not mentioned by Tarr. Leaf curl, which attacks *G. barbadense* cotton in the Sudan, is the most fully documented of the virus diseases, and is transmitted by the whitefly, *Bemisia tabaci*. It causes curling of the leaves and twisting of the petioles, accompanied by a thickening of the veins in the leaves and bracts, and sometimes by distortion of the internodes of the stem. The disease reached the Sudan Gezira from West Africa, where it is present in the quasi-indigenous *barbadense* of the region (Hutchinson *et al.* 1950); although these were introduced into Africa from the New World, the disease is not known there, and it was probably acquired from local malvaceous species in West Africa. Resistant selections were found in the Lambert varieties of the Sudan, which were replaced by the first resistant variety, Bar XL 1, in the early fifties; susceptible varieties are still grown in the northern Gezira where the disease is less troublesome. El Nur and Abu Salih (1969) give a full account of the history and status of leaf curl disease in the Sudan, and Hughes (1969) and Siddig (1969) discuss the resistance breeding programme.

Halliwell (1966) lists leaf curl, anthocyanosis, leaf crumple and perhaps leaf roll as economically important virus diseases of cotton. Anthocyanosis occurs in Brazil (Costa 1956), where it is found on cotton and numerous weed hosts, primarily *Sida* spp.; it is transmitted by aphids. The symptoms are similar to those of magnesium deficiency, a deepening of the anthocyanin pigmentation of the middle and lower leaves. It does not affect the growth habit of the plant, but similar aphid-transmitted diseases, blue disease in West Africa (Cauquil 1977) and leaf roll in Thailand (Nonglak and Sompark 1980), produce stunted growth and shedding of buds and young bolls.

Leaf crumple is found only in the United States, and is transmitted by whitefly. It does not greatly reduce the current season's growth, but distorts the new growth of ratooned cotton; since the production of ratooned cotton has been discouraged, the disease is no longer a serious problem.

A different leaf roll is reported as a disease of cotton in Russia, and appears to be distinct from the other leaf distorting viruses.

Non-parasitic disorders

Psyllose is transmitted by the larva of a psyllid *Paurocephala gossypii* Russell, and can lead to complete sterility, but the nature of the disease is not known (Dineur 1957); it is reported chiefly from the Congo (Soyer 1947), but may occur occasionally in Malawi (McKinley 1965). It is characterized by a purple colouration of the older leaves and stalks; internodes developing from the terminal buds of the stem and branches are severely shortened and rosettes of stunted scale-like leaves are formed, olive green with purple margins.

'Bunchy top' and 'crazy top' are names given to various disorders of cotton about which little is known except the symptoms. Pearson (in Hutchinson *et al.* 1950) describes these as follows:

Bunchy top is characterised by a progressive reduction in the size of all foliar and floral parts, and of the internodes... The stipules remain normal in size... The disorder somewhat resembles that known as 'Psyllose'... but the symptoms are less striking and the reddish purple tinge of stems, petioles and leaves on plants affected by Psyllose is absent.

O. F. Cook described the crazy top of cotton found in Arizona in similar terms (Tharp 1953); the disease occurs only on calcareous soils, and is closely related to irrigation practice, although water shortage may not be the only factor involved. The disease is controlled by applying irrigation frequently enough to prevent a check to plant growth during the summer months.

Boron deficiency in Zambia (see Ch. 8) was first suspected to be Psyllose; it causes distortion of the leaves and flowers, a bushy rosetted appearance of the plant, and characteristic necrosis of the pith of the petioles (Rothwell *et al.* 1967).

2,4-D weedkiller is used against broadleaved weeds in cereal crops and sugar cane; minute traces of the weedkiller drifting from applications to neighbouring crops are sufficient to distort the leaves of a cotton crop, which grow narrow and strap-like with exaggerated points to the lobes (see Fig. 7.1). Cotton can recover quickly if the symptoms are not too severe.

Lightning is a rare hazard in the cotton crop, but where it does strike the symptoms suggest an attack of wilt or blight. It produces a round patch 20 or more metres across, in which the plants in the centre are killed, the damage tapering off towards the circumference where only the tops of the stems are affected.

Chapter 14

Cotton breeding

Genetic basis

From the standpoint of the plant breeder, cotton is regarded as a self-pollinating crop, although some natural cross-pollination occurs in most environments. The main point is that it does not suffer from inbreeding depression, and the methods of breeding self-pollinating crops described in the textbooks (Allard 1960, and others) are those which are most commonly used.

Some control of outcrossing is usually necessary, however, if the purity of genetic lines is to be maintained. Natural cross-pollination is mainly dependent on the activity of honey bees, and varies from nil up to 25 per cent or more; average values for Uganda range from 7 to 11 per cent between adjacent plants (Arnold and Innes 1976). A figure of 5 per cent of outcrossing between adjacent rows of different varieties is quoted for Egypt by Brown (1953), while Balls (1912) states that 'a value of 5 per cent to 10 per cent for natural crossing under field conditions in Egypt has been confirmed by numerous, though non-systematic, pieces of evidence'. The bee population is affected by the use of insecticides, and their increased use in India appears to have reduced natural crossing to very low levels (Innes 1971).

The Old World species of cultivated cottons are diploids ($2n = 26$), and self-pollination rapidly produces a heterogeneous population of homozygous plants. When two varieties are crossed, segregation starts in the second filial generation F_2, and the majority of individuals will still be heterozygous; by the time that the F_5 is reached (assuming that each generation is selfed), individual plants are likely to be homozygous or nearly so for most characters, and selections made at this stage will mostly breed true.

The New World species are tetraploids ($2n = 52$), containing two complete sets of chromosomes, the A and D genomes, from different species; such tetraploids are amphidiploids (Manning 1955, Fryxell 1979), although Hutchinson (1959) uses the more general term allopolyploid. These amphidiploids might be expected to function as diploids, but Manning (1955) quotes evidence to show that the reduction in genetic variance with inbreeding in tetraploid cotton is surprisingly slow, and discusses four possible explanations, any or all of which may be contributary sources of enhanced genetic variance. Very briefly, these are: (1) certain heterozygotes may show hybrid vigour and be selected at singling; (2) a high mutation rate of major genes; (3) irregular pairing of homologous chromosomes from the different genomes; (4) interaction between duplicate genes in the two genomes.

In a breeding programme the breeder must continually ask himself whether there is sufficient variability in his material, and if not, how best he should go about increasing it. Hutchinson (1959) states that 'the prospect of improvement in breeding work lies in the selection and propagation of superior material out of a heterogeneous population. There can be no breeding progress without selection, and selection becomes ineffective when variability is exhausted, or falls below the level at which it can be effectively detected'. The programme will generally proceed from primary selection and introduction to secondary selection and hybridization. Each of these phases will be discussed in turn, but they all have one factor in common – selection.

Selection

Selection is the act of choosing a plant which is different from its neighbour, and harvesting the seed of that plant separately to produce progeny in the next generation; the progeny may breed true to type or may segregate. Selection was practised successfully long before the genetic mechanisms were understood, but we now know that self-pollinated crops tend to produce homozygous individuals, and that selection within selfed progenies will soon cease to show improvement, unless a sudden heritable change, or mutation, occurs. Cross-pollinated crops are naturally heterozygous, and selection produces unstable lines which continue to segregate; its effect is to change the number and frequency of the genes which make up the gene pool in the population, a subject studied in population genetics (Falconer 1960).

In selecting apparently superior plants in a population, it is important to distinguish between those whose superiority is caused by their environment and those which have a superior genotype. This is particularly important in cotton, which responds more than many other crops to small differences in the environment such as spacing, soil, weather and insect attack. Progeny testing is the most effective way of distinguishing genetic from phenotypic variation, and it is informative to take 100 random plants and compare their progenies with those of 100 deliberate selections; it will often be found that there is little to choose between the two sets of progenies, especially if selections are taken from plots which are uneven due to variation in soil and stand. Selection efficiency can sometimes be improved by dividing such plots into smaller units, and taking a fixed number of selections from each unit. It can be taken as a general rule in cotton that the performance of the progeny is more important than the appearance and behaviour of the single parent plant. It not only helps to distinguish genetic from phenotypic variation, but by its uniformity or otherwise may distinguish between homozygous and heterozygous parents.

In the first stages of selection, the accurate judgement of visual characters requires that the breeder knows his crop thoroughly. He must be familiar with the morphological and physiological characteristics of acceptable varieties, so that he can make a quick visual evaluation of the hundreds of progenies in the programme, and reduce them to a number which can be tested with greater precision and for less obvious characters.

Selection may well be directed towards a single character which has been identified as the primary objective; at the same time it is important that the other good qualities of the material are not lost. In so far as these qualities are indicated by visual characters, they can be maintained by ensuring that the selected plants conform to the desired type, and this is where a thorough knowledge of the crop is essential. While selecting for a single character such as jassid resistance, the experienced breeder will note such things as plant height, vegetative branching, where the bulk of the crop is set, internode length, boll size and so on, and use these observations in discarding plants or progenies which are off-type and undesirable. The number of selections should be as large as possible considering the staff and facilities available, and it is often a major problem to reduce this number to a level which can be incorporated into replicated trials without discarding potentially good material.

Special techniques may be used to provide the conditions necessary for the selection of some characters, particularly those relating to resistance. Disease and pest resistance can only be judged in the presence of the pathogen or pest, and the ability to withstand drought, lodging and low fertility requires the appropriate conditions before they can be assessed. These conditions must strike a happy mean between levels of adversity too low to ensure that susceptible plants are affected and levels so high that any differences in susceptibility are obscured.

When selecting for disease resistance, artificial inoculation can ensure the presence of the disease (Knight 1946); it may be more convenient to do the screening for resistance on seedlings in the laboratory or greenhouse (Arnold and Innes 1976). To test resistance to insect pests, the trial plots may be interspersed with susceptible varieties to ensure adequate numbers of the pest (Parnell *et al.* 1949). Duplicate variety trials are often carried out at sites of high and low fertility to ensure that new selections are suitable to both, or to measure any variety × fertility interaction. Skip-row planting can be used to encourage lodging and prevent interference between lodged and resistant varieties.

If conditions are unsuitable for selection, progenies can be harvested and resown each year until suitable conditions are encountered.

Selfing

Although cotton is generally speaking a self-pollinating crop, a certain amount of cross-pollination takes place, and the breeder has to decide whether it is desirable to exclude foreign pollen by selfing, as artificial self-pollination is usually called. There are four stages in cotton breeding when selfing may be desirable: in maintaining pure stocks of imported varieties; during a programme of selection and breeding; in maintaining seed of new lines while they are being tested in variety trials; and in maintaining a supply of breeder's seed to supply a seed renewal scheme.

It is first necessary to establish how much natural crossing occurs in the field (Innes 1961); if this is already very rare, there is no need for selfing. Outcrossing is generally measured between adjacent rows of cotton, but for practical purposes it is important to know how many rows constitute an adequate barrier to crossing; it is usually easier to plant a large plot and discard the outer rows than to self an adequate number of flowers in a small plot with insufficient discard rows. Cotton itself constitutes a better barrier than surrounds left unplanted or sown with maize or other crops.

In the breeding programme there is often a conflict between the maintenance of purity and the maintenance of variability. In some countries, especially those growing long-staple cotton, a very high standard of uniformity is demanded; in others, more heterogeneity is acceptable, and may even be encouraged with multi-line seed issues. To the breeder the greater the purity the less scope there is for improvement by selection, and he is conscious that purification may result in the loss of potentially valuable characters.

Selfing is very strict in Egypt, where much of the breeding material is grown in bee-proof cages, and if grown outside the later flowers are bagged to provide selfed seed (Brown 1953). In Manning's BP52 programme in Uganda single plants for selection were selfed every year; from 1962 Arnold reduced this to selfing every second generation, growing unselfed progeny rows in alternate years (Arnold and Innes 1976). In Tanzania no attempt was made to self all the material, and many of the selections stemmed from open pollinated plants (Peat and Brown 1961). In Malawi no selfing was done in the breeding programme.

When selections reach the stage of testing in replicated yield trials, it is not advisable to depend on seed taken from the trials to maintain the lines, as the layout usually favours intercrossing; separate multiplication plots should be sown to provide a source of pure seed, and the size of these can be adjusted according to the amount of seed likely to be required for extended trials and bulking up if the variety proves successful.

If seed has to be taken from the plots of a replicated trial without artificial selfing, the plots should be large enough to allow discards. The varieties to be included in the trial must also be considered: the entries should preferably be fairly closely related, so that crossing between them will be innocuous; the controls should not differ markedly from the progenies being tested.

In Ecuador plots specially designed for selfing were sown each year to provide a supply of pure seed. The entries consisted of three groups:

1. every entry in the variety trials;
2. varieties of historical interest, previously used in the breeding programme;
3. varieties with special characters available for use in the crossing programme.

The plots (2×15 m) were sown on the skip-row pattern, two rows wide with an unplanted row between; this arrangement allowed easy access and visibility for selfing, the skip-rows reduced the shedding of selfed bolls compared with solid planting, buds and selfed bolls were easy to see and the risk of mistakes in picking was minimized (see Fig. 15.4).

Primary selection

With very diverse material, a large initial advance may be expected from a general survey of a large amount of material, a technique which Mason (1938) called 'primary selection'. This was the situation in many of the African territories when systematic improvement was started; it arose from the uncontrolled introduction and cultivation of a range of varieties, which soon became mixed and lost their identity. At the same time natural selection had reduced or eliminated types unsuitable to the particular environment. Primary selection in these 'land races' or 'local bulks' quickly produced results. Mz 561 was selected from Mwanza Local in Tanganyika (Peat and Brown 1961); BP 52 was selected from Uganda Local in 1928 (Nye and Hosking 1940); U4 was selected from Nyasaland Upland in 1925

(Parnell 1935); Samaru 26C was selected from Nigerian Allen in 1940 (Freeman 1946).

It is becoming rare nowadays for the breeder to have access to such diverse local material. Breeding programmes were established about 50 years ago on the eastern side of Africa from the Sudan to South Africa under the Cotton Research Corporation (CRC, formerly Empire Cotton Growing Corporation) and in West Africa under the Institut de Recherches du Coton et de Textiles Exotiques (IRCT); Egypt, India, the West Indies, Mexico, Brazil, Peru – countries which produced cotton long before the advent of modern insecticides – have an even longer history of cotton improvement. Even a well-established programme may benefit from breeding work carried out in other parts of the world, and at some stage in the breeding programme it will be desirable to introduce new material from outside. Cotton expansion is now taking place in areas where it had not been grown before, and is based on insect pest control and the importation of improved varieties from well-established sources.

Plant introduction

Commercial production of cotton at the present time bears little relation to the centres of origin of the cultivated species. Even Pakistan, the centre of origin of the Old World diploid species, is expanding its production of tetraploid Upland varieties at the expense of diploid 'desi' cotton. The spread of Upland cotton into the United States cotton belt from Mexico began in the early seventeenth century (Kearney 1930). Sea Island cotton in the West Indies and in South Carolina was first established towards the end of the eighteenth century. Sea Island cotton was introduced into Egypt after 1820, where, crossed with Jumel's perennial tree cotton, it gave rise eventually to the present commercial varieties (Balls 1912). The Upland cottons of India and Pakistan are mainly derived from two sources: the East India Company first imported seed from the USA about the middle of the nineteenth century (Burt 1913) and in 1905 Upland types were imported from Cambodia, Hanoi and the Philippines, where they had become acclimatized (Main 1912). Most of the countries of Latin America now grow varieties bred in the USA; Peruvian Tanguis is an exception, being derived from a natural cross between the sporadic village cotton of the Indian tribes and an Upland variety (Hutchinson *et al.* 1947, Harland 1949). American varieties of Upland cotton were introduced into Africa south of the Sahara in quantity at the beginning of this century, although the Portuguese navigators in the sixteenth century had brought New World species with them which replaced the indigenous diploids for local consumption.

More recently, U4 bred at Barberton in South Africa by Parnell and his associates (Parnell 1935) was so great an advance on current stocks that it replaced them in South Africa and spread to what are now known as Zimbabwe, Malawi, Zaïre and Mozambique. Nigerian Allen spread to francophone West Africa, where after local selection it became the commercial variety in Chad and Cameroon and was used in the breeding programmes of other countries (IRCT 1978). Albar, selected from Nigerian Allen, became the commercial variety of Malawi, Mozambique, Zimbabwe and Zambia, and was used in the breeding programmes of Uganda, Tanzania and the Sudan. The Reba varieties from West Africa have been widely used as a source of resistance to bacterial blight, and have been grown commercially in Thailand, Zaïre, Burundi, Paraguay and Bolivia; they are derived from a series of crosses made in the Central African Republic between Allen selections and varieties from the USA about 1956. Varieties from the USA have been sent all over the world, and are grown commercially in many Latin American countries, also in Syria, Indonesia, the Philippines, and Australia.

It is not possible to predict the behaviour of an imported variety in its new environment. For example, Reba B50 was bred for the Central African Republic, and was grown for a time in India and Thailand; it was crossed with Deltapine Smooth Leaf from the USA, giving rise to Reba P279; this in turn is now being grown on a large scale in Paraguay. On the other hand, no imported Upland variety has performed well *per se* in India, and the commercial varieties are Indo-American, derived from crosses with local varieties. In Egypt several Sea Island types were introduced in the early 1920s, but the only selections actually used were two off-types from St Kitts, called Sakha 10 and Sakha 11. Introductions were made again in the 1940s, and crossed with Egyptian varieties; the crosses gave weaker yarn and were later-maturing than comparable Egyptian varieties, and showed little commercial promise (Brown 1948). The Pima cotton of Arizona and New Mexico, however, is derived from Mit Afifi seed imported from Egypt; Pima selections were brought back to Egypt in 1918 under the name Maarad, an ancestor of Karnak; they were crossed with Sakel and Giza 7 in the USA, giving rise to the high-quality varieties Amsak and Pima 32 (Brown 1953, Feaster and Turcotte 1962).

In any new cotton development, as wide a range of varieties as possible should be tested; this testing should be carried out over a number of sites and seasons before deciding on the variety to be used as the basis for a breeding programme. At the same time, the characteristics of the imported varieties should be noted for possible future use as crossing parents. In an established breeding programme, one section can well be devoted to regular importation and testing of the latest products of breeding in other countries.

Some breeders maintain that an introduced variety needs a period of acclimatization to allow it to 'settle down' in its new environment (Hutchinson 1938). The author does not know of any evidence for this; Allard (1960) lists 'Adaptation' in the glossary, but makes no reference to it in the text. Gene frequencies can be altered over a number of generations by natural selection, but such changes occur much too slowly to require consideration in testing imported varieties.

Varieties which are phenotypically uniform in their own environment have been known to segregate into different types when grown elsewhere (Brown 1963). This means that they are in fact mixtures of genotypes which are phenotypically similar in their own environment, but react differently to a different environment; they may provide sufficient variability to make selection worth while in their adopted home.

Secondary selection

Hybridization

All breeding systems depend on a supply of varied material in which to practise selection, and this is generally provided by the F_2 of a cross between two varieties. There are no hard and fast rules about the choice of parents for such a cross, and the results are rather unpredictable. Brown (1948) cites several examples of unexpected results in Egypt, among them Amon (Giza 39), a cross between two high-quality varieties, which was found to have finer lint than either of the parents; the Albar × Acala crosses made in Uganda have done unexpectedly well in Swaziland, Malawi and Zambia (Munro 1974); the Reba B50 × Deltapine Smooth Leaf cross made in Thailand, Reba P279, is particularly favoured in Paraguay, Argentina and Bolivia (IRCT 1978). Most cotton breeders will have had similar experiences.

Crossing or hybridization only becomes necessary when useful variation is exhausted: it may be noted that both the Uganda BP52 and Tanzania Mz561 programmes included no crosses until the introduction of Albar. The parents chosen for a cross may be widely different or closely related. In Egypt Brown (1948) built up a stock of parents with extreme qualities in any direction for crossing in different combinations. On the other hand, Feaster and Turcotte (1970) made crosses between closely related lines of Pima in Arizona, and were able to show significant advances among the progeny, when straight selections showed no improvement.

No single variety is ideal in all its characters, and breeders usually select as parents a locally proven variety and one with qualities which are likely to strengthen the weaknesses in the local variety, or two local varieties with complementary characteristics. The charts compiled by Ramey (1966) of the pedigrees of American cotton varieties show how crosses have been made at various stages, followed by selection among the progenies.

Crossing techniques are described at the end of this chapter.

Panmixis

To provide a highly variable gene pool for plant breeding improvement in East Africa, J. B. Hutchinson (1959) chose 19 major types of *G. hirsutum* race *latifolium* from different parts of the world; seed of these was mixed and sown in 1955 as a randomly mating population, or panmixis, and interbreeding was ensured by transferring pollen from flower to flower with a camel-hair brush. 'As a long-term plant breeding reserve, maintained at very little cost, a bulk of diverse origin maintained under a system of interpollination offers attractive possibilities.'

After three generations the distinguishing features of the original strains had completely disappeared, and selections were made in the highly variable population. Arnold (1970), however, reported that 'after a further three generations, the whole population had declined to such a low level of productivity that it seemed doubtful if successful types could easily be produced from it. Although selection lines had by this time been established, their performance was poor when tested in replicated trials in the 1961–62 season, and after one more season the material was finally discarded.' Four different panmixis populations established in Tanzania in 1956–57 were equally unsuccessful and were discarded after the 1959–60 season.

Arnold discusses the possible reasons for the failure of the method, and concludes that the use of a small number of genotypes, with selfing introduced after one or two generations of crossing, would have given better results; and that the systematic plan of crossing, intercrossing and backcrossing used in Uganda has produced more interesting material than the panmixis.

Two commercial varieties have been obtained from panmixis bulks at Bebedjia in Chad – Pan 575 and Pan F3–52. The panmixis included 34 lines and varieties originating from the basic stocks of N'Kourala, Triumph and Allen. Selections were taken in 1965–66 after one or two seasons of panmixis, and underwent mass selection and selfing for a year.

Breeding methods

Having established a variable population in which to exercise secondary selection, it remains to be decided how to exploit this population to the best advantage. The methods used in breeding self-pollinated crops have been classified broadly by Allard (1960) into three groups: 'pedigree', 'bulk population' and 'backcross' breeding. For cross-pollinated crops, the selection procedures take much the same outward form, but the results are not the same; they are used to identify superior genotypes, but new varieties are seldom developed from the offspring of a single plant. Recurrent selection, synthetic varieties and hybrid seed are common in cross-pollinated crops, but are rarely used in cotton breeding; nevertheless, the genetic basis of these procedures is worth studying and may contribute fresh ideas to a cotton-breeding programme.

Pedigree breeding

Pure line selection as a method of breeding had become well organized by the late nineteenth century, notably by Vilmorin in France, who combined selection with progeny testing. Johannsen in the early years of the twentieth century defined 'pure lines' and described the genetic mechanism by which they are established (Allard 1960). Up to 1933 cotton breeding in India was dominated by Johannsen's pure line concept; segregating material was regarded as exceptional and undesirable, and was either 'purified' or discarded. In 1933 Hutchinson and Panse at Indore started to use the recently evolved statistical techniques of field experimentation in their progeny row breeding programmes, and emphasized the direct relation between genetic variance and the prospect of advance in plant breeding (Hutchinson and Panse 1937, J. B. Hutchinson 1959).

The progeny row system used by cotton breeders in India and Pakistan is illustrated in Fig. 14.1. Starting with a cross, single plant selections are taken in the F_2 and sown as progeny rows; three out of every four of these progenies are discarded in the F_3, and four single plants selected in each of those remaining; the progenies of these plants are sown as the F_4, and the process is repeated to form an F_5 and F_6 if segregation is still taking place.

Once uniformity has been achieved, the four progenies of each of the best families are mixed and sown in 'microtrials' at two sites; the best strains are then tested in 'progeny row trials' at five sites, the remaining seed being sown in multiplication plots. The best strains are now entered for the National Variety Trials, where they are tested against entries from other cotton breeders at experimental stations spread over the whole country. Entries are tested in these trials over three seasons, while at the same time they are tested locally in growers' fields. In the F_3 and in the final stage of progeny testing, plots are left unsprayed to assess resistance to jassid and other insect pests.

This system meets the theoretical requirements for a self-pollinated

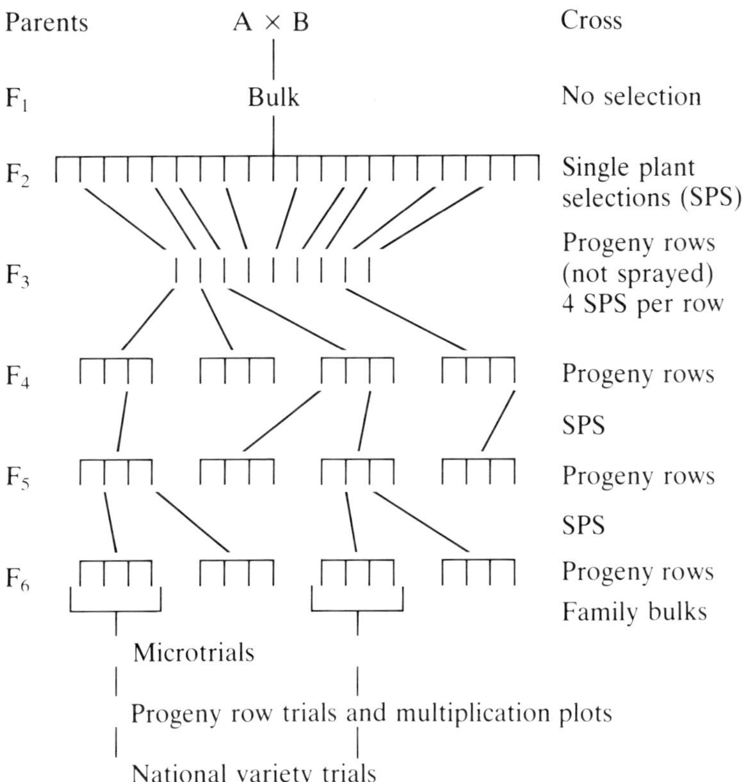

Fig. 14.1   Pedigree breeding system.

crop: a cross between two homozygous parents, single plant selection in the F_2 and subsequent generations as the plants become more homozygous, bulking of selections within a family when no further segregation is apparent, and finally testing the new families in replicated trials at a number of sites. With three to five generations of selection, starting in the F_2, and five years in variety trials, there is a minimum of 10 years between making the cross and the registration of a new variety. During this time records must be kept of the relationship and main characteristics of a large number of progenies; there is obviously a limit to the number of crosses which a plant breeder can handle, and there will be a conflict between the number of crosses which can be tested and the number of plants which can be selected in each cross.

A characteristic of the progeny row system is the dependence on eye judgement in the early stages of selection. This in turn depends on the ability of the breeder to recognize the most promising lines and to avoid discarding potential winners. He can be helped by measurements of characters which are not greatly influenced by the environment, such as seed weight, ginning percentage, jassid resistance, disease resistance, and possibly lint length. The yields from unreplicated plots are notoriously unreliable, but may have some value in eliminating the poorest yielders, provided they are used with caution: growth should be fairly uniform, with no abrupt changes in fertility; the field notes should be checked to see that the poor yield was not caused by outside influences or poor stands; and yields should be compared with those of adjacent plots rather than with a general mean.

The number of generations of progeny row selection can be varied to suit the circumstances. For example, some countries attach more importance than others to phenotypic uniformity, which is achieved by increasing the number of selfed generations to increase the homozygosity of the progenies. But the importance of uniformity can be exaggerated, as has been shown by the success of the multi-line seed issues in Tanzania: besides giving better yields, there was a marked improvement in the market reputation of the crop (Peat and Brown 1961). Hutchinson and Ghose (1937a) quote the case of the Indian cotton Malwa, which yielded a very uniform and acceptable lint, but was in fact a mixture of a diploid (*G. arboreum*) and a tetraploid (*G. hirsutum*). Seed mixtures of this kind are much more acceptable than mixtures which occur during marketing and ginning; the latter are likely to vary in quality from one lot to another, or even within the same bale, because the mixing is much less thorough. On the other hand, a variable population may be an indicator of careless handling of the planting seed, with the possibility of deterioration in quality if a variety becomes mixed or crossed with inferior types.

Homozygosity can be achieved with less effort by growing the hybrid population for a number of generations before selection starts; these hybrid bulks are much less time-consuming to maintain than progeny rows, which can be limited to one or two generations of selection.

The breeder may prefer to sacrifice some uniformity to get quicker results, if, for instance, the introduction of a new variety is urgently required. If this is so, the progeny row selection will have to be reduced to a minimum, and the early introduction of formal trials is essential; an actual example from Malawi is given in Table 14.1. Others may prefer to maintain variability in the population, offering a wide range for further selection and allowing hybridization between the components. This can lead to a two-stage programme of selection: the first selections from the F_2 of a cross may undergo progeny selection for only one or two generations before going into formal trials, and a second stage of selection is applied to later generations of the best progenies, aimed at improving uniformity

Table 14.1   *Progeny row programme, Malawi, 1970–73*

Cross No.	Cross	1970–71	1971–72	1972–73
	Group A (American)	(Original)		
1	Albar × Deltapine SL*	19	11 (1)	5
2	× Acala 4–42*	14	8 (2)	5
3	× Deltapine SL	14	11 (3)	—
4	× Coker 100 AWR	19	2 (1)	1
5	× Rex	17	14 (4)	6
6	× Stoneville 7A	19	4 (1)	2
7	× Austin	18	6 (—)	—
8	× Empire 61 WR	17	6 (3)	1
	Sub-total	137	62 (15)	20
	Group B (West African)			
9	Albar × A333–57*	13	6 (—)	3
10	× HG 9*	16	12 (2)	6
11	× Reba B50	40	9 (4)	7
12	× BJA 592	19	7 (3)	3
13	× HG 9	18	14 (2)	4
	Sub-total	106	48 (11)	23
	Group C (Europe) early			
14	Albar × T59–501[†]	21	14 (10)	4
15	× SR 1054–2[†]	13	—	—
16	× SR 1054–2	49	22 (11)	7
17	× K3811	15	5 (4)	—
18	× K4219	41	14 (13)	2
19	× K4245 (1245)	17	3 (3)	2
20	× K4255	36	14 (9)	6
21	× K4371	36	27 (7)	14
22	× K4614	32	13 (13)	1
	Sub-total	260	112 (70)	36
	Total	503	222 (96)	79
	% Selection	100	44.1 (19.1)	15.7

Notes

1970–71 – The original progenies were single plants selected from the 1969–70 F_2 stock.

1971–72 – '11 (3)' means that a total of 11 progenies were selected for further screening and that of these 3 were in unreplicated progeny rows.

1972–73 – All progenies grown in replicated trials.

In 1971–72 one cross A1-SR1054-2[†], was completely discarded. This number rose to four in 1972–73, in which Al-Del, Al-Austin, and Al-K3811 were also discarded.

* Crossed material obtained from Zambia.

[†] Crossed material obtained from Rhodesia.

Source: Munro 1974.

and correcting any weakness which may have shown up; material from the first stage may show sufficient improvement to be worth issuing as a commercial variety.

This two-stage testing may be applied in other ways. Some

crosses are likely to prove more successful than others, and an early test of the combining ability of the parents may be desirable. Composites can be made up from the selections to represent each cross and compared in replicated trials; this allows some crosses to be eliminated, and a greater number of selections to be retained in the crosses with the best performance. In the same way, unselected hybrid bulks in the F_2 and later generations can be compared in replicated trials before the first selections are taken, to give a guide to their potential. How good a guide it may be is uncertain, as there do not appear to be any measurements of the correlation in cotton between early generation trials and the success of the final selections; Allard (1960) reviews the work on this subject in other crops.

The stage at which replicated trials should be introduced depends among other things on the number of progenies which can be accommodated in the trials. While there are lattice designs for testing large numbers of varieties, they require great care and accuracy at all stages, and the consequences of a mistake may be serious. The number of progenies reaching the trial stage will generally be less than 100. To obtain full value from the trials, observations and measurements will be made on every plot, which means the work involved in making them is multiplied by the number of replications; it is of course possible in the measurements of seed cotton to mix the produce from all replications of each treatment and take, say, the ginning percentage of the mixed bulk, but this does not give any measure of variability.

Selection index

Manning (1956b, 1963) developed the selection index as a breeding system in Uganda. He had seen that the systems used in India and Africa depended to a great extent on the experience and judgement of the breeders, and wanted a more objective system which could be used by an inexperienced breeder; for this he needed replicated trials at an early stage. He was dealing with a rather specialized crop, the long-staple BP52, and aimed at a refinement of the selection process to exploit the genetic variance still present, even after many years of selection (Manning 1955). For this he used estimates of heritability, genetic advance and variability, which were calculated from the variance and covariance among progenies, strains and lines (Walker 1960).

To obtain the necessary data he divided the seed from each selfed single plant into two portions, one of which was used to sow a single row for selfing and selection, and the other to sow a replicated progeny trial. He took lint per plant as his measure of the economic worth of the selection, and divided it into three component parts – bolls per plant, seeds per boll and lint per seed – each of which was measured and recorded in the plots of the trial. The relative importance of these components was calculated from a set of equations in the same way as Smith (1936) calculated his discriminant function.

The selection index has not been widely adopted by cotton breeders for a variety of reasons. One reason has been an unwillingness to embark on the collection and analysis of such a mass of data, another the difficulty in understanding the statistics involved, when simpler and more direct methods are still giving good results. The very small plots (10 plants) used by Manning in his progeny trials were distrusted by some breeders, especially under raingrown conditions when losses in stand could reach 4 plants out of 10, on the grounds that the conclusions drawn were unreliable. The most important reason, however, is that the advances claimed by Manning were far greater than those actually achieved (Arnold and Innes 1976); the interaction of the environment (site and season) with yield is such that the advances in fact ranged from an increase of 35 per cent measured over 17 trials in 1961–62 to a decrease of 8 per cent measured over 28 trials in 1968–69.

The number of single plant selections is limited by the size to which the progeny trial can be stretched; many breeders prefer to take a much larger number of selections and whittle them down by admittedly rather crude criteria before testing them in a formal trial. Nevertheless, Manning's work on the selection index and modal bulks has led to a much better understanding of heritability in cotton and of the genotype-environment interaction.

Testing new varieties

In the final evaluation of promising strains and lines Allard (1960) stated that 'the compromise reached by many plant breeders is to release when a variety proves to be superior in five years of trials at each of five representative locations in the intended area of use'. In many developing countries this is considered to be too rigorous a test; the tendency in Africa has been to have district trials at 10 or more sites, and to be satisfied with two or three years testing in these trials.

In Pakistan there are several one-variety zones, and within each zone the breeders claim that their varieties are adapted to that zone and outyield all others. The National Variety Trials were introduced in 1978 as an impartial test of these claims; they correspond to the Regional Cotton Variety Tests in the USA, and those of the National Institute of Agricultural Botany in Britain. In a country with a small production, say less than 100,000 bales, a single authority is usually responsible for all the breeding work: it may be the government, a research institute or a specialized agency like

IRCT or the former CRC. This authority runs its own series of district variety trials, and decides when an existing commercial variety should be replaced; the evidence is usually submitted to an independent body, including representatives of government, growers, ginners and the textile industry, for final approval. A promising variety may even be issued for multiplication while testing continues, with the possibility of withdrawing it in the early stages of multiplication if it fails to live up to its promise. Two seasons of formal testing at 10 different sites is recommended as the minimum requirement. If a strong genotype–environment interaction renders the results inconclusive, testing will have to continue further, and consideration should be given to dividing the area into two or more zones, each with its own variety.

A pedigree breeding system

It will be seen that there is considerable scope for individual initiative in choosing a breeding programme to suit the circumstances. In the absence of an established programme, the following is suggested as a pedigree breeding system suitable for most circumstances.

Starting with a mixed population, or the F_2 following a cross, about 1000 single plant selections (SPS) are marked early in the season; some of these may be discarded before picking, and some after harvest on seed weight, ginning percentage and yield. The remainder (say 800) are sown as progeny rows, without selfing, in the next season (F_3). Discard about half of these in the field, taking one or more SPS in each of the remainder; discard about half of the selected progenies on ginning percentage, seed weight and yield, along with any SPS taken from them. Sow the remaining selections as progeny rows without selfing (F_4), and reduce the number of progenies to below 100 by selection in the field and in the laboratory.

The surviving progenies now enter the testing stage. Keeping aside about 50 g of seed for selfing plots, the remainder is used to sow two 7×7 simple lattice trials (2 replications each); the 98 entries are reduced by reselection to between 40 and 45. These proceed to a further trial, using either a 7×7 lattice with 3 or more replications, or two 5×5 balanced lattices with 6 replications. The seed for these trials is made up from any surplus selfed seed, unselfed seed from the selfing plots and if necessary seed from the previous trials. Three or four commercial control varieties must be included in each trial, and selfed seed used to plant selfing plots and small multiplication plots.

Anything equal to or better than the commercial controls now goes to a variety trial including selected material from other sections of the breeding programme, using unselfed seed for the trial and

selfed seed for small multiplication plots (say 10×10 m); watch out
for outstanding individuals in these plots – poor plants should be cut
out and good plants selected for progeny row testing.

Bulk population breeding

Bulk population breeding in its simplest form starts with a mixed
population or the segregating F_2 of a cross. This is planted in a plot
of several hundreds of plants, the plot is harvested in bulk and the
seed is used to plant a similar plot the following season; this process
is repeated as many times as desired.

In a self-pollinating crop this will have two effects on the population. The percentage of homozygous individuals will increase,
approaching 100 per cent by the 10th generation even when a very
large number of genes are segregating (Allard 1960); at the same
time natural selection will alter the gene frequencies in the bulk
population. Natural selection can be expected to exert selective
pressure on intangible characteristics that provide for good adaption,
and it may be more discerning than the plant breeder in this respect.
On the other hand, natural selection may operate in a direction
opposed to that desired by the plant breeder, when valuable
characters are not those which promote survival in a mixed population.

Certain characters which might logically be supposed to have
survival value, such as resistance to some diseases, are apparently
neutral in survival. Artificial selection can, however, always be
brought in to complement natural selection and shift the population
toward agriculturally desirable types; in fact the earliest recorded
use of bulk population methods, by Nilsson-Ehle, was to select
winter hardiness in wheat, and he discarded plants that had suffered
winter damage even though they were capable of setting seed.

Mass selection

The most common use of mass selection in cotton is in maintaining
the purity and characteristics of a commercial variety, sometimes
called filtering. In a field of the variety concerned, several hundred
typical plants are selected; the seed from these plants is mixed and
planted the following season to produce a purified bulk. The procedure can be elaborated in various ways if the plants are harvested
individually. If the seed is planted as progeny rows in separate but
contiguous plots the next season, progenies which appear to be
mixed or do not conform to type can be eliminated, and the
remainder bulked to constitute a new lot of purified seed. The seed
cotton from selected plants can be tested for such characters as

ginning percentage and seed weight, and atypical plants rejected; this is the basis of the modal bulk system described later. The selection of plants in the field or in the laboratory can be biased deliberately to apply selection pressure in favour of any desired character. The mirror image of mass selection is roguing, where atypical plants are removed from an otherwise uniform bulk before harvest. Mass selection can also be combined with pedigree breeding in a variety of ways.

The 'mass-pedigree' method proposed by Harrington (1937) can be useful to cotton breeders. Characters such as lodging and disease resistance depend for selection on conditions conducive to their occurrence, which may only occur in certain seasons. As pedigree breeding for these characters is obviously a waste of time if no lodging or disease occurs, he proposed carrying the material as bulks until the necessary conditions for selection occurred; single plant selections can then be taken and handled subsequently by pedigree breeding methods.

Harland (1944, 1949) used a different mass-pedigree system to restore the original quality of the Tanguis crop in Peru, when it had deteriorated since its origin as a single plant in 1908. He decided that pure line selection and selfing were unsuitable for the improvement of the very heterogeneous commercial crop with which he had to deal; it would take too long, and selfing would tend to stabilize a proportion of undesirable characters, while desirable characters not capable of accurate measurement would be lost. Instead he established a series of norms for six main characters, and discarded all progenies which were below these norms. The norm for any character was set according to its importance, and might be the mean, the lower quartile or the upper quartile of the frequency curve. Starting with a sample of about 22,000 loculi from the commercial crop in 1940, each locule was examined for colour, lint length and ginning percentage; discarding those below the norm set for each character, 2863 locules were planted in progeny rows. At the end of the 1940–41 season, each progeny row was tested for the same three characters, plus boll weight, fineness of fibre and yield, leaving 41 progenies which passed the tests. From these progenies 200 élite plants were selected for progeny testing the following year; in this second generation Harland was able to demonstrate substantial improvement, and the best 43 progenies in 1941–42 were mixed for distribution as a new variety SNA 242.

Modal bulks

Manning (1955, 1956b) started his modal bulks in Uganda in 1947 as a standard for comparison to measure yield improvement from his selection index technique; he hoped to obtain 'a relatively stable check variety' by successive bulking of material which was 'modal' for several traits. The system is described in detail by Walker (1964).

Between 200 and 300 reasonably productive single plants are selected in a quarter-acre plot, and harvested individually. They are analysed in turn for lint per seed, seed weight and lint length, and in each case all plants which deviate from the population mean by more than one standard deviation in either direction are eliminated. Seed from the remaining plants is mixed to provide the modal bulk of the following season, when the whole procedure is repeated.

After seven generations (7 MB) Manning concluded that, contrary to his expectations, there had been an average yield improvement in the modal bulk of 3.4 per cent per annum. He suggested that the modal bulk might provide an 'effective mechanism of positive selection for yield'.

After 15 generations Walker (1964) reviewed the data that had accumulated. He concluded that a definite increase in lint length had occurred in the first five seasons, after which length stabilized around a value of 35 mm; but that no significant changes had occurred in the other two characters, lint per seed and seed weight. He showed that Manning's data on yield were not sufficiently extensive to provide a reliable estimate of progress, and that when subsequent trials were included, the overall advance was of the order of 1 per cent per annum. A reassessment of the data by Arnold (1972) after 22 generations showed that differences in yield between modal bulks and BP52 local were associated with earliness in the modal material. He concluded that 'results of yield trials do not support the inference of a progressive yield advance from modal selection, and it is suggested that the positive advances . . . probably arose mainly from environmental effects'.

Meanwhile, modal bulks had been established in Malawi (Gausi 1967), in eastern Uganda (Low 1968, Riggs 1967b) and in Nigeria (Lee 1962) but little, if any, change in yield or quality was found to occur. The BP52 modal bulks were continued for 22 successive generations at Namulonge until 1969–70, when the replacement of BP52 with Albar derivatives as the commercial crop of Uganda had been completed.

Backcross breeding

The backcross method is used to improve a variety (A) that excels in a large number of attributes but is deficient in a few characters, by transferring the deficient character or characters from another variety (B). After the first cross between the two varieties (A × B), the F_1 is crossed back to the recurrent parent with the large number of desirable attributes ($A^2 \times B$); this is repeated as many times as may seem desirable. After each cross, only those individuals in the F_1 showing the character being transferred are used in the backcross, and it is essential that this character can be identified readily in hybrid populations by visual inspection or by simple tests. Recessive characters are more difficult to transfer by backcrossing, as they require a much more elaborate system to identify them. The character being transferred may be controlled by a single gene or a number of genes, and may be qualitative or quantitative, provided it has a high heritability and is not tightly linked to other, undesirable, characters.

In the system outlined by Allard (1960) the F_1 of the first, third and sixth backcross is selfed two or three times before the next backcross is made. These F_2 and F_3 generations are selected for plants with the general features of the recurrent parent and the character required from the donor parent. The reasons for this are, firstly, that selection for the type of the recurrent parent is more efficient in selfed material than in the same number of backcrosses; secondly, it provides the breeder with much larger populations in which to select the desired combination of characters. The method is illustrated diagrammatically in Fig. 14.2.

In cotton the classical example of backcross breeding is the transfer of bacterial blight resistance to the commercial cotton varieties of the Sudan by Knight (1945). It was based on extensive research into the genetics of backarm resistance, and the identification of the major genes B_2 and B_3. In adding both these genes to a susceptible type, they were handled separately as parallel transferences, and combined after backcrossing was completed. The method of transferring the factor B_2 is fully described by Knight (1946):

The resistant and susceptible parents should be crossed in bulk (not single plants). Bulk crossing is a means of preserving the genetic variability of the recurrent parent. The F_1 plants of this cross, being all of B_2b_2 composition, will be resistant and this resistant F_1 should be bulk backcrossed to the susceptible parent, using the latter as female. The progeny of this backcross, which should comprise at least 20 plants, will consist half and half of resistant (B_2b_2) and susceptible (b_2b_2) plants. The susceptibles should be removed and the remainder backcrossed in bulk to the susceptible variety (backcross parent). This process should be continued backcross by backcross until variety tests of the type described elsewhere (Knight 1945) show the resistant and susceptible components of the backcross progenies to be identical in spinning quality, yield, etc.

With most American varieties not more than 5 backcrosses would be required in transferring B_2 from any of the Upland control types listed. Where the susceptible backcross parent strain is of particularly long staple and high quality, 6 backcrosses might be required, whereas 4 (or even 3) would probably be adequate for the lower quality bread-and-butter cottons.

When the end-point is reached, all the resistant plants in the backcross

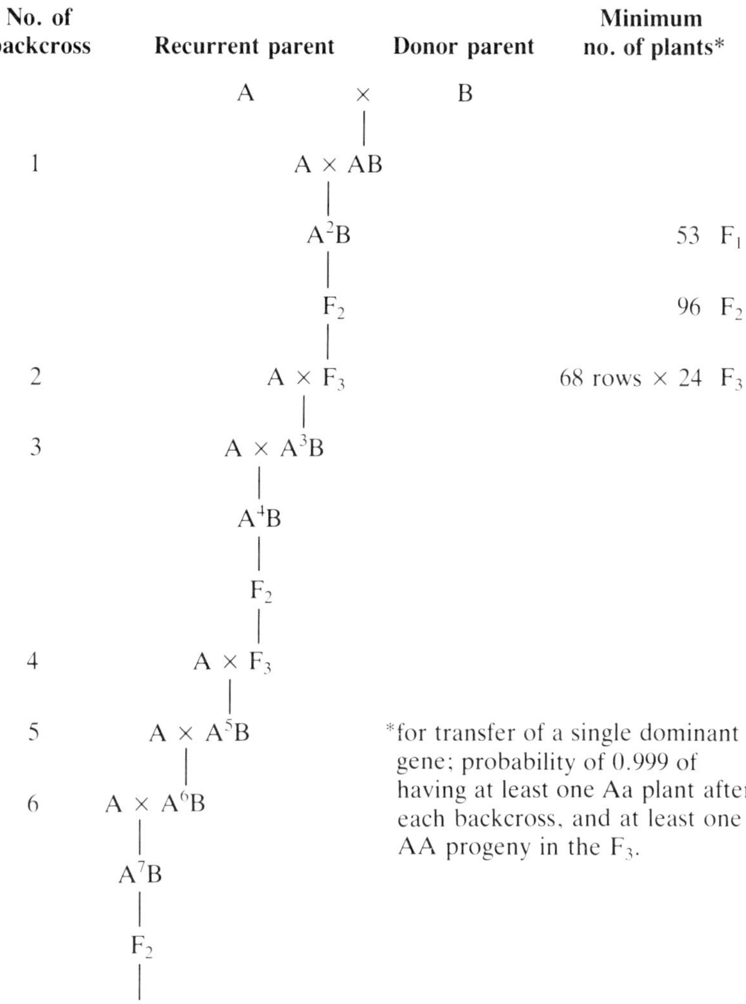

Fig. 14.2 Backcross breeding (after Allard 1960).

progeny should be selfed and their seed bulked and sown as a single
"family". This family would consist of 75 per cent of resistant plants (25 per
cent B_2B_2 and 50 per cent B_2b_2) and 25 per cent susceptible (b_2b_2) plants.
Progenies (each consisting of at least 25 plants) from all resistant plants
should be sown in rows in an isolated area, sprayed with inoculum and
those rows 'splitting' for resistance pulled up. The remaining rows then
become the first propagation plot of the new resistant strain of the original
susceptible commercial variety.

The transfer of B_3 is similar, but since this factor shows only partial dominance it is advisable to include reference plots of the susceptible parent (b_3b_3) and of the F_1 (B_3b_3) when measuring susceptibility.

In combining the resistance genes B_2 and B_3 there are two alternative methods: in one the F_1 is selfed and the heterozygotes are eliminated in two stages in the F_2 and F_3; in the other the F_1 is backcrossed to the B_3B_3 parent, and again two stages of selfing are needed to eliminate all except the homozygous progenies.

Backcrossing can be combined with other breeding methods. Imported varieties may be backcrossed once or twice to a local variety with no specific objective other than to improve their adaptation to the local environment before the start of a selection programme. In interspecific crosses, backcrossing is usually necessary in the early stages to restore fertility, and in the later stages to produce a commercially acceptable type. Backcrossing is also used as a tool in genetic studies, for example in distinguishing between major gene segregation and additive inheritance in resistance to bacterial blight (Arnold, Innes and Brown 1976); or in breaking the linkage between desirable and undesirable characters (Allard 1960).

Specific characters which have been introduced by backcrossing into local stocks are frego bract (Jones and Andries 1969); nectariless (Meyer and Meyer 1961); okra-leaf (Andries *et al.* 1968); storm-proof (Niles and Richmond 1962); high gossypol (Sappenfield and Dilday 1980).

Many breeders have tried to combine the long staple of *G. barbadense* with the productivity of Upland cotton (*G. hirsutum*), usually without success, but a new variety with this origin, Varalaxmi, was released in India in 1972. These two tetraploid species cross freely and give a vigorous F_1; the F_2 and subsequent generations segregate into a welter of types, the majority of which are inferior in vigour and productivity and often morphologically abnormal (Harland 1936, Hutchinson 1940); types intermediate between the parents continue to segregate and are impossible to stabilize. Harland showed that the genetic balance in these intermediate selections had been upset, and only those similar to one of the two parents would be stable. Kammacher (1965) therefore suggested that the establishment of hybrids would be accelerated by backcrossing two or three times to one or other of the parents, and selecting for a few of the desired characters from the other parent; in effect, the aim should not be an intermediate type, but the transfer by backcrossing of a few characters to the genetic background of the other species.

Mutation breeding

The great increase in mutation rates in maize and barley caused by irradiation with X-rays encouraged plant breeders in the 1930s to direct considerable effort towards using the new discovery in practical breeding. The results were discouraging, as it was found that for one favourable mutation about 800 were deleterious. Plant breeders at that time had more variability at their disposal in existing germplasm collections than they could conveniently handle, and the use of induced mutations dwindled. Since the Second World War atomic energy research has discovered many new forms of radiation, and radioactive chemicals are much more widely available. This has resulted in a revival of interest in mutation breeding (Allard 1960).

The agents, called mutagens, which produce artificial mutations, include chemicals as well as radiations and may be grouped as follows:

1. Densely ionizing radiations:
   alpha particles, neutrons.
2. Sparsely ionizing radiations:
   X-rays, gamma rays.
3. Ultraviolet light.
4. Alkylating chemical mutagens.
5. Mutagenic nucleosides.

These groups are in decreasing order of the energy they dissipate; those with greater energy tend to produce a high proportion of mechanical changes in the chromosomes, such as fractures, segmental rearrangements, losses and so on; those with less energy tend to produce point mutations, that is, changes in specific genes or loci (Allard 1960).

The dosage necessary to produce the greatest number of mutations is critical, and very close to the lethal dose, and will usually cause the death of a number of the treated plants. It is usual to treat germinating seeds, which are easy to handle and contain actively multiplying cells at an early stage of differentiation. Mutations induced early in the life of the plant ensure that they affect the reproductive organs, and are transmitted to the next generation.

The disadvantages of mutation breeding are largely associated with the necessity for testing large second generation (M_2) populations. With the chances of a beneficial mutation at 1 in 800, and the probability that it will be a recessive character, a population of at least 3000 has to be screened to find one useful mutation; much greater numbers are desirable. As a corrolary to this, an efficient screening technique is necessary to select the useful mutant; such techniques have been developed for disease resistance, in which large numbers of germinating seeds are sprayed with cultures of the disease organism (Konzak 1956, Singleton 1955).

Cotton breeders in Pakistan, India and Mexico have induced mutations in cotton seeds, but so far have had little success in incorporating favourable mutations into commercial varieties. It is probable that the results which may be achieved compared with the work involved could be realized more economically using conventional methods.

Natural mutations occur in commercial crops, and the breeder should always be on the look-out for any unusual plants in the field, whatever their origin; it is easy to bring back some seed and plant a progeny row to see what develops.

Hybrid vigour

Although cotton does not suffer from inbreeding depression, hybrid vigour or heterosis is found in the F_1 of crosses within species, even between fairly closely related lines, and in interspecific crosses within the cultivated diploid and tetraploid groups (Hutchinson *et al.* 1947).

The direct exploitation of hybrid vigour in cotton on a commercial scale has had limited success because of the difficulty of mass emasculation of the female parent (Davis 1979). In southern India, however, F_1 hybrid seed has proved popular, even though its cost is 20 times that of conventional seed; all the emasculation and hybridization is done by hand. The most successful hybrid to date is H4, a cross between a late-maturing variety, Gujerat 67, and a nectariless variety of American Upland. The hybrid variety Suguna (CPH 2), issued in Tamil Nadu in 1978, was the first in which a male-sterile line was used as the female parent. A wide range of crosses is being tested for hybrid vigour, and experimental plots show increases in yield ranging from zero to 100 per cent above the parental mean. In Pakistan a small quantity of hybrid seed has been distributed from the central cotton research station at Multan, and work on cytoplasmic male sterility and restorer genes is being carried out using the techniques developed by Weaver and Weaver (1979) in the United States; these techniques have reached the stage when the commercial feasibility of hybrid cotton is being tested (Davis and Palomo 1980). The hybrid NX-1, released in New Mexico in 1979, has shown improvements in earliness, yield and fibre qualities over Acala 1517–75 (Palomo and Davis 1983), but 12 hybrids tested in Arkansas showed no significant differences from the standard controls (Smith and Varvil, 1984).

Multi-line seed issues

From 1946 to 1958 the commercial seed issues in the Lake Province of Tanzania were mechanical mixtures of from two to seven selected lines; such mixtures were called multi-line issues, and had the prefix UK, derived from the name of the research station, Ukiriguru (Peat and Brown 1961). In Uganda the NC seed issues (Walker 1963) were mixtures of lines selected from BP52, and formed the basis of the BP52 crop from 1954 to 1966; BPA was issued in that year, and was derived from a single plant progeny, A(61)6, but from 1969 onwards the seed issues were again multi-lines (Arnold and Innes 1976).

The main reasons for issuing multi-lines rather than single lines were either a shortage of seed or the fact that no single line was satisfactory in all respects. The theory of multi-lines was not considered at the time, but various theoretical advantages have since been suggested to explain their success.

One suggestion is that the component lines may respond differently to season and locality, and while line A yields well under one set of conditions which do not suit line B, another set of conditions will suit line B rather than A. When the seed is distributed, one or more of the components will be able to perform well under any set of conditions which are likely to occur, and the multi-line will on average be more stable than a single line would be. This implies that the interactions of the different components with the environment were markedly different; Walker (1963) showed that this was not so with the components of the NC issues of BP52; a similar study of the components of the UK issue has not been made, but from UK55 onwards the components were so closely related that they were unlikely to differ in their reaction (Lee *et al.* 1974).

Another suggestion is that a mixture of types may have an intrinsic advantage over a pure culture; differences in growth habit, root range, or other less obvious characters might complement each other, making more efficient use of the available resources. Trials in Uganda (Riggs 1970, Jones 1973, Arnold and Innes 1976) have given some indication that mixtures can outyield the mean of their components, but other comparisons show no difference at all. Arnold and Innes (1976) conclude that 'evidence on the possible advantages of heterogeneous varieties, with respect either to greater yielding potential or to greater stability, is inconclusive'.

Cotton seedlings are usually thinned out after germination, and where this is done by hand there is a natural tendency to thin out the weaker plants. This results in unconscious selection of any F_1 hybrids which may be present, and Brown (1953) suggested this as the reason for the persistence of Hindi Weed (*G. hirsutum*, race *punctatum*) in the Ashmouni variety in Egypt. Walker (1963) demonstrated that the size of the cotyledons in an F_1 hybrid was greater than that of the parents, which could be traced back to a common origin seven generations before; he also quoted an experiment showing the extent of the selection pressure exerted at thinning in a BP52 multi-line issue (NC 58 M2); the mean dry weight of the thinned plants was greater than that of the one left in only 6 per cent of the hills.

In this paper Walker put forward a new theory of multi-lines depending on the release of hybrid vigour by pollinating insects. Having shown that hybrids have a selective advantage at thinning, he calculated that this could result in an entirely hybrid crop within a few years if natural outcrossing was sufficient to give 20 per cent of hybrid seed in any one season. If crosses between the components of the multi-line show hybrid vigour, its yield should increase until the crop consisted entirely of hybrids. From a regression analysis of yield on years of multiplication, he concluded that NC 54 did show progressive increases in yield each year up to five years from the constitution of the multi-line reaching 27 per cent in the fifth year. Arnold (1970) points out that critical tests of this claim would need the simultaneous testing of successive generations of the multi-line over a range of localities and seasons, which was not in fact carried out.

An important consideration in this and subsequent multi-lines was probably the inclusion of line C(50)20, which was later shown to possess high general combining ability.

Interspecific crossing

The use of interspecific crosses in sorting out the taxonomy of cotton has been discussed in Chapter 3; the identity of species and their relationships are indicated by the fertility of crosses between them and by the study of chromosome pairing in the hybrids (Fig. 14.3). Between cottons in the same genome group, crosses usually are fully fertile, and even show hybrid vigour in the F_1, but they then segregate and lose their vigour and competitiveness. Species maintain their identity when grown together; while vigorous hybrids may occur in the first generation, the later generation hybrids disappear from the mixture. The treatment of such crosses in a breeding programme has been discussed under backcrossing.

The kind of changes which wider interspecific crossing can bring about in the cultivated crop can be grouped under two broad headings: firstly the transference of unit genes or chromosome segments from the wild to the cultivated species, and secondly quantitative changes due to the integration of an indeterminate amount of foreign chromatin.

In the first group, a desirable character is found to exist in the wild diploid, but not in the cultivated tetraploid. The transfer entails the progressive elimination of foreign chromatin from the hybrid by repeated backcrossing to the cultivated cotton, the new character being retained in each generation. The ease with which this can be done will depend on the fertility of the hybrid and on whether the character in question is dominant or recessive in its new environ-

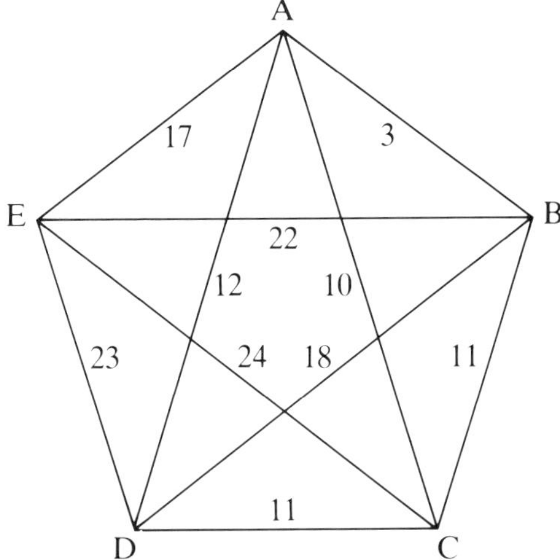

Fig. 14.3 Summary of average univalent frequencies for intergenomic hybrids of *Gossypium* (Phillips 1966, in Prentice 1972).

ment. Successful transferences of this kind have been the integration of 'smooth leaf' from the D genome of *G. armourianum* into *G. hirsutum* (Lukefahr *et al.* 1976); the transfer of the blight resistance gene B_6 to Sudan Sakel, *G. barbadense*, from the A genome in *G. arboreum* (Saunders 1972); the transfer of nectariless from *G. tomentosum* and lint strength from *G. thurberi* (Meyer 1974); a restorer gene for cytoplasmic male sterility transferred from *G. harknessii* (Weaver and Weaver 1977); and hairiness from *G. anomalum* to *G. barbadense* (Saunders 1964).

The theory behind the second group is that the original cross between Asiatic and American diploids and the doubling of the chromosome number to give the tetraploid species can only have occurred a very few times; the gene complement is therefore limited to what was carried by the few individuals involved. This complement will be increased by repeating the original cross using other diploid individuals or species carrying the A and D genomes, and this offers tremendous possibilities of releasing variability and producing hitherto unknown characters. Although the wild species do not produce spinnable lint, crosses with cultivated cottons have produced a full range from very coarse to very fine lint, with staple lengths in all groups and even exceeding that of the commercial standards. On the Ivory Coast, French workers have produced the HAR varieties from synthetic tetraploids *G. raimondii* × *G. arboreum* crossed with Acala varieties (*G. hirsutum*), and similarly the ATH varieties from *G. thurberi* × *G. arboreum* (Goebel 1968).

These interspecific crosses, especially the former, have produced lines with particularly high fibre strength and ginning outturn, as well as substantial yield increases (IRCT 1978); also a strain with greatly reduced bracteoles, a character unknown in any of the parent material.

The fact that the wild species are nearly all diploids, and most of the commercial cotton varieties are tetraploids, presents one problem in interspecies crossing; another problem is the fact that there are at least five distinct genomes with varying degrees of compatability, two of them with affinities to the tetraploid genomes. Many of the crosses which have been made between species have been studied cytologically, and the degree of chromosome association reported; the occurrence of univalent chromosomes indicates that certain chromosomes have no partner, and the greater the number of univalents present, the wider the relationship between the species concerned. This information has been summarized by Phillips (1966) in the polygraph of the interrelationships of the genomes which is reproduced in Fig. 14.3; the numbers on the lines connecting any two genomes represent the average number of univalents which are found in hybrids between species of the genomes in question. Lower numbers indicate closer affinities.

In crosses between species that have been effectively isolated from one another, the delicate balance of the genetic mechanism is rudely disturbed. The first concern is whether or not a viable F_1 plant can be obtained; in this plant the pollen and ovules will probably be infertile and such plants rarely if ever set seed, even when pollinated with normal pollen from one of the parent species. Some of the natural barriers between species may be circumvented by culturing the ovules and embryos artificially *in vitro* (Stewart and Hsu 1981). There are no rules laying down the best procedure to adopt in overcoming incompatability and sterility, and each worker must find by experience and by his particular needs what approach is most rewarding. When a few viable plants have been obtained in an interspecies cross, the next step is usually the doubling of the chromosome number with colchicine. Subsequent procedures outlined in the following paragraphs are largely derived from the chapter on cytogenetics by J. H. Saunders in the earlier edition of this book (Prentice 1972).

The cross may be between a diploid and a diploid, or between diploid and tetraploid, giving respectively a diploid or a triploid hybrid; fertility can usually be restored by doubling the chromosome number, giving a tetraploid or a hexaploid (Fig. 14.4 (a) and (b)). It is usual to try to synthesize such a tetraploid from diploids with the A and D genomes, so that it will produce fertile seed when crossed with one of the cultivated tetraploid species (AD). The fertility may not be complete, and may require several backcrosses to the natural tetraploid before full fertility is restored. The B genome is not

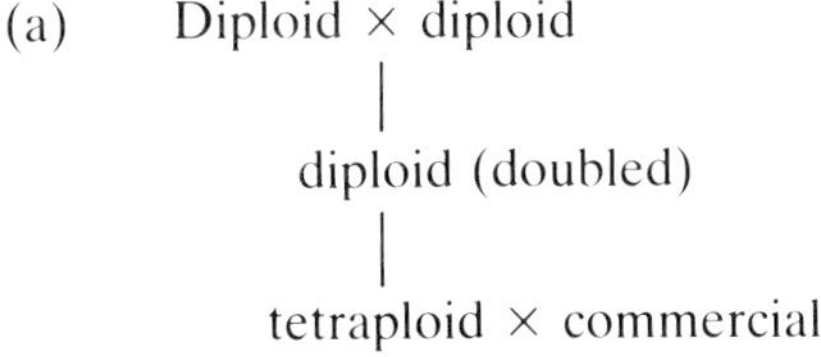

(a)   Diploid × diploid
            |
      diploid (doubled)
            |
      tetraploid × commercial

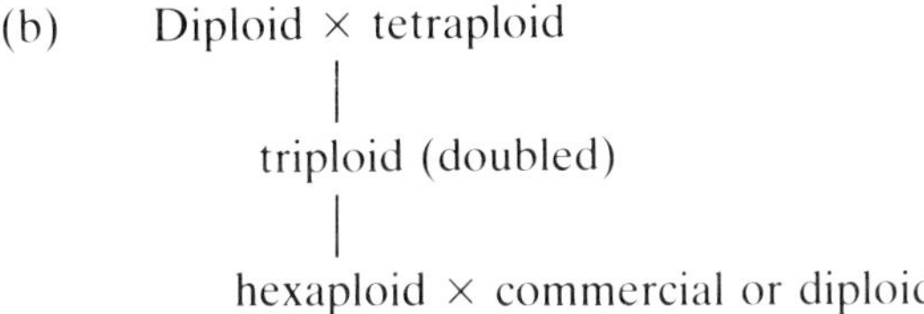

(b)   Diploid × tetraploid
            |
      triploid (doubled)
            |
      hexaploid × commercial or diploid

(c)   Hexaploid × diploid
            |
      tetraploid × commercial

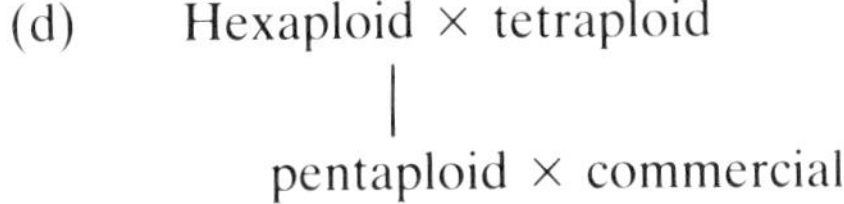

(d)   Hexaploid × tetraploid
            |
      pentaploid × commercial

Fig. 14.4 Systems of interspecific crossing.

widely differentiated from the A, and synthetic tetraploids BD can be crossed successfully with the tetraploid AD species.

The use of the hexaploid is more likely to make possible very wide hybridization, when combinations other than AD or BD are being attempted. When one or more hexaploid plants have been synthesized, there are three possible procedures to follow: the hexaploid may be selfed; it may be crossed with a diploid giving a tetraploid F_1; or it may be crossed with a tetraploid giving a pentaploid F_1 (Fig. 14.4 (c) and (d)).

In selfing the hexaploid, each chromosome will have its precise counterpart or homologue due to the doubling of the chromosome complement of the triploid. Pairing will normally take place between homologous chromosomes, i.e. genome A will pair with A and genome D with D. However, the diploid genome may have some affinity with one or other of the AD subgenomes of the tetraploid parent; if this is the case, then strict bivalent formation will not occur and multivalent associations will be seen because the chromosomes of the diploid parent are attracted to those of one of the subgenomes as well as to each other. To illustrate the point, if an Asiatic A genome cotton were the diploid that had been hybridized with an (AD) tetraploid to make the hexaploid, the genome complement could be written 2(AD)A′. The A′ is so written to distinguish the Asiatic diploid genome from the A subgenome of the tetraploid. At meiosis the following associations could occur:

$$
\begin{array}{lll}
\text{bivalents} & \text{AA} \quad \text{A}'\text{A}' \quad \text{A}'\text{A} \\
\text{trivalents} & \text{AAA}' \quad\quad \text{A}'\text{A}'\text{A} \\
\text{quadrivalent} \quad \text{AAA}'\text{A}' \\
\text{univalents} & \text{A} \quad \text{A}'
\end{array}
$$

This illustration is an extreme case where all the 13 chromosomes of the haploid complement of the Asiatic could have paired with the 13 chromosomes of the A subgenome. In such a hybrid there would be a high degree of sterility. However, if the diploid genome had not been so closely similar to the (AD) of the tetraploid but had displayed limited affinity only, then occasional or rare aberrations would be observed. This would mean that on self-breeding the hexaploid would not necessarily breed absolutely true. Some segregation due to reassortment of genes might occur, and the hexaploid is often unstable, gradually losing chromosomes with successive generations of inbreeding until it has reverted to a tetraploid complement.

The fact that irregular pairing can occur is of considerable importance to the breeder who is trying to incorporate part of the genic material from a diploid species into a tetraploid. Bivalent formation between the chromosomes of different genomes means that there is an opportunity for chiasma formation and the exchange of chromatin. The synthesis of the hexaploid is the first step in the integration of new gene material into tetraploid genotypes. No hexaploid that has been synthesized to date can match the agronomic performance of the tetraploid cultivated types, and it is not profitable to consider it as a potential variety in its own right. Whether there is an inherent optimum performance in the tetraploid genotype is not known; perhaps it is due to a gene balance that has been acquired over countless generations, but at present any improved cultivated cotton must be a tetraploid.

Crossing the hexaploid with a diploid will result in a tetraploid, but unless the F_1 contains two pairs of homologous genomes, meiosis will not be possible and it will be sterile. This possibility therefore suffers from the same limitations as a doubled cross between two diploids – the genomes involved must be A or B and D. Success in this kind of hybridization is usually greater if the plant with the higher number of chromosomes is used as the female parent.

The most usual way of treating the hexaploid is to cross it with pollen from the tetraploid commercial variety which it is wished to improve. The resulting pentaploids will have two AD sets of chromosomes and one set of 13 from the diploid, and are backcrossed to the commercial tetraploid. At meiosis, one set of AD is available for each cell at the reduction division, but the 13 chromosomes from the diploid genome do not pair and can only divide at random between the cells. What is hoped for is that some of them will then pair with an A or D chromosome from the male parent; at the next meiosis chiasmata will be formed and an exchange of chromatin will take place. No progress will be made unless the diploid chromosomes can be integrated into the tetraploid genomes.

It is very necessary at this stage to grow the largest possible populations of plants in order to ensure that as many different combinations in crossing over can take place. Following the second or third backcross, it is advisable to self every second generation to consolidate gains in structural hybridity, because repeated back-crossing tends to eliminate introduced chromatin. Any plants different from the standard type are indicative of genetic change; phenotypic change is not desirable for its own sake, but acts as a marker. Only field tests on material in which full fertility has been restored will show whether real changes in cropping characteristics have been achieved.

Selfing techniques

Insects are responsible for any natural cross-pollination that occurs between cotton flowers, so the problem of ensuring self-pollination is simply that of excluding insects from the reproductive organs of the flower. This can be done in many of the ways used in other crops, such as bags, cages or other artificial barriers; in cotton the easiest and most common method is to prevent the flower opening. The bursting of the anthers takes place perfectly normally inside the unopened flower, with no discernible loss of efficiency in the setting of seed.

The petals can be prevented from opening by sealing the tip of the flower bud on the day before it is due to open; this stage is easily recognized because the bud elongates on the day before flowering in a way that is unmistakable once it has been seen. The bud can be sealed by tying the tip with fine string or wool, by gluing the petals together with locally available materials such as mud, gum arabic or shellac, by fitting a perforated card, a cone of paper or a spiral of wire over the bud, or by enclosing the bud in a bag made of paper or mosquito netting; clear plastic bags have been found to increase shedding, but light paper bags, 3 × 4 in., are satisfactory, and protect the bud against bollworm damage. This protection is

Fig. 14.5 Self-pollination. The flower bud was tied with cotton yarn before the flower was due to open. The photograph shows the flower on the following day, when it would otherwise have been fully open (S. J. Brown).

enhanced if the bags are impregnated with insecticide (Saunders 1956). In wet weather the use of gum or clay is risky, but it is effective in dry climates. The size of the hole in a perforated card is critical – it must be the right size to fit over the bud at the right distance from the tip, sufficiently firmly to prevent it sliding off as the petals bulge outwards.

String or wool is cut into 80 mm lengths by winding it round a piece of wood 30 × 10 mm and cutting along one edge, which may have a slot cut along it. The string is tied with a single knot about 6 mm below the tip of the bud, lightly enough not to cut into the petals (Fig. 14.5). The selfed flowers should be checked the following morning for careless tying; a loop made too low on the bud means that the tip of the stigma can be exposed, while a tie made too near the tip of the bud can allow one or more petals to slip out, providing an opening for insects.

The petals are shed within a few days, along with the selfing mechanism; to identify the selfed bolls, a marker must be attached to the flower stem or peduncle. This may be a piece of wool or string similar to that used in selfing, or a label or small tag; the marker should be brightly coloured so that the selfed bolls are easily identified at harvest time. The marker may be attached at the time of selfing, or when the selfing is checked the following day. The best time for selfing is the afternoon or evening, as flowers usually open early in the morning, but this is not always so, and should be verified.

A good worker – women and children are often defter than men – can tie buds at the rate of three or four a minute if they are plentiful. If they are scarce, more time may be spent searching for suitable buds than in selfing, so the work should be concentrated at the peak of flowering, over a period of perhaps a month.

Crossing techniques

When making a controlled cross, the flower of the female parent is emasculated at the same large bud stage as in selfing; emasculation is usually done about 12 hours before the flowers are due to open: in the afternoon or evening when the flowers open early in the morning, or in the early morning if the flowers open late in the day. To gain access to the anthers, scissors may be used to cut round the base of the corolla, which can then be lifted away bodily to expose the staminal column (Fig. 14.6(a)); this is in the form of a tube surrounding the style, with the stigma protruding from the top. The anthers are pulled off in ones and twos by gripping the stalks of the stamens with forceps (Fig. 14.6(b)), which should be dipped periodically in a tube of methylated spirits to kill any pollen grains which may have been released accidentally. Rather than remove the whole corolla, some workers prefer to slit up the side of the bud with scissors and cut around the base of the corolla only enough to open it up and remove the anthers as before. It is also possible to remove the petals and the staminal tube in one piece by inserting the thumbnail near the base of the bud and peeling away both petals and anthers; this needs a delicate touch to leave the ovary and style undamaged. Less tedious methods of emasculation have been tried, but so far no satisfactory alternative to manipulation by hand has been found; the use of alcohol or other chemicals does not guarantee the destruction of all pollen when used at concentrations which

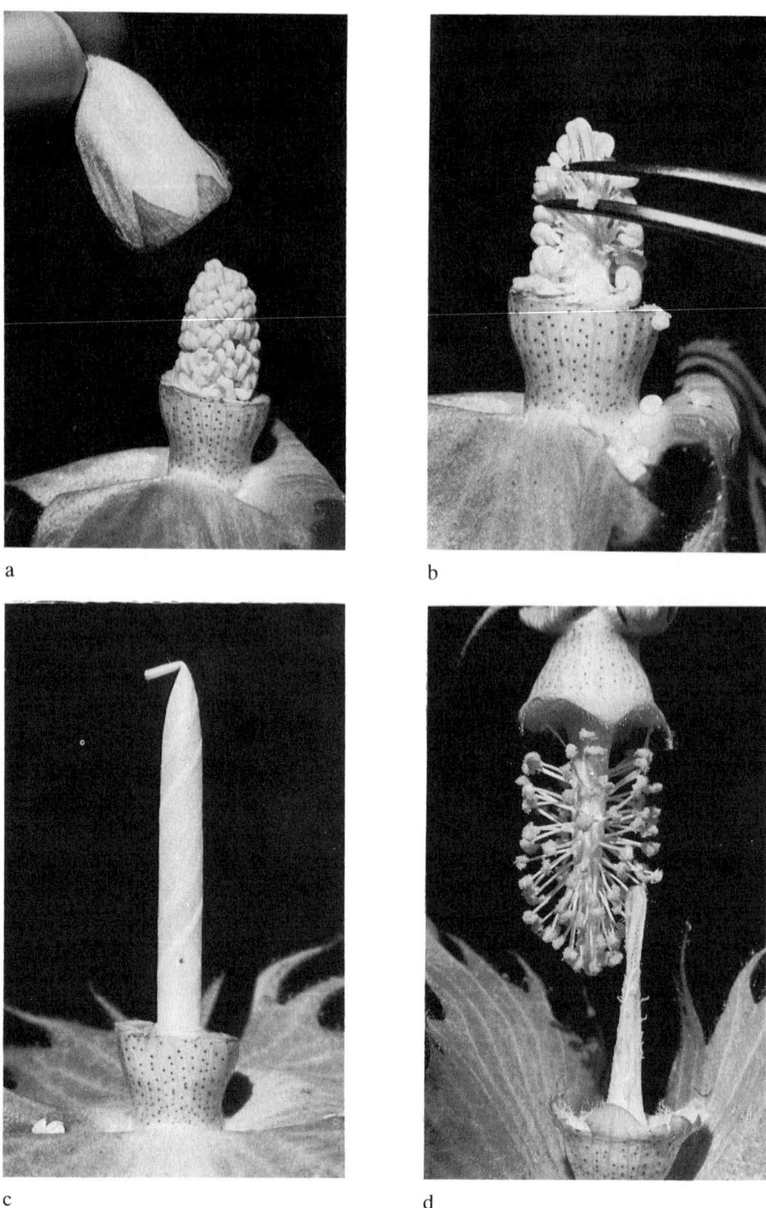

Fig. 14.6 Controlled cross-pollination (S. J. Brown). On the day before the flowers were pollinated, the corolla and part of the calyx were cut from the female flower (a), and the unripe anthers removed (b). A drinking straw was then placed over the stigma (c). The male flower was tied as if for self-pollination (Fig. 14.5) and, on the following day, the ripe pollen was transferred to the female stigma (d). The straw was then replaced to prevent uncontrolled cross-pollination.

leave the stigma undamaged. The chances of a stray pollen grain landing on the stigma are very small, but a recommended precaution for the beginner is to carry a hand lens and examine the stigma for pollen; the grains are fairly large and spiny, and if any are found the flower should not be used for crossing.

The exposed stigma has to be covered as soon as possible to prevent pollination by insects; this is conveniently done by slipping a short length of drinking straw (about 5 cm) over the pistil and bending the top end of the straw to close it (Fig. 14.6(c)). A little cotton lint may be wound around the base of the pistil and the straw pushed over it to provide a firmer grip. Alternatively, the emasculated flower can be covered with a paper or plastic bag of suitable size, folded or clipped round the flower stalk to prevent it being blown away. Mosquito netting has also been used, but unless the stigma is protected by the petals pollen grains from an insect which lands on the net can still pollinate it.

The flowers of the male parent are tied up as for selfing, at about the same time as the female flowers are being emasculated. After a suitable interval, when the stigma is receptive and the anthers of the male parent have burst to release the pollen, the pollination is carried out. A few selfed flowers with ripe pollen are picked from the male parent, and the petals cut away or folded back. The cover is removed from the female bud, pollen from the male flower is dusted copiously on the now sticky stigma (Fig. 14.6(d)), and the straw or bag replaced; one male flower is sufficient for at least half a dozen pollinations, and can be carried round for several hours in a box without obvious detriment to the pollen. A small tag with details of the cross and the date is tied to the stalk of the newly pollinated flower.

Crosses should be made early in the season, when physiological shedding is at a minimum; the chances of shedding may be reduced by cutting off any other bolls which have set on the plant, so that the available nutrients are channelled to the crossed boll. The recovery of crossed bolls varies with the conditions following crossing, and anything over 50 per cent may be considered satisfactory. It is worth while recording the number of emasculations made by each operator and their success for future reference, as operators vary considerably in their skill.

Chapter 15

Seed supply

Organization

Cotton is peculiar in that the grower seldom retains his own planting seed on the farm, the whole of the seed cotton being sold either directly or indirectly to the ginneries. The seed thus passes out of the farmer's hands, and each season a fresh supply of seed has to be passed back from the ginneries to the growers. As a result, it is possible to exercise considerable control over the seed supply and the variety being grown, and to a large extent the quality of lint produced. An adequate seed distribution scheme is necessary to exercise this control.

Another difference between cotton and most other crops makes it essential that the organization of seed distribution works smoothly and efficiently. This is the fact that varieties are often practically indistinguishable on inspection, and it is extremely difficult to verify their purity in the field. In a potato crop, for example, it is quite possible with practice not only to spot individual plants as rogues, but also to name the variety of the rogue plant. In cotton the plant breeder himself is often unable to name the varieties in his experimental plots, far less identify an individual plant as a rogue. The appearance of a cotton plant depends very much on its environment, and seed from apparently off-type plants often produces progenies which are true to type. Under tropical conditions lush growth tends to mask differences in growth habits which may be obvious in other environments. Field inspection does not therefore provide a complete check on the efficiency of a seed multiplication scheme.

In most cotton-growing territories in the tropics there are no commercial seedsmen who deal in cotton, and government departments or agencies control the distribution of cotton seed. The large number of small-scale farmers typical of developing countries makes the establishment of one-variety zones almost essential if varietal purity is to be maintained; the advantages of such zones have been recognized by the growers of Acala cotton in California.

In some countries seed renewal is a continuous process, with a routine progression through successive stages of registered, certified and commercial seed from a nucleus stock maintained by the plant breeders (Lewis 1970). Such a process makes it easy to introduce a new variety at any time. In other countries seed is only replaced when there are signs of deterioration in the yield or quality of the commercial variety, or when an improved variety is being introduced.

The question of paying a premium for the production of pure seed for planting is a vexed one. Where seed is produced by a farmer on contract to a seed merchant, a premium is paid to compensate for the conditions laid down to ensure purity and quality, such as isolation from other cotton crops and care in handling and processing the seed. A premium is less easy to justify if a farmer finds his farm included in a seed multiplication zone, with no obligations other than to collect and sow the seed issued to him, and sell his produce at a designated market. It is also argued that he receives a hidden premium from the better yields expected from any improved variety which may be introduced, which benefit him in advance of other farmers.

Rate of multiplication

The time taken to introduce a new variety depends on the quantity of seed to start with, the multiplication rate of the seed, and the area to be covered. The cotton breeder will almost certainly have started to bulk up his promising varieties before a final choice is made. The multiplication rate under close supervision in the early years is high, but usually drops as the crop gets further from the centre and depends more on the commercial grower; in the early years the yield per plant is maximized, but naturally enough the emphasis swings more towards yield per hectare as the commercial grower takes up the running. The history of the U4 variety in Barberton in South Africa has been described by Parnell (1935); starting with a promising single plant selection labelled U4 in 1925, the 1929 crop gave 200 tons of seed in spite of drastic rogueing in the first two years of bulking. At the other end of the scale, in parts of East Africa the ratio of seed recovered to seed issued has been as low as 2 : 1 or 3 : 1 in a bad season under peasant conditions, and cases are known of the return of seed being less than the amount issued. With a seed rate of 20 kg per ha and a ginning outturn of 40 per cent, an average yield of 1000 kg per ha of seed cotton (600 kg of seed) will give a multiplication rate of 30 : 1.

The areas chosen for the early stages of a multiplication scheme should have a reputation for good farming and good yields; they should be compact and so situated as to reduce the risk of contamination of the crop to a minimum. In addition, a single buyer should operate in the area, and competitive buying should not be permitted (King 1957, Faulkner 1972).

A typical seed multiplication scheme from Nigeria is shown in Table 15.1; the Uganda scheme is described by Thomas (1970).

Provincial Seed Corporations have been set up in Pakistan under a scheme financed by the World Bank. The Khaniwal Seed Farm in the Punjab has been operating successfully since 1975, and the buildings at Sakrand in Sind were completed in 1982. Certified seed of cotton and other crops is produced on the farm directly and by tenants, and by growers under contract in the surrounding areas. The seed is processed in a plant equipped with cotton gins, flame delinters, a gravity table and seed dressing machine, besides equipment to deal with seed of wheat, maize and berseem (World Bank 1976).

Estimation of seed requirements

The quantity of seed required for planting should be estimated with some care; although an under-supply is the more serious, and a comfortable safety margin should be allowed for, no grower,

Table 15.1   *Southern Katsina (Nigeria) cotton seed multiplication scheme*

Year	Approx. area (Acres)	Multi-plication rate	Seed required Per acre	Total	Production Seed cotton per acre	Total seed
0	½ × 45		—	4 lb	580 lb	180 lb
1	20 × 32		9 lb	180 lb	465 lb	5,760 lb
2	360 × 16		16 lb	5,760 lb	410 lb	41 tons
3	4,000 × 9		23 lb	41 tons	330 lb	370 tons
4	27,000 × 6		28 lb	340 tons	270 lb	2,000 tons
5	108,000 × 4½		28 lb	1,350 tons	200 lb	6,000 tons

Year	Stage	Location
0	A	Samaru research station
1,2	B,C	Department of Agriculture, Daudawa
3	D	Settlers around Daudawa
4	E	4 market areas near Daudawa
5	F	Remainder of Southern Katsina

Note: Areas and seed rates beyond Stage C are estimates only. Wide variations in yield occur, and in a good season these production figures will be exceeded. Stage D production allows for 30 tons to be sent to the Gombe seed multiplication area. Stage E production allows for 650 tons to be used for planting a 'protective zone'.
Source: After King 1957.

department or seed merchant wants to be left with a large surplus of ageing seed. Records of previous seed requirements are the most reliable guide to the needs of an area, but if these are not available the requirements may be estimated from the average seed rate and the area to be sown. Estimates for raingrown cotton should include an additional allowance for replanting of about 50 per cent of the estimate for the first planting; in a bad season whole areas may have to be replanted two or three times.

Seed should be issued to the growers at least a month before the planting rains are expected.

The seed rate for cotton planted by hand in Africa is normally around 20 kg per ha. For machine planting a rather heavier rate is used; in the USA the average allowance is 35–45 kg per ha for ordinary drill sowing, 25 kg per ha for hill drop sowing and 12–16 kg where acid delinted seed is used (Munro 1966). Seed rates in Egypt are very high, almost certainly the highest in the world; spacing is about 60 cm between ridges and 15–20 cm between holes, and 15–20 seeds are sown per hole, requiring 100–130 kg of seed per ha (Brown 1953).

Special care is needed when a new variety is under multiplication. If the supply of seed runs out, the only recourse is to use seed of the old variety to refill gaps or to replant; it is practically impossible to separate such mixtures when the cotton is bought at the end of the season. On the other hand, a surplus of seed will mean that multiplication of the new variety will not be as fast as it might have been. Where a scheme for continuous replacement has been established, successive waves of seed following each other will flush out the previous seed, and some contamination of the new variety can be tolerated.

Soundness of seed

Besides varietal purity, a multiplication scheme must produce seed of good planting quality. This may be defined as the ability 'to germinate and produce seedlings that rapidly emerge from the soil, under both favourable and adverse environmental conditions' (Leffler 1981). Peacock and Hawkins (1970) and others have shown that the quality and vigour of the seed can affect plant growth, development and yield.

The quality of the seed depends on environmental conditions during development; Gipson and Joham (1969) found a positive relationship between germination and night temperature during seed development; the number of bolls developing on the same plant also affects seed quality by competing for the products of photosynthesis. Other factors are more easily controlled by the seed producer. Seed quality can be improved by genetic selection, as is shown by the multi-adversity resistance (MAR) breeding at Texas A & M (Bird 1981, Bird *et al.* 1981). Leffler (1981) quotes a number of references showing that the best seed is produced in the middle of the harvest period, and that wider spacing enhances seed quality; high rates of fertilizer nitrogen can lower the quality of the seed produced; the crop should be as free as possible from pests and diseases.

Weathering of cotton seed in the open bolls decreases the germination rate and the vigour of the seedlings. The main factor influencing weathering is moisture, and cotton seed exposed to rain or high humidity deteriorates much faster than seed ripened during dry weather. In extreme conditions cotton seed can be found germinating in the boll, but moisture itself is not usually the direct cause of the deterioration; it provides conditions favourable to the development of fungi and bacteria, leading to bollrot. 'Avoidance of exposure of cotton seed to moist conditions through timely harvesting has long been known, and continues to be the best way to circumvent weathering damage of planting seed.' (Halloin 1981).

Quality of cotton seed can be affected by the methods used in harvesting, ginning, delinting, handling, seed treatment and storage (Noggle 1973). Harvesting by hand is unlikely to cause any mechanical damage, but mechanical harvesters (spindle pickers) of certain designs can cause quite severe damage by cracking the seed coat (Douglas *et al.* 1967). It is not surprising that ginning and machine delinting can cause mechanical damage and reduce germination; Watson and Helmer (1964) found that harvesting and ginning each contributed about 5 per cent of visible damage to the seeds; increasing the feed rate to the gin and higher moisture content of the seed cotton tended to increase the percentage of damaged seed. In handling, improperly adjusted conveyers can cause substantial damage, and pneumatic conveyers should not be used for moving seed. All of the acid-delinting systems can reduce seed quality if not properly controlled and managed.

Seed storage

Given good storage conditions, cotton seed viability can be maintained from harvest to the following planting season without significant deterioration. The principal factors affecting viability are humidity and temperature, and of these humidity is the more important; a high moisture content provides ideal conditions for the growth of fungi and bacteria, which not only damage the seed directly, but can cause bulks of seed cotton or seed to heat in the store, occasionally causing fires.

Sorenson and Wilkes (1973) quote the following safe storage periods for seed cotton at various moisture contents:

Seed cotton moisture (%)	*Safe storage period (days)*
8–10	30
10–12	20
12–14	10
14–15	3

It follows that cotton should be picked as dry as possible, and if it has to be picked wet it should be spread out and dried as quickly as possible before it is finally packed into bags.

Ginning is most efficient at a moisture content of $6\frac{1}{2}$ to 8 per cent for saw gins and 5 to 6 per cent for roller gins (USDA 1977), and tower·dryers at the ginnery may be needed to condition wet seed cotton. After ginning the moisture content of the seed should not be allowed to rise above 12 per cent; a safe level is below 10 per cent, and the seed should be kept as cool as possible. This is not a problem in most cotton-growing areas, as the storage period usually coincides with the dry season in the tropics and subtropics; seed is sown within about six months of ginning, and can be stored satisfactorily at ambient temperatures from one season to the next provided it is kept dry. In Ecuador advantage is taken of the cooler climate at higher altitudes to store planting seed at Cuenca (2540 m), some 200 km from the cotton-growing areas on the coast.

If warm and humid conditions are expected during the storage period, special precautions will have to be taken to keep the seed cool and dry. Ramachandran (1983) has described a method used in India for storing bags of cotton seed satisfactorily for nearly eight months; 20–25 cm open channels criss-cross the floor of the seed store every 1.2 m or so, and the bags are stacked over these in such a way that a vertical hollow space is left in the centre of each stack (Fig. 15.1). Any incipient heating in the stack will warm the air in these spaces, and draw cooler air in through the channels in the floor, providing automatic air circulation through the godown. Additional ventilation may have to be provided by the use of fans, with or without heaters to dry the incoming air, depending on how humid it is. Dry seed can be packed in sealed polythene bags, preventing contact with moist air, for quite long periods without deterioration. Fully air-conditioned storage is only justified if seed is to be stored for long periods, as when it has to be carried over to a second season or in maintaining seed stocks for reference. For really long-term storage, a temperature of 5 °C and a relative humidity of 50 per cent is recommended (Holman and Snitzler 1961).

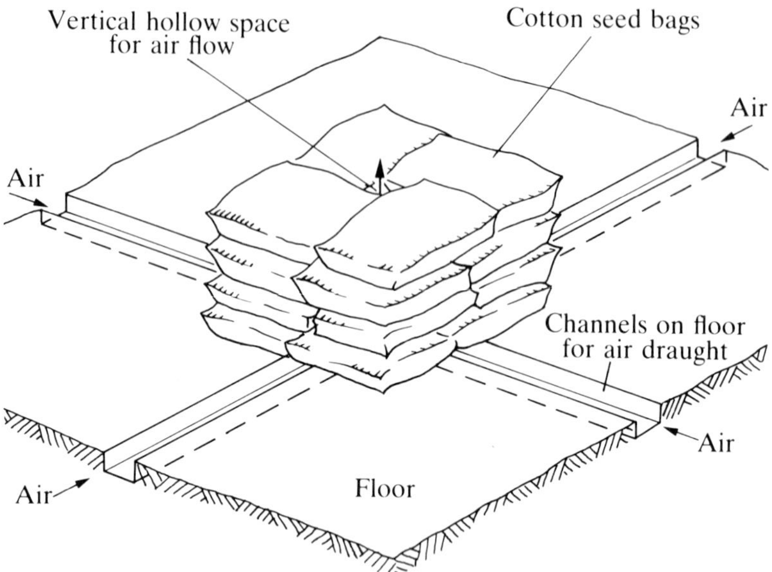

Fig. 15.1   Method of storing cotton seed bags in godown (Ramachandran 1983).

Seed stores should be inspected periodically for damage by rodents and insects pests, and suitable measures taken against them if they are found.

Germination tests

The simplest and most direct method of testing the quality of seed for planting is a germination test; this can give reliable counts of germination, judged by the percentage emergence of healthy radicles, in about three days. In the standard germination test, a sample of 100 seeds chosen at random is soaked in water for a few hours, then rolled in wet sacking or blotting paper and placed in an incubator at 30 °C; this gives slightly quicker germination than the alternative regime of 8 hours per day at 30 °C and 16 hours at 20 °C (Association of Official Seed Analysts 1978). If an incubator is not available, the rolled seed can be placed in a plastic bag to retain moisture and germinated at room temperature; alternatively the seed can be sown in boxes of sand kept moist by watering, counting the seedlings that emerge; this takes a little longer, as the cotyledons emerge a day or two later than the radicle. Tests made within a few weeks of boll opening may give lower figures than normal; this is the result of seed dormancy (Simpson 1935), which disappears within a month or so, and is thought to be caused by the presence of abscisic acid (Halloin 1976).

The standard germination test takes three to five days, and it is often necessary to obtain a quicker estimate of seed quality. The simplest method is to cut a sample of individual seeds, lengthwise, and rate them for the fullness and colour of the contents; a high proportion with the contents stained or shrivelled indicates poor quality; good seed is hard and brittle, and soft seeds indicate too high a moisture content. There are chemical tests which depend on a colour reaction on the cut surface of the seed to show its viability. By applying tetrazolium salt (2,3,5-triphenyl tetrazolium chloride) to the cut surfaces, the viable respiring tissue is stained red, while non-viable tissue remains colourless; the location and intensity of the staining and the amount of dead tissue are used to classify the seed into categories for germination capacity and vigour (Baskin 1981c). Other tests of seed quality are based on free fatty acid content and electrical conductivity (Presley 1958).

Conditions for the standard germination test are chosen to give the optimum germination, but conditions in the field are seldom optimal. Rate of emergence and the stand achieved in the field depend not only on germination capacity, but also on the ability of the seedling to withstand the stresses it encounters in the seed bed. The deficiencies of the germination test are illustrated in Table 15.2: 50 commercial lots of cotton seed labelled 80 per cent germination were collected in Mississippi in 1967, of which 34 lots showing 80–85 per cent germination were selected for field emergence tests; percentage emergence in the field ranged from below 40 to over 80 per cent (Delouche 1981). A variety of tests of cotton seed vigour have been suggested, the most widely used being the tetrazolium test and the cool germination test. 'The cool germination test is similar to the standard germination test except that a temperature of 18 °C is used, and the percentage of normal seedlings $1\frac{1}{2}''$ long or longer is determined after 6 and 7 days respectively for acid delinted and mechanically delinted seed' (Delouche 1981).

Table 15.2 *Field emergence of 34 lots of cotton seed with germination percentages from 80–85 per cent*

Date of sowing	Number of lots with emergence % of:					
	80+	70–79	60–69	50–59	40–49	40–
Mid-April	1	8	11	6	3	5
Mid-May	6	10	13	2	2	1

Emergence tests made at Mississippi State, Miss., in 1967.
Source: Reproduced from Delouche 1981.

Delinting of cotton seed

Seed fuzz

The ginning process which separates the cotton lint from the seed does not remove the short hairs, called fuzz or linters, which are characteristic of most commercial varieties. Generally speaking the Upland cottons (*G. hirsutum*) and the diploids (*G. arboreum* and *G. herbaceum*) produce fuzzy seed, while the Egyptian cottons (*G. barbadense*) have no fuzz or only a small tuft at the distal end, known as naked and tufted seed respectively. Various grades of fuzziness (Fig. 15.2) have been recognized (Low 1968, Ware 1940). Any grade from fully fuzzed to naked seed can germinate successfully, but the degree of fuzz affects its behaviour and how it can be handled. Naked seed requires less power for ginning, and the seed is easier and cheaper to handle; fuzzy seed sticks together and presents difficulties in hand sowing, while it is practically impossible to sow by machine without delinting.

The fuzz impedes the imbibition of water by the seed and delays germination. This is probably a safety mechanism developed under arid growing conditions to ensure that germination does not occur until sufficient rain has fallen to maintain the growth of the seedlings. When the rainy season starts late, cotton is often sown in dry soil – dry sowing – to get the quickest possible germination when the rains eventually arrive; the use of fuzzy seed will prevent premature germination after an isolated shower of rain.

Naked or acid-delinted seed has many advantages. The seeds do not stick together and the seed rate, whether sown by hand or by machine, can be controlled much more effectively. Mechanical seeders designed for other crops such as maize and beans can be

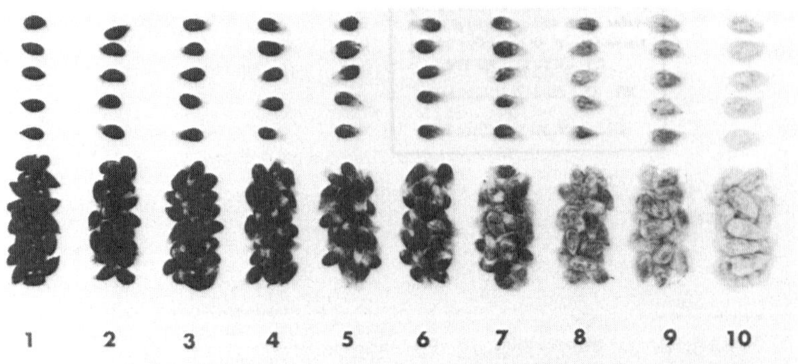

Fig. 15.2 Grades of seed fuzz (Low 1968).

used without any modification other than adjustment for seed size. Light and damaged seed can be separated out during acid treatment or on a gravity table to improve germination. In the field, given adequate moisture, germination is quicker and more uniform, allowing more exact timing of subsequent operations, and providing more uniform stands in field experiments.

Black seeds, either naked or tufted, are occasionally found in fuzzy-seeded varieties. This is sometimes taken as an indication of deterioration in the variety, but except in extreme cases this is probably not so. Low (1968) found that selection in Albar could reduce the seed fuzz grade (Fig. 15.2) from the typical grade of seven or eight to grade four or five without any loss of quality or yield, but beyond that point reduced fuzz was associated with reduced ginning percentage and coarser lint (Costelloe 1970).

Mechanical delinting

Mechanical delinting is the simplest and cheapest way of reducing the amount of fuzz to a level acceptable for sowing. Delinters are similar in action to saw gins, but the saws and teeth are set more closely together; the seed may be passed through two delinters in succession, giving a first and second cut. Fuzzy seed may yield as much as 10 per cent of its weight as linters; the product is not suitable for normal spinning, but is used in the manufacture of such commodities as surgical cotton wool, paper, packing materials and felt. It is a source of pure natural cellulose, and was originally used in the manufacture of man-made cellulosic fibres. The cost of delinting is largely recouped by the sale of the linters, which may even show a small profit. Delinters for treating planting seed should be installed at the ginnery; seed for processing into oil and cake is usually delinted mechanically to increase the percentage of oil extracted and the protein content of the cake; for this purpose the delinting machines may be installed in the ginnery or in the oil mill.

Mechanical delinting improves the handling quality of the seed, allowing more accurate sowing by hand and the use of mechanical seeders fitted with a stirring rod to maintain the flow of seed. The remaining fuzz allows the seeds to take up a significant load of fungicide, and up to 1 part in 150 by weight can be added to the seed in the form of fungicidal or insecticidal dust without the use of detergents or stickers. On the other hand, mechanical delinting has been shown to aggravate the problem of seed-borne bacterial blight by distributing the spores from a few infected seeds to the majority of uninfected seeds. Mechanical delinting, with or without the addition of dusts, is the usual form of seed preparation in developing countries.

Acid delinting

Delinting by acid removes all the fuzz to produce naked seed; machine delinting before the acid treatment reduces the quantity of acid needed in the process. Commercially, the method most commonly practised is the wet acid treatment, using 98 per cent commercial sulphuric acid; a gallon of acid will treat about a hundredweight of fuzzy seed (1 litre to 12 kg seed). The seed and acid are stirred together until all the fuzz is removed, giving a black treacle-like mass, which is then washed off in successive changes of water, the last wash containing some lime or caustic soda to neutralize any acid residue. Less dense seed floats in water, and this seed (floaters) can be separated from the denser seed (sinkers) during the washing process. The floaters include damaged and poorly developed seed, and their elimination improves the quality of the seed sample. Not all the floaters are of poor quality, however, and the percentage varies with growing conditions; in South Africa the percentage is usually between 10 and 20 per cent, while in Nigeria around 50 per cent of the seed floats in water.

Sulphuric acid treatment is carried out on a small scale at many cotton research stations in Africa (Costelloe 1968); it allows an economical and uniform seed rate to be used in trials, ensures quick and even germination, and the process kills any disease organisms, such as those of bacterial blight, which are carried on the seed coat. The germination of sound seed is not affected by the application of concentrated acid for the time required for treatment. The floaters derived from an acid-treated bulk of a scarce new strain may be held as an emergency reserve; although rarely reaching the standard of germination required in field trials, such seed can be used for strain multiplication should the need arise.

Strong sulphuric acid must always be handled with care. Operators should wear gumboots and overalls, and plenty of clean water must be immediately available to wash off any acid spilt on the skin or clothes. If a small quantity of water is added to concentrated sulphuric acid, sufficient heat can be generated to vaporize the water explosively, splashing the acid in all directions; in dilution, the acid should always be added to an adequate volume of water. The heat can also damage cotton seed in the first washing, unless cold water is poured on quickly and in sufficient volume to disperse the heat.

Acid treatment plants have operated for many years in Zimbabwe, Zambia and South Africa. Watson (1975) describes a small installation based on an ordinary cement mixer which will treat 1.25 tons of seed in 4 hours; drying the seed in the open air occupies the rest of the day, and limits the capacity of the plant. The disadvantage of the wet acid treatment is the difficulty of arranging for the supply and disposal of the large quantities of water required for washing the seed. A system requiring less water and acid has been described by Jones and Slater (1976), in which dilute acid (about 12 per cent) is used. In this process the seed is first soaked in the dilute acid, surplus liquid is removed and the moist seed dried, leaving an increasingly concentrated residue of acid which eventually burns off the fuzz; by adjusting the concentration of acid according to the amount of fuzz, the quantity of free acid left to be removed by washing is relatively small. This method is theoretically very attractive, but has run into problems when used on a commercial scale.

Another new process is the treatment of seed with an acid foam, which requires less acid than the usual wet acid treatment. The acid has to be diluted to a concentration of about 60 per cent before the foaming and wetting agent is added; this creates problems in remote areas, as the diluted acid is more bulky and corrosive than the commercial grade of acid, and dilution with water at the treatment site can be hazardous.

The dry acid process is another method of treating large quantities of seed. The seed is exposed to hydrochloric acid gas in sealed tanks, or to a mixture of sulphuric and hydrochloric acids, and the charred remains of the fuzz are removed from the seed by abrasion. This process is superseding the wet acid process, but it requires skilled supervision and a high capital expenditure, and is only economic when large quantities of seed are to be treated.

Flame delinting

In this process the seed is dropped through a bank of burning gas jets, which singe off the bulk of the fuzz but do not leave the seed completely smooth. The adjustment of the burners and the flow of seed is critical to prevent damage to the seed, which must be separated from the hot ash and cooled quickly after flaming.

Seed conditioning

Seed conditioning is the preparation of seed for marketing, including cleaning, grading and treatment with insecticides and fungicides. Acid-delinted seed can be graded on a gravity table to select high-density seed. Mechanically delinted seed cannot be separated in this way as it clings together and other methods of separation have not proved adequate for commercial use. In the wet acid process seed which floats in water can be separated during washing.

Packaging of planting seed does not normally receive much attention in the less developed countries, but suitably labelled packages can make the distribution of good quality seed easier and more efficient. The size of the package may be a full-sized sack, or smaller quantities suitable for the individual farmer; a 10 kg packet of seed is sufficient to plant half a hectare or one more. Where seed is on offer from a number of unregulated sources, attractive labelling and packaging can help to increase the market share of registered or certified seed of guaranteed quality.

Seed may be treated with chemicals to protect it from diseases and insect pests (see Ch. 13). This is usually in the form of a dust at ratios between 1 : 150 and 1 : 500 by weight. Mechanically delinted seed retains enough fuzz to take up this dust without difficulty, but some form of sticker is needed to help the dust adhere to the surface of naked seeds. Small quantities of seed can be treated in batches, being shaken up with the dust in a sealed container of suitable size; an oil-drum (44 gallons, 200 litres) can be mounted on a stand and rotated end over end with a handle. Large quantities can be treated in batches in a Booth drum mixer, or as a continuous flow, the dust being dispensed from a hopper at the start of a screw conveyer; seed and dust are mixed as they pass along the screw, which must be enclosed to prevent pollution of the atmosphere. It is advisable when treating seed with any kind of dust that the operators wear masks to prevent them inhaling it, and wash any dust off the skin immediately work is finished.

Maintenance of varietal standards

Cotton-growing countries differ in the standard of varietal purity that they wish to maintain, but once the standard is set it should be maintained against the many possible sources of contamination. The first of these and probably the most common, is mechanical mixing caused by careless handling; this may occur during seed distribution, at planting, at harvesting and marketing of seed cotton, and at ginning, seed preparation and storage between seasons. A well-established routine is a great help in avoiding such mistakes, as it leaves no doubt about procedure and avoids hurried decisions and possible misunderstandings.

Apart from mechanical mixing, a slow deterioration tends to take place in an established variety due to segregation of heterozygotes remaining even in the most highly bred varieties. If natural selection is left to work unhindered, those segregates which possess some advantage in reproduction, such as smaller or more numerous seeds, will gradually alter the modal type; this is often associated with a deterioration in lint quality. The best known example of a prolific contaminant is the continuing presence of Hindi Weed (*G. puncta-tum*) in the Ashmouni variety in Egypt; organized work on purifying the main stock started in 1920, but Brown (1953) states that 'Hindi-free nucleus stocks have now been continuously distributed for thirty years, but it as regularly reappears in the bulk crop'.

The Egyptian system is an example of the maintenance of a nucleus or élite stock to correct any tendency to change or deterioration. It is kept free of off-types by roguing, mass selection or at the extreme by single plant selection, and is grown in a plot physically surrounded by a crop sown with seed of the previous nucleus from which it was selected. Reselected seed can be issued annually or at short intervals from a centre, reaching out to the periphery of the area like waves or ripples in a pond. An alternative to this wave system is to flood the areas as quickly as possible with a new variety when the original stock needs replacement, but this calls for concentrated effort at irregular intervals; a steady annual replacement from the centre is preferable (see Ch. 14).

Seed supplies for the breeding programme

A breeding programme involves the handling of a large number of selections, and maintaining their purity and identity from season to season through harvesting, weighing, ginning, sampling, seed treatment, storage and sowing. This requires the setting up and maintenance of a routine as simple and foolproof as possible; mistakes will occur, but they must be kept to the minimum. It is disconcerting, to say the least, to find red or okra-leaf plants during the multiplication of a new variety which should not possess these characters; it means that the routine is faulty, and that similar mistakes which are not so obvious are being made.

Stores

The store for breeding material should be dry, well-lit and rat-proof, and it should be conveniently placed with regard to ginning and other facilities. Seed cotton samples coming in from the field need a large store, giving access to the ginning and weighing room; smaller stores for ginned lint and seed are best located at the other end of this room, giving a one-way flow. The ginning room should open on to a concrete drying floor, spacious enough to spread out a large number of small individual lots of seed or seed cotton (Fig. 15.3). Shelving in the stores makes access to individual lots much easier, and avoids having to turn over a heap of sacks on the floor every time a particular sample has to be found. If sacks have to be piled up, they should be arranged in such a way that the labels identifying them are all easily visible.

Bags

A range of bag sizes is needed to accommodate seed cotton, seed

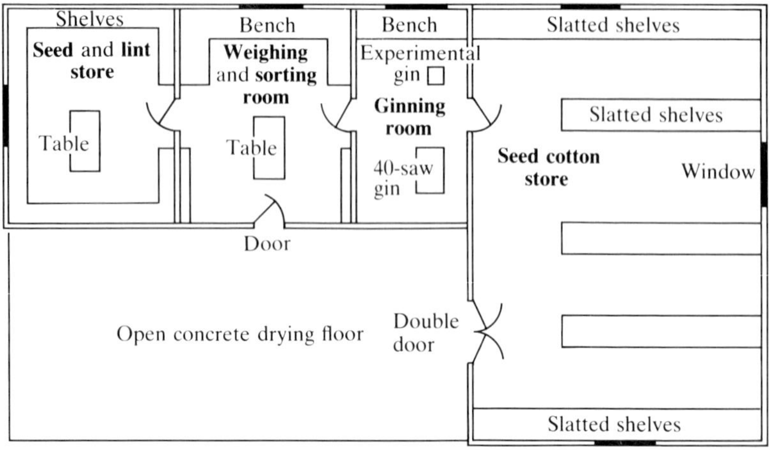

Fig. 15.3   Suggested layout of cotton experimental unit; scale 1:100.

and lint from single plants, progeny rows, yield trials and multiplication plots. Paper bags are cheap and easily handled, and have the great advantage that the identification can be written on the bag itself; their disadvantage is that they are fragile, and liable to burst if handled carelessly or overfilled. Plastic bags are more robust, and a label placed inside the bag can be read without having to open it. They are also weatherproof, and may be tied to a plant or stake in the field; a succession of pickings can then be made without having to sort out the bags each time. Cloth bags can often be made quite cheaply from locally woven cotton cloth; they can be used over and over again, washed after use, and turned inside out to make sure that no cotton has been left in them.

Paper and plastic bags containing a relatively small amount of material can be closed by simply rolling them up; if fairly full, they should be tied with string or closed with paper clips or staples. Cloth bags can also be rolled up, but are usually tied with string; as they require a tie-on label, the same string can be used to tie the mouth of the bag. An alternative is to sew tapes to the bag just below the mouth, on which a label can be threaded before the bag is tied up. A small curtain ring can also be sewn in, so that the bags can be strung in order on to a wire or bamboo cane. Grain sacks or gunny bags are suitable for the produce of large plots, and for keeping groups of small samples together during storage.

Labels

Every lot must be clearly labelled, either on the bag itself or on a separate label tied to the bag. Strong paper (manila) labels are commonly used, and black or coloured pencil is more resistant to moisture than ink. A second label placed inside the bag is a useful insurance against the loss or illegibility of the outside label. Small labels are less vulnerable to damage than large ones, but they must be big enough to take the necessary details clearly written.

Numbering systems

The numbering system in a breeding programme should be simple and concise, as the more complex it becomes the greater the possibility of mistakes. In a small programme, each selection may be identified by the year of selection and a serial number, e.g. 66/1, 2, 3, etc. With larger numbers of selections, different groups may each have their own serial numbers, the groups being distinguished by names or letters, e.g. Albar 66/1 (or A 66/1, or A (66) 1) and UPA 66/167, etc. When the programme involves a series of reselections, some breeders substitute the number of the parent for the year of selection, e.g. A/25/3 is the third selection from A/25.

When progenies reach the stage of formal trials, they may be identified by the treatment number allocated in the trial, e.g. Lankart × Stoneville (68) 257 becomes LS 4. It is important to maintain a key to such numbers, so that all the data can be linked together when summarizing the record of a particular line over a number of seasons. For ease of reference, the annual report can quote the previous year's number in brackets, along with the current season's number.

A variety entering commercial production may well be given a different name or number. This may relate to the pedigree, the research station, the district, or may be arbitrarily chosen. The long-staple Uganda variety BP 52 was named after the Bukalasa station where plant 52 was selected; its successor, BPA, indicated its selection from Albar, but linked it with the commercially well-known BP cotton. In Tanzania the early seed issue MZ 561 was called after the Mwanza district where it was grown; later issues, such as UK 63, indicate the 1963 multi-line issued from the Ukiriguru research station. In the Sudan the Lambert (L) varieties have quality characteristics similar to X 1730 A, and the first one which incorporated resistance to bacterial blight was called Bar XL 1; the Sakel (S) varieties have the better quality of Domains Sakel, and include the improved variety VS 1 issued in 1960. In Egypt new varieties were given individual names, such as Sakel, Ashmouni, Karnak and Dendera, as well as serial numbers from the Giza Research Farm, e.g. Dendera = Giza 31, but most varieties are now identified by the Giza number only, e.g. Giza 70, Giza 67, etc. Some of the older varieties from the United States were introduced into Africa before 1914, and the American names still persist, e.g.

Allen, Bancroft, Acala. Varieties in India and Pakistan are usually identified by letters and numbers, e.g. MCU 4, Hybrid 4, AC 134; but they may also receive a name, e.g. Mahalaxmi (1301–DD), Bharathi (MCU 6), Krishna (AC 122).

Seed source

Seed supplies must be planned so that there is enough seed to meet the requirements of the programme. These needs range from sufficient to plant a progeny row to a hectare or more to start the multiplication of a new variety. Where natural crossing is negligible, this only involves a decision on the size of the plot needed at each stage of breeding and testing, but if there is appreciable natural crossing there must be a balance between selfing and other forms of isolation. A typical programme will include selfing plots producing selfed and unselfed seed, multiplication plots without selfing in which the outer rows are picked separately from the inner plot, and seed from variety trials; this gives rise to five grades of seed with respect to purity:

Source	Order of purity
Selfed seed	1
Multiplication plots: central	2
borders	3
Variety trials with guard rows	4
Unselfed seed from selfing plots	5

Selfed seed should be used for planting the selfing plots and multiplication plots in the following season, supplemented if necessary with seed from the central areas of the multiplication plots. Only after this has been set aside should seed be prepared for variety trials; a small proportion of hybrids is unlikely to affect the results of the trials, and it is not advisable to depend on seed taken from the trials for future planting as the layout usually favours intercrossing. The seed used for variety trials should nevertheless be as pure as possible, following the order of purity shown above. As this may comprise different degrees of purity, all the trial seed should be thoroughly mixed before the seed for individual trials is weighed out.

If seed has to be taken from the plots of a replicated trial without artificial selfing, the plots would be large enough to allow discards, and the varieties in the trial should be fairly closely related (see p. 235).

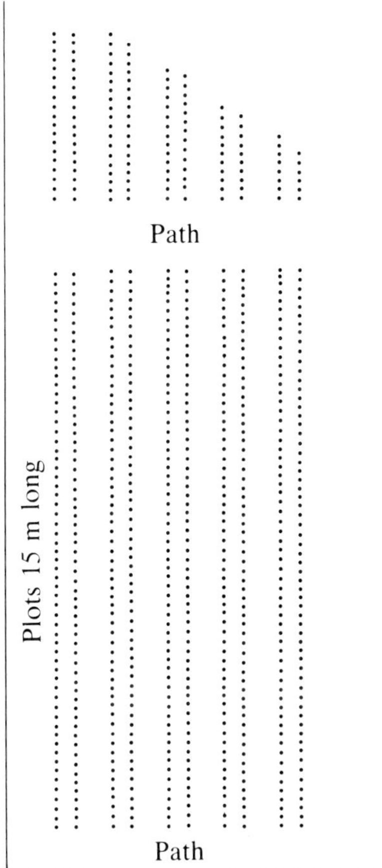

Fig. 15.4 Layout for producing selfed seed; two-row plots, one unplanted row separating plots.

Selfing plots

The layout of the selfing plots is worth special attention to ensure the best return from the labour involved. It is known that border rows yield over 50 per cent more than the interior rows of a plot of cotton, and some of this is due to a higher percentage of bolls set. One- or two-row plots with unplanted rows between them ensure that every row is a border row, and the proportion of selfed buds which produce bolls will be increased. Paths between the plots make it easier to spot buds at the right stage for selfing, and to locate the selfed bolls at harvest. They also reduce the risk of mistaken identification, as adjacent rows may intermingle their branches to such an extent that it is not clear to which row a boll belongs (see Ch. 14). The selfing area can therefore be laid out as a series of plots (2 × 15 m) sown in a skip-row pattern, two rows wide with an unplanted row between each (Fig. 15.4). Fifty grams of seed is ample to sow each plot.

Chapter 16

Field experimentation

The use of statistics

Snedecor (1946) in his introduction to *Statistical Methods* states that 'the experiment is the distinctive tool of science'. Like any other tool, it is a means to an end, and can be very useful when the job requires it; but the scientist has to steer a middle course between those who consider an elegant statistical analysis showing significant differences as an end in itself, and those who analyse their data as a matter of routine, but do not use the analysis to assess the validity of their conclusions.

'The object of all experiments, whether of a field or laboratory character, is to provide the research worker with an answer to a question he has put.' (Wishart 1940). In formulating these questions and designing experiments to answer them, there are two aspects of the problem to be considered: the size of the difference and its reliability. Where an estimate is needed of the difference between a new variety or treatment and existing practices, in terms of yield, quality or other characters, the estimated difference must be qualified by the extent to which it can be relied on to represent the true difference between the treatments.

The estimate of the difference will be based on experimental results. In the experiment itself, every effort is made to produce an unbiased result by choosing a suitable design, by randomization and by adequate replication. But an experiment is carried out at a particular site in a particular season, and there is no certainty that the circumstances were typical of the area or zone as a whole. This can only be determined by repeating the experiment at other sites and in other seasons, and averaging out the results (Munro 1968). The sites should be as representative as possible of the area in which the new treatment is to be applied. There is a tendency in annual reports to ignore the results of previous years, but it is essential that all the available evidence, including trials in which the differences are not significant, should be taken into account before a realistic estimate can be made.

The reliability of the estimate is tested by setting up a null hypothesis, postulating that there is no real difference between the treatment means, and that the observed differences are chance variations from a general mean. Statistical analysis enables us to calculate the probability that the differences are compatible with this hypothesis; if it is found that such a difference would occur by chance only once in 20 experiments, it can be regarded as a rare event and there is a good chance that it is a true difference. In statistical terms, the difference is said to be significant at the 5 per cent level, or the probability, P, of the difference being due to chance is only 1 in 20: this is abbreviated to $P = 0.05$.

This value ($P = 0.05$) has been chosen arbitrarily as the criterion for significance, along with a stricter level of 1 per cent ($P = 0.01$). The significance is determined from the analysis of variance by the ratio of the treatment variance to the error variance; this is called the variance ratio, F, and is tested against tables showing the 5 per cent and 1 per cent points for the distribution of F. But there is no magic about these figures, and results which do not reach the 5 per cent level must not be ignored; if the F value for signifiance is shown in published tables as, say, 3.01 and the analysis of variance finds $F = 2.96$, the differences are much more likely to be real than if the analysis showed $F = 1.43$. Only if $F = 1.00$ or less can we say with some confidence that the null hypothesis has been proved, and that there is no real difference between the treatments.

There is another class of experiment 'in which neither significance tests nor estimation of means and differences need chief emphasis' (Finney 1963). This is in preliminary trials of new selections in a breeding programme, where a more or less fixed proportion is discarded at each stage of screening.

Cotton as test material

There are two characteristics of the cotton plant which must always be borne in mind when conducting field trials. One is the indeterminate fruiting habit, which enables it to compensate in later life for setbacks it may encounter in the early stages. Sometimes this compensation is complete, as Brown (1965) demonstrated in Tanzania: the loss of early flowers in some cases actually increased the final yield. On the other hand, Balls (1953) in Egypt and Crowther (1941) in the Sudan showed that with irrigated cotton a setback to the seedling was carried through to yield. This emphasizes the importance of making and recording observations on an experiment at regular intervals throughout the growing season.

The second characteristic is that vegetative vigour is not necessarily an indication of yield. The harvest comes from the reproductive phase, the boll and seed, which may be in competition with the vegetative phase for the nutrients available (Hearn 1972b). The plant responds to an attack of bollworm by producing more and more buds to compensate for those shed or damaged; this means more leaves and taller plants, ending up with the tall vigorous plants carrying little or no crop. The application of nitrogen late in the season can cause an obvious change in the colour of the leaves, but no measurable increase in yield. As a general rule, however, vegetative vigour is essential for good yields, and both vigour and yield are strongly affected by the nutrient status of the crop.

Soil variation

Variation due to soil heterogeneity is probably the most important component of experimental error. 'It is usual in designing experiments to consider it separately owing to its magnitude and systematic nature, whereas most of the other factors affecting performance in a given year are small in size and random in incidence.' (Wishart and Sanders 1955). This variation is not always obvious, and subsoil irregularities may be worth investigating by digging holes at regular intervals before placing an experiment on an unknown piece of land. Slope of the land and drainage must be taken into account, and previous cropping history.

The extent of soil variation was illustrated at Rothamsted by Mercer and Hall (1911), who harvested a 'very uniform area' of wheat in plots of $\frac{1}{500}$ acre; the results are shown in Fig. 16.1, and range from 2.7 to 5.1 lb per plot. Uniformity trials of this kind emphasize the range in yield caused by soil heterogeneity, but can also provide other useful information. The data may be grouped to give yields from plots of various shapes and sizes to choose the most convenient and efficient plot layout for future trials; if an experiment is laid out on the same plots in the following year, the experimental error may be reduced by covariance analysis with the yields of the previous year; a series of uniformity trials may be used to pick the most uniform land for future experiments, although Wishart and Sanders, (1955) warn that correlation between seasons is generally low for annual crops; possibly the most interesting application is to get an estimate of what the crop can do at its best. King (1953) harvested an acre of cotton at Samaru in Nigeria in plots of $\frac{1}{880}$ acre (4.6 m^2), and found that the yield of seed cotton varied from 157 to 2550 kg per hectare, about a mean of 755 kg; in 1957/58 an attempt was made to achieve a successful combination of all the main factors affecting the yield of cotton, and the result was a yield of just over 1900 lb per acre (2130 kg per hectare) of seed cotton on a half-acre (0.2 ha) plot of land, realizing very nearly the potential shown in the uniformity trial.

Less formal ways of reducing the effect of soil heterogeneity can

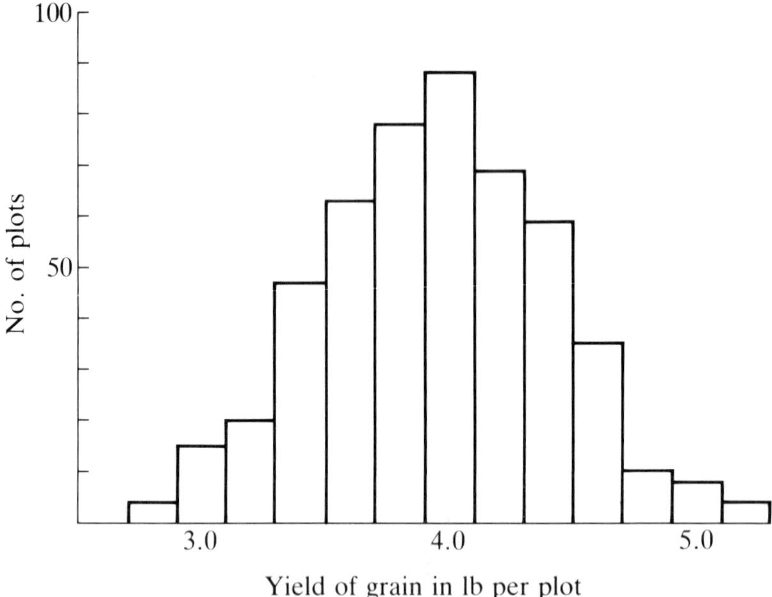

Yield of grain in lb per plot

Fig. 16.1   Yield of wheat from one acre divided into 500 plots of equal size (after Mercer and Hall 1911).

be used. In the tropics it is standard practice to avoid the sites of termite mounds and low-laying areas liable to waterlogging; a walk over the land to be used for experiments before the previous crop is harvested can indicate what areas should be excluded. When laying out the trials, changes in colour and texture of the soil can indicate where boundaries between blocks should be located. Such considerations should take precedence over a neat and tidy rectangular design.

Climatic variation

The experimenter has even less control over the weather than over the soil; all that he can do is to repeat his trials as often as possible at different sites and in different seasons. Here the importance of recording meteorological data cannot be overstressed, so that results can be related to the climate; at the very least, daily rainfall measurements at or near the site of each experiment should be recorded.

Good husbandry

'The first essential of good experimentation is sound husbandry, preferably based on specialist experience of a crop' (Paterson 1939). Uniformity in all operations, other than the treatments being tested, is the only way to increased precision and reliable results. Wishart and Sanders (1955) are equally emphatic:

All work on a formal trial must be done at the right time, and any one operation should not occupy more than a day or two. It is beside the point to argue that a normal crop would not receive such precedence; a normal crop is not required to give reliable comparative figures, and an experiment will not be made more normal by haphazard attention which may affect treatments differentially.

This does not mean that all experiments should be carried out under conditions of high fertility or sown at the optimum time. Trials at different fertility levels and sowing dates may show up important interactions, but at whatever level they are conducted there must be uniformity in all operations. Economics in the usual agricultural sense should be ignored until the worth of a treatment has been established.

Good quality seed is necessary to produce a uniform stand; acid-treated seed usually gives quicker and more even germination than fuzzy seed. For a variety trial all the seed should, if possible, have been grown and harvested under similar conditions; the yield of any variety which gives poor stands throughout the trial will be suspect, as poor stands indicate poor quality seed, which may be reflected in yield.

Any operation which takes more than one day to carry out should be organized so that work stops at the end of a block or replication; any effect which it may have will then appear as block variation rather than experimental error.

The same standards of good husbandry should be aimed at in trials on farmers' fields. Selection of a uniform area of land and supply of good quality seed will present no problems, and subsequent operations should be of the highest standard consistent with the resources available to the farmer.

Observation plots

In spite of all that has been said about soil heterogeneity and the need for replication, single unreplicated observation plots are still a useful tool in experimental work. The emphasis must be on observation; the experimenter should visit his plots periodically during growth, walk through them, observe and make notes. Even though the notes may never be used in detail, they help to fix attention on differences which might otherwise be passed over and left undefined. They can also act as a checklist of the points to be observed – insect pests to be looked for, or details of plant habit to be observed.

Whether the produce from the observation plots should be weighed is debatable. Wishart and Sanders (1955) state categorically that 'it is generally best not to weigh the produce of observation plots . . .; if the produce is not weighed there is no temptation to give figures for yield credence to which they are not entitled'. Bearing in mind that such figures can be misleading, there are still some circumstances in which they can be useful. For example, if a plant breeder is faced with some hundreds of progeny rows and wishes to reduce them to a more manageable number of 50 or so, yield can be a useful criterion for elimination; not the yield compared with a general average, but compared with that of neighbouring plots, making use of the fact that there is some correlation between the yields of adjacent plots, and only discarding the very worst. If the observation plots are repeated at more than one location, the significance of yield differences can be assessed (see p. 293).

Observation plots are much less expensive to run than formal trials and are often used when a new line of enquiry is initiated. The new method may prove impractical under ordinary farming conditions, unforeseen difficulties may occur, a new variety may be susceptible to the local pests or physiologically unsuited to the environment; it is much better to learn about this with little expenditure of time and money. Occasionally the new method may be so much better that the difference is obvious, such as the use of basic slag on pasture at Rothamsted and the jassid resistance of U4 cotton at Barberton.

Observation plots also have considerable value throughout an investigation. When laying down an experiment, it is good practice to apply some or all of the treatments to observation plots much larger than those in the trials, and subject them to normal farm-scale cultivation. If these plots are sited in the same field as the experiment, they can provide a very impressive demonstration for visiting farmers, who will often show more interest in the large plots than in the experiment. Observation plots of new varieties under test can also provide a source of seed for multiplication; looked at the other way, seed multiplication plots, accompanying variety trials, provide very useful plots for observations supplementing those made in the trial itself.

Design of trials

Statistical designs

In choosing the best design to use in an experiment, it is necessary to have a clear idea of the objectives of the trial and a knowledge of the alternative designs available and their strengths and weaknesses. If the experimenter himself lacks sufficient experience, the time to seek advice is when the experiment is being planned; the statistician can advise on the design and analysis most suited to the problem, but 'finds repeatedly that he makes a much more valuable contribution simply by getting the investigator to explain clearly why he is doing the experiment, to justify the experimental treatments whose effects he proposes to compare, and to defend his claim that the completed experiment will enable its objective to be realised' (Cochran and Cox 1957). Whether skilled advice is available or not, the proposals for an experiment should always be written down, and the following headings are suggested:

1. Objectives.
2. Treatments and controls.
3. Design and replication.
4. Plot size, discards (if any) and guard rows.
5. Basic treatments – fertilizers and insecticides.
6. Measurements to be made.
7. Form of analysis of the results.

The majority of field experiments are of the randomized block design. The number of treatments and replications can be varied at will, and the statistical analysis is straightforward. Missing plots can be estimated and complete replications omitted from the analysis if mishaps occur to any of the plots. Arrangement of the treatments in blocks, often called local control, is always preferable to a fully randomized design unless there is conclusive evidence that site variation is unimportant; this may be so in some greenhouse experiments. There should be at least 10 degrees of freedom (d.f.) for error; with fewer than 10 d.f., the F value required for significance is so large that it will rarely be achieved. If only two treatments are being tested, there should therefore be at least 11 replications. If the blocks become too large, the advantage of within-block uniformity is lost; in general if the number of treatments is more than 10 or 12, designs which reduce block size by the use of incomplete blocks should be considered.

In agronomic trials, two or more groups of treatments are often included in the same trial, providing tests of any interaction between them; for example, different levels of nitrogen may be tested along with different levels of phosphorus and different cultivation methods, to find the best combination of treatments from the three groups. Trials of this kind are usually designed either as factorials or as split-plot experiments. A factorial design includes all combinations of the groups or factors, each combination allocated to a separate plot: a 3 × 2 × 2 factorial might have three rates of nitrogen, two of phosphate and two cultivation treatments, requiring 12 treatments in all, which are randomized in the usual way. In a split-plot experiment, one of the factors is allocated to the main plots, forming a randomized block experiment on its own; these main plots are then subdivided into sub-plots to accommodate the remaining treatments. Where all the comparisons are of equal interest, factorial designs are more suitable, and block size may be reduced by confounding. In split-plot trials, the comparison between main plot treatments is less precise than that between sub-plot treatments; this is convenient when already established treatments can be allocated to the main plots, and newer treatments or finer differences are tested in the sub-plots. Practical considerations may also affect the design; many cultural treatments, if carried out by tractor, cannot conveniently be applied to small plots and will be assigned to the main plots for practical reasons.

In variety trials it is often necessary to test large numbers of varieties or selections, and it is difficult to find blocks of uniform soil sufficiently large to accommodate them all in a randomized block layout. They are best handled by the use of lattice designs, in which the treatments are divided into incomplete blocks. In a lattice design, the number of treatments must be an exact square ($x^2 = 9$, 16, 25, etc.), and each replication is divided into x blocks each containing x treatments. If the treatments are written down in the form of a square, the blocks making up the two replications of a simple lattice are formed from the rows and columns of this square; in a balanced lattice ($x + 1$ replications), every pair of treatments occurs once and only once in the same block, and all comparisons between treatments have the same degree of precision. The treatment totals are adjusted mathematically according to the totals of the blocks in which they occur (Cochran and Cox 1957).

There is much to be said for the use of these designs even if the lattice analysis is not always carried out; they involve no additional work in the field, they can legitimately be analysed as randomized blocks, and the lattice analysis can always be carried out if the block differences within replications appear to warrant it. The material can be divided between two or more trials; with about 100 selections to test, they may be sown as one 10×10 lattice, two 7×7 lattices or four 5×5 lattices, with the same control varieties included in each trial.

Additional controls may be added to make up the exact number of treatments required for these designs; the controls are treated exactly as if they were additional treatments until the statistical analysis is completed. After analysis, the control treatments are averaged to give a mean value, which will have a lower standard error (S.E.) than the other treatments (S.E./$\sqrt{x}$, where x is the number of controls).

When the control treatment represents the current practice, such as the commercial variety or the recommended fertilizer treatment, a comparison with the control is often the principal object of the experiment, and the inclusion of several identical controls will give a more reliable standard of comparison. The control suffers from the same random variation as the other treatments, and with a single control it may happen that all the treatments under test appear to be better than the control in one trial, but worse in another, although the trials were grown side by side in the same season. Cochran and Cox (1957) cite a Rothamsted experiment in which eight treatments and four controls were included in each block of a randomized block experiment. D. R. Cox (1958) recommends that the number of control treatments should be approximately the square root of the number of alternative treatments with which it is to be compared; in balanced incomplete block designs, the control treatment can be included one or more times in each block. Incidentally, the differences between identical control treatments can bring home very clearly the magnitude of the experimental error.

Many other designs are available to suit a variety of requirements, and are described by Cochran and Cox (1957) in *Experimental Designs*, without which no agricultural library is complete.

Observation plots and unreplicated progeny rows cannot be analysed to find a standard error, and even large differences must be accepted with caution until they are confirmed by replicated experiments. If, however, the same treatments are repeated at two or more sites or in different seasons, the results can be combined for analysis by treating the different sites as replications of a randomized block experiment. Another possibility is to sow duplicate plots of some of the progenies, and calculate a standard error for these as representative of the trial as a whole; as there is only one d.f. for replications, between 10 and 20 pairs are needed to give a reasonable number of d.f. for error.

The standard error (S.E.) is the basic value in all tests of significance, and should always be quoted in reporting results, whether or not the differences are significant; it is often necessary to combine the results from a series of experiments, and 'erroneous conclusions are likely if inadequate individual reporting leads to omission of experiments in which a difference was not significant' (Finney 1963).

Statistical analysis

The analysis of variance is a laborious business, even with the advent of electronic desk calculators, and there is an increasing tendency to send all the raw data for analysis by computer. Where several factors have to be analysed from a series of similar trials, the use of a computer programme is justified, and may be the only way in which the results can be presented within a reasonable time; but reliance on a computer has its dangers. There may be an error in the computer programme, and this should be checked by a few sample analyses made by the experimenter himself; there may be errors in the data fed to the computer caused by mistakes in copying, and the means, standard errors and any other statistics should be studied closely to see that they are within the normal range for that statistic. Any result that appears abnormal should be thoroughly checked, and a full print-out of the raw data enables this to be done directly. Another danger is that the experimenter may miss some useful information; he is bound to take a closer interest in data analysed by himself than in those from a computer which may arrive many months after the trial is completed; modern minicomputers are less remote.

The experimenter often makes the calculations more difficult for himself. It is rarely necessary to measure plot data to more than three significant figures, yet the author has often seen a lattice analysis carried out on plot yields with five figures. In one case the total yield was made up of the first and second picks, and a sample of 30 bolls per plot; the figures in grams were like this: 1450 + 820 + 158.3 = 2428.3 g. The field balance weighed to the nearest 10 g, and only the last and smallest component was measured to one decimal point. The accuracy of the total can only be to the nearest 10 g, and the extra two figures in any case add little to the accuracy of the analysis.

Another way to make things more difficult is to include the decimal point when doing the calculations: ginning percentage may be 37.3, boll weight 6.55 g or plant height 87.4 cm, and the decimal point can be ignored during the calculations and replaced only in presenting the results. This arises from the fact that the plot data may be divided or multiplied by a constant figure without affecting the analysis. Similarly a constant figure may be added to or subtracted from the plot data; ginning percentages usually lie between 30.0 and 40.0 or more, and subtracting 30.0 from each figure leaves data much more easily managed. While rounding reduces the accuracy of the figures, subtraction does not; if all the yield data from a trial lie between 1000 and 2000 g per plot, the analysis will be more accurate if 1000 g is subtracted from each value than if the figures were rounded to the nearest 10 g, leaving three significant figures in both cases. Care must be taken to replace the decimal point at the right place in the mean and standard error, especially if they themselves have been calculated to one or more decimal places; when a constant is added or subtracted, only the mean needs to be corrected.

Shape of plot

The plots in an experiment are arranged in blocks, each of which should provide soil conditions which are as uniform as possible. A compact block will in general include less soil heterogeneity than a dispersed one, and the most compact rectangular shape is a square. This square shape may be achieved in different ways. Suppose nine treatments are being tested in a randomized block experiment; square plots may be arranged in three rows of three plots each, while long narrow plots (9 × 1 in any units) may be placed side by side (Fig. 16.2). While it may not be possible to achieve a perfectly square block, this should always be borne in mind as the ideal.

Square plots have certain advantages. For a given size, they have the shortest perimeter of any rectangle, and thus suffer least from edge effects and interplot competition. Natural crossing between such plots is at a minimum, and differences in plant habit are easy to see. On the other hand, access to the centre plot of a 3 × 3 square is difficult, and the exposure to edge effects is not the same for each plot: if the block is surrounded by a path, as it usually is, the centre plot has no exposed borders, four plots have one side and four have two sides bordering the path. To even out possible border effects, all four sides of each plot must be discarded (Fig. 16.3), or internal paths must be provided, sometimes amounting to two-thirds of the experimental area. Square plots surrounded by access paths are ideal for seed multiplication and for observation plots.

Long narrow plots eliminate most of the disadvantages of grouping square plots into a compact block. Every plot can be observed from both ends. If guard rows are planted at the ends of each block, every plot is exposed to the same extent to edge effects (Fig. 16.4(b)), and such plots can often be harvested without discards. The use of guard rows and discards is discussed more fully in a later section.

If there is a fertility trend across the area of the block, the length

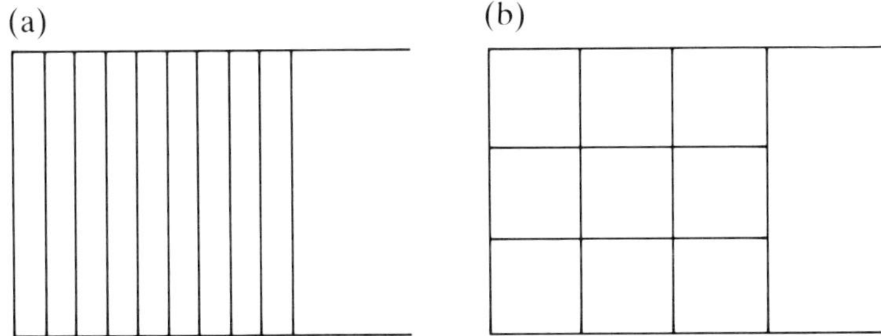

Fig. 16.2 Shape of plot: (a) long narrow plots; (b) square plots.

of the plots should be in the direction of the trend, so that each plot includes both low and high fertility. When machinery is used in cultivation trials, or for sowing variety trials, the plot width should be a multiple of the width of the implements used, and plot size can be adjusted by varying the length. Wishart and Sanders (1955) claim that 'long narrow plots usually give lower experimental errors than square plots'. Differences in plant habit may be difficult to judge in one- or two-row plots because adjacent rows intermingle their branches, and at least four rows per plot are desirable where the observation of such differences is important.

Size of plot

The classical uniformity trials of Mercer and Hall (1911) at Rothamsted with wheat and mangolds showed that plot to plot variation was reduced by increasing the plot size up to $\frac{1}{40}$ acre (100 m^2), but beyond that there was little further reduction. In the appendix by 'Student' it was shown that smaller plots with increased replication required much less space to produce the same accuracy as the large plots. While cotton in the tropics might not give exactly the same results, Prentice (1972) stated that plots of 100 m^2 have been found by experience to be of a good general utility size for trials with cotton. Plot sizes in fact differ considerably between different workers, and even between experiments in the same programme. It is convenient to consider three types of trial separately: variety trials, agronomic trials and insecticide trials.

In a breeding programme, small plots tend to be used in the early stages of selection, when a large number of progenies is being tested. Progeny row plots are often between 10 and 20 m^2 in size. Manning (1955) used plots of 10–12 plants at a spacing of 3×2 ft (6 m^2 per plot) because the quantity of seed for his replicated progeny rows was limited. At the same time, the district trials in Uganda were 5×5 latin squares with plots of 60 m^2, and larger plots are commonly used in the final stages of testing, e.g. Egypt 200 – 300 m^2, Sudan 100 m^2, Pakistan 85 m^2. In 1962 the Uganda latin squares were replaced by 4×4 balanced lattices so that more varieties could be included in each test, and the plot size was reduced to 25 m^2. As might be expected, the average standard error reported in the four seasons after the change was about 20 per cent greater than that in the four previous seasons, but the greater flexibility of the layout and the increase in the number of varieties which could be tested more than compensated for the loss in precision. Lattice designs with small plots are commonly used for variety trials, plots of 2×8 m and 2×10 m giving square blocks in 4×4 and 5×5 lattices respectively.

Agronomic trials usually need much bigger plots, from 50 to

200 m^2 or more. They often have to accommodate tractor-mounted implements; the plot width should be a multiple of the width covered by the implement, and the plots long enough for the implement to bed in at the start of the treated area. The treatment boundaries may not be sharply defined, especially if the trial is continued for a number of years, and the area of transition is usually planted with discard rows.

Interference between adjacent treatments is the main problem to be faced in insecticide trials. The size of plot required depends to some extent on the mobility of the main pests; Pearson *et al.* (1952) noted that plots of $\frac{1}{40}$ acre (100 m^2) showed only small differences in average population of lygus between treated and untreated plots. 'These differences were larger the larger the plot, and resulted in contrasts in plant appearance and crop formation which were detectable by the eye in plots of $\frac{1}{17}$ acre (240 m^2) and well marked in plots of $\frac{1}{5}$ acre (800 m^2).' Bowden and Ingram (1958) used plots of approximately half an acre (2000 m^2) for each treatment to obtain evidence of the value of DDT spraying in Uganda. Joyce and Roberts (1959) in the Sudan showed that very large plots of up to 3 ha had to be separated by more than 150 m before the 'interplot effect' could be overcome.

The use of large plots is costly in time, materials and labour, and McKinlay (1953) used a much simpler single-row spraying technique at the Uganda Cotton Research Station, Namulonge. The single rows were sprayed to run-off every week, and were separated by seven or more rows of unsprayed cotton, one of which was chosen as the control. McKinlay claimed that the spray not only controlled the lygus nymphs and bollworm, but repelled the lygus adults, had little effect on the population of lygus in the field, and did not suffer from the interplot effect. Using this technique in trials for the next seven years, it was found that the sprayed cotton yielded little more than the unsprayed (11 per cent), and in two of those years it actually yielded less. This was explained by McKinlay *et al.* (1957) by the powers of recovery of the cotton plant, a late crop compensating for the earlier losses. Similar results were obtained from trials carried out between 1965 and 1970, using plots of 1000 m^2, in which the central 400 m^2 were harvested for yield.

Reed (1972) is highly critical of using unsprayed controls in a spraying trial to measure the yield losses caused by insect pests. He compared the yields from farm bulks with those of small plot spraying trials on research stations over 20 years in Uganda and 9 years in Tanzania, and concluded that the 'yields of unsprayed control plots within spraying trials may be greater, or less, than the yields which would have been obtained from larger blocks of unsprayed cotton'. He recommends that a large sprayed area should be compared with a large unsprayed area.

Such an approach would have forsaken that of formal statistics, but the difference between the yield of sprayed and unsprayed cotton at Namulonge is so large that formal statistical comparisons are unnecessary. In this case, the quest for a statistical measure of the effects of spraying appeared to hold back the development of crop protection at Namulonge for over ten years. (Reed 1976).

Discards and guard rows

Treatments in adjacent plots are likely to affect each other to a greater or less extent, and this interplot effect must be minimized if the effect of a treatment on the crop is to be measured accurately. Competition is likely if the height and vigour of the treatments in a trial are markedly different. An exceptionally vigorous plot may gain at the expense of a less vigorous one next door, and yield better than it would have done if it had to compete with others of its own kind, while the less vigorous treatment yields less than its true potential. Competition may occur both above and below ground; it is known that the roots of adjacent cotton rows intermingle, and penetrate into each other's root space at least as far as the next row and probably further. The boundary between treatments is not always well defined: for example, in a comparison between flat and ridge cultivation, the boundary row of a ridged plot may have a flat row on one side. Fertilizer applied to one plot may be accessible to the roots of the boundary rows of an adjacent plot. When a fertilizer trial is carried through to a second season, the preplanting cultivations will mix some soil from one plot with another along the plot boundary. An implement such as a plough or subsoiler does not reach its full depth for some distance after it is lowered to the ground, so that the start and finish of the treatment is imprecise. Access and boundary paths and unplanted areas adjacent to a plot will affect the growth of cotton growing alongside them.

The purpose of discards and guard rows is to eliminate these interplot effects as far as possible from the recorded area, so it is important to know how far these effects extend. Work on skip-row planting has shown that the row of cotton bordering an unplanted area can yield as much as 50 per cent more than an interior row, and that this effect extends on a diminishing scale about two metres into the plot (see Ch. 6). To eliminate this effect all plants within two metres of a path have to be discarded, that is, they are excluded from the plot yield at harvest; the plants may be cut off at ground level or harvested separately. If they are cut out, this should be done just before the first picking; if it is done too soon, this in itself may affect the plot yield. If they are left standing to be harvested separately, the discards should be harvested before the plots, when supervision is not distracted by problems of bags and labels.

The effect of one treatment on another is usually less severe than the complete absence of competition, except when dealing with pest movement, which has already been discussed; for internal discards, i.e. between the plots in a block, it is usually sufficient to discard all plants within one metre of the plot boundary. These discard rows receive the same treatment as the rest of the plot up until harvest. The terms guard rows and discards are practically synonymous, but strictly speaking a guard row is an extra row outside the plot, which may or may not receive the plot treatment; thus a trial may be laid out with one-metre discards round each plot, but an extra guard row is planted round each block to give a two-metre discard alongside a surrounding access area (Fig. 16.3).

Discards add considerably to the size of an experiment, and consequently to the cost, and by increasing the distance between the harvested plots they may increase the experimental error. Discarding one metre all round a 10×10 m (100 m^2) plot (Fig. 16.3) will reduce it to 64 m^2, while discarding two metres leaves a net size of 36 m^2; in long narrow plots the proportion of discards is even greater (Fig. 16.4 (a)). It is therefore worth while considering the circumstances under which discards can be dispensed with. These are found when the competition between and within plots is similar, and when any edge effects are the same for all treatments. In a breeding programme the selections are usually fairly similar in size and vigour, and in trials the whole plot is often harvested without discards. Each replication is surrounded by access paths, and two guard rows are added at each end so that the first and last plots are subject to the same competition as the others (Fig. 16.4(b)). All of the plots have the same exposure to the unplanted area, and it is assumed that they all derive the same benefit from the edge effect; with long narrow plots the exposed plants constitute quite a small proportion of the total.

These assumptions were tested in trials in Uganda in 1970 and 1971 (Table 16.1). Six selections with different plant habits were grown in a variety trial using six-row plots; yields were recorded from the two centre rows of each plot, from the four central rows and from all six rows. Increase in the plot size led to a reduction in the standard error, but the relative position of the six varieties was similar, and it was concluded that 'for the range in plant habits at present available in the breeding programme at Namulonge, there is no need to discard the outside rows' (Innes *et al.* 1973). Similar results were obtained using four-row plots in Ecuador, but in Malawi measurable competition effects were found in trials where the differences in plant habit were greater than those in Uganda (unpublished data).

Discard rows are usually necessary in agronomic trials, and it is important that they should be adequate. This was illustrated in a

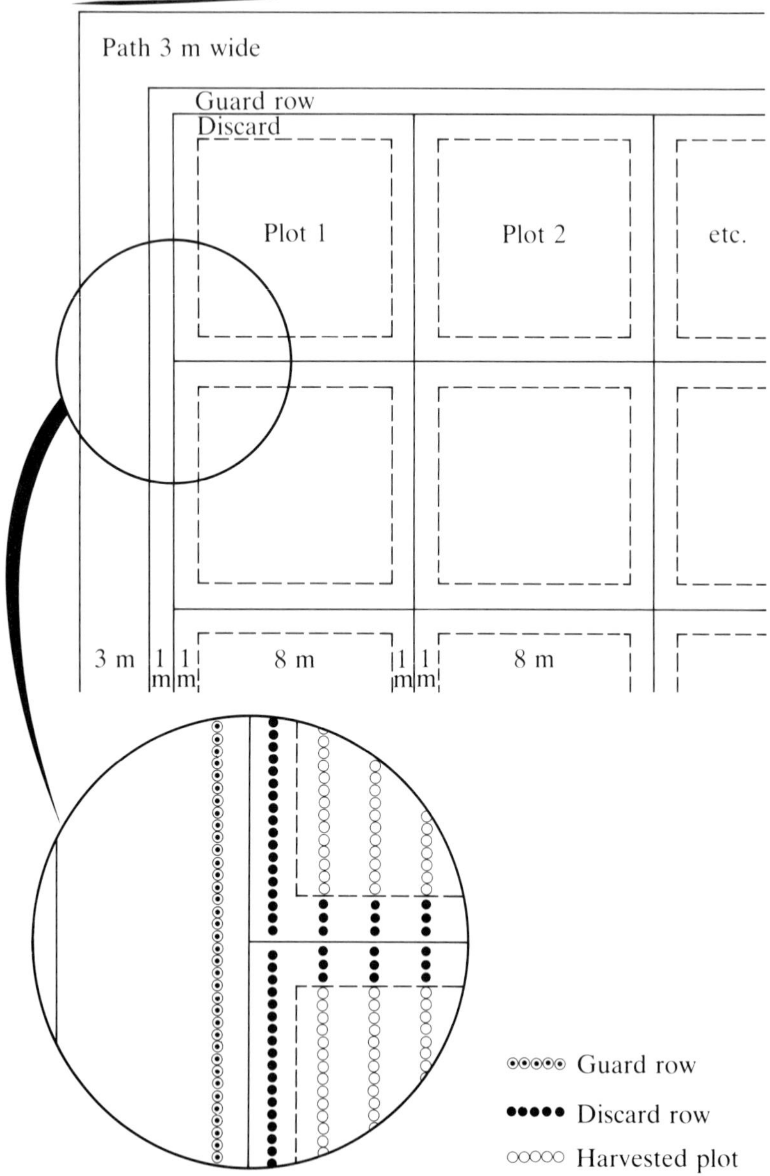

Fig. 16.3 Square plots – discards and guard rows.

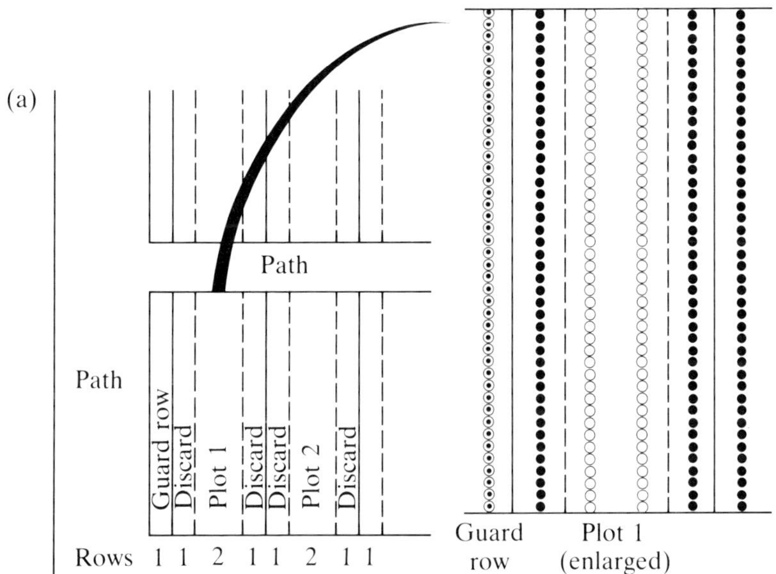

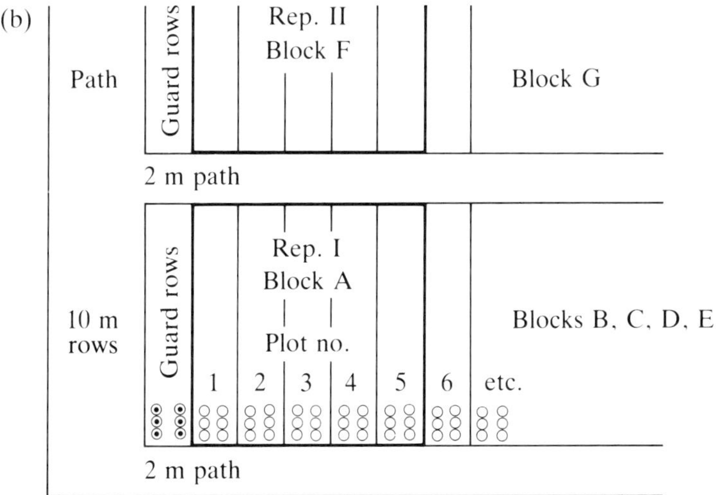

Fig. 16.4 Layout of variety trials, with and without discards: (a) four-row plots with discard rows; (b) two-row plots without discards, part of 5 × 5 lattice.

Table 16.1   *Plot size trials in Uganda*

| Variety | Yield of centre rows of six-row plots Seed cotton in kg per ha | | | | | |
| | 1970–71 | | | 1971–72 | | |
	Two rows	Four rows	Six rows	Two rows	Four rows	Six rows
BPA 68	2315 (5)	2270 (6)	2289 (6)	2073 (3)	2090 (3)	2109 (3)
Satu 65	2458 (2)	2445 (2)	2483 (1)	2045 (4)	2012 (5)	2004 (5)
A (66) 131	2424 (3)	2441 (3)	2479 (2)	2121 (2)	2158 (2)	2139 (2)
BP52 NC63	2267 (6)	2307 (5)	2388 (5)	1993 (5)	2158 (4)	2065 (4)
CA (68) 36	2376 (4)	2417 (4)	2398 (4)	1947 (6)	1978 (6)	1957 (6)
CA (68) 41	2489 (1)	2481 (1)	2469 (3)	2179 (1)	2165 (1)	2145 (1)
Mean	2388	2394	2418	2060	2073	2170
S.E. ±	69.6	43.1	38.1	47.7	46.5	37.7

Note: Rank order in parenthesis.
Source: Cott. Res. Rep., Uganda, 1971–72.

series of spacing trials in Ecuador, in which for three successive seasons narrow row spacings consistently outyielded the standard spacing of 1 m between rows; the population density was the same for all treatments. This aroused interest in other departments, and narrow rows were included in some population × fertilizer trials run by the soils section; these showed no difference between the row spacings, and the discrepancy led to a reappraisal of the original trials. In these trials the plots were four rows wide, yields being recorded from the two centre rows, and it was noted that the narrow row plots had correspondingly narrow discards – 40 cm in the closest treatments compared with 1 m for the standard controls; at the same time a dry season had demonstrated border effects extending more than 1 m from the access paths, and a skip-row trial had shown enhanced yields up to 1.5 m from the skip-row. The conclusion, confirmed by subsequent trials, was that the narrow row plots had benefited substantially from the wider spacings in the adjacent plots, and that the apparent benefit was due not to the treatment but to inadequate discards in the design (INIAP 1980).

In special circumstances it may be desirable to leave unplanted paths round every plot in an experiment. Provided these are all of the same width, the edge effects will be the same for all plots, and it may be decided to dispense with discards. The purpose of the paths may be to give better access to the plots for spraying, observations, insect counts, etc., especially when square plots are grouped as in Fig. 16.2(b); to provide space for manoeuvring agricultural machinery; for selfing (see Fig. 15.4); or to encourage lodging when testing selections for resistance to this weakness.

Parnell *et al.* (1949) used discard rows for a different purpose.

They were selecting for resistance to jassid, and inter-planted their progeny rows and yield trials with discard rows of a susceptible variety to build up a high and uniform population of jassid throughout the area.

Sampling

It is not always possible or convenient to take measurements on all the plants in an experiment or on all the produce from it. It is then necessary to take representative samples of a size than can be measured. The aim of sampling is to get as representative a fraction of the whole as possible, so that conclusions drawn from the samples can be applied with confidence to the population as a whole. Wishart and Sanders (1955) discuss various ways in which the samples may be chosen and the size of the sampling unit, while Snedecor (1946) has a chapter on the design and analysis of samplings.

Observations on plant habit and development can often be made on whole plots rather than on samples, assessing characters such as the incidence of pests and diseases visually as a percentage, or on an arbitrary scale of 0–5 or 0–10: a system of grading for resistance to bacterial blight on a scale in which 0 represents immunity and 12 full susceptibility is described by Knight and Clouston (1939) (see Fig. 13.1). Differences in height and vigour can often be measured by classing each plot into one of five grades: very tall, tall, medium, short, very short; if figures are assigned to each grade (1–5), statistical tests can be applied to them, although theoretically this may not be strictly valid. The alternative of actually measuring the height of a sample of plants is much more laborious and may not give any greater accuracy. Visual grading should always be done without reference to the field plan; if the treatments are not known, there will be no unconscious bias.

Developmental studies help to explain subsequent differences in yield, but this sort of study is only feasible on a small sample of the plant population; the size of the sample will depend on the measurements to be taken and the facilities available. The sample may be chosen by any method of random sampling, and a single cotton plant is a suitable unit. Cotton growth is strongly influenced by spacing, and if the choice falls on a plant which is not surrounded by a full stand, it should be discarded and replaced by the nearest plant which fulfils the required conditions.

When several numerical observations are obtained by sampling during the growth of a crop the question arises as to whether the samples should be the same, or whether a fresh randomization should be made, at each count. The usual procedure is to keep to the same positions, and experience has shown that if it is proposed to use covariance in the statistical analysis the units must be precisely the same, for a shift of a few inches along the row greatly decreases the correlation between the figures at different counts, thereby diminishing the accuracy of correction. (Wishart and Sanders 1955).

The sample plants should be marked with brightly coloured labels, string or wool, so that they can be easily identified for subsequent measurements.

The handling of the plants and the local trampling when taking such records can affect plant growth, sometimes to a visible extent. Sometimes plank bridges are used to avoid trampling, but usually such effects have to be accepted, and should always be noted for future reference if the plants are visibly affected.

It is common practice in cotton variety trials to take boll samples from each plot just before the first pick, and use them to determine boll weight, seed weight, ginning percentage and to test fibre qualities. Open, well-developed, healthy bolls are chosen at random along the plot, not more than one or two from any one plant. Such samples may not be representative if the trial has a high proportion of diseased or badly opened bolls, but does represent the genetic potential of the varieties concerned. Usually some number between 20 and 40 bolls is decided upon, and the measurement of average boll weight depends on taking exactly the same number of bolls from each plot; this can be checked either by laying out the bolls separately at the end of the plot before they are put in bags, or by using egg trays with the correct number of spaces to be filled by individual bolls. Similar samples can be taken from agronomic trials, but it must be borne in mind that boll damage can sometimes be caused by the treatments applied, and this will only be shown by a truly representative sample which is more easily obtained by sampling the whole produce after harvest.

Sampling of the harvested produce presents different problems. It is generally better to harvest and weigh the produce of the whole plot and then take samples from it than to demarcate a sample area for harvesting; the latter may be necessary in the case of very large plots, or in comparing plots of irregular size and shape in farmers' fields. The seed cotton from a plot should be thoroughly mixed before sampling, and then separated into four quarters; these may be treated in two different ways to provide the final sample. One way is to discard two opposite quarters; the two remaining quarters are mixed, and the process repeated until the bulk is reduced to the size required. Alternatively, each quarter may be quartered again and again, and a random sample taken of the final small lots; or a little seed cotton may be taken from each of the small lots to make up the final sample.

If statistical tests are to be applied to the various measurements, the produce from each plot must be harvested, weighed and sampled separately, but such tests are not always required. Commercial seed cotton is usually marketed in two or more grades, clean cotton being separated from stained or dirty cotton; if it is certain that the grading is not affected by the treatments, as is usually the case in variety, cultivation or spacing trials, the percentage of each grade may be determined as a matter of general interest after bulking the plots of each treatment or after each picking. Similarly, it may be desirable to have a record of the ginning percentage of trials carried out in different areas and different seasons, quite apart from the results of the particular experiments. Samples of lint for spinning tests are commonly taken after the produce of each treatment has been bulked and ginned, as the cost of testing each plot individually soon becomes prohibitive.

Attention to detail

No amount of statistical juggling can make up for careless management of an experiment; like a computer, the results can only be as good as the data put into it. Mistakes occur even in the best-run experiments, and 'it is much better to aim for a standard verging on the ridiculous than to be content with one which, it is hoped, will just avoid inaccuracies large enough to have appreciable effect on experimental error' (Wishart and Sanders 1955).

In the field, it always pays to be meticulous in the layout of the plots and in checking the length of rows; if planting by hand, to see that each 6 m length of row with plants spaced at 0.3 m contains 20 plants, and not 21 because both extremities have been planted. Seed and fertilizer should be prepared in convenient lots beforehand; if each plot is labelled before planting, the layout can be checked; if the bags of seed, fertilizer, etc., labelled with plot number and treatment, are laid out beside the appropriate plot, this again provides an opportunity for checking.

Small plots will usually be planted by hand, and lines of the appropriate length and marked with the appropriate spacing are often used. These lines may be wire, rope or chain, with coloured tape threaded through and knotted firmly to them; two lines are moved forward alternately after each row is planted. Time spent in teaching the planting squad to plant to a uniform depth and to sow approximately the right number of seeds per hill is never wasted, and attention to detail saves not only time but mistakes.

The use of machinery presents special problems, particularly in choosing the most convenient plot size and shape to suit the implements used in applying the treatments and in harvesting the produce. The systems used at the Norfolk Agricultural Station, Sprowston, are typical of those used both on experimental farms and in trials on commercial farms in England; they are described in *The Practice of Arable Crop Experimentation* by Harvey (1952).

When treatments such as insecticide sprays have to be repeated at intervals during the season, coloured pegs indicating the different treatments are very useful; searching for and reading the plot labels, or consulting a plan of the layout every time the treatment is repeated, are time-consuming and prone to mistakes.

Whether the plot labels should show only the plot number, or the plot number and the treatment, is a matter of debate. Field observations during the growing season can be biased by a knowledge of the treatment and some people think the only safe way to prevent this is to show only the plot number; others find it useful to be able to check the treatment without having to consult a field plan every time. As a compromise, it is suggested that the labels should be fairly small, so that they cannot be read from a distance, and can be ignored in making plot by plot assessments.

The first plant of a row should be some distance from the stake carrying the plot label, so that the label can still be seen when the plants reach full size. All experiments should follow a standard practice regarding the position of the plot label; it should be on the left-hand side of the plot, or the right-hand side, or in the centre; to have it on the left in some trials and on the right in others will inevitably lead to confusion and mistakes at harvesting.

Labels for marking plots should be durable, weather-proof and legible. There are many labelling methods and materials, some expensive in first cost and more durable, others relying on local materials such as bamboo. Many hundreds of stakes and labels are needed even on a small experimental station, and a sufficient number must be ready for use well before planting starts.

Packaging of seed and fertilizer must be strong enough to withstand the handling it will undergo; two or more burst bags in a parcel of seed of a variety trial will vitiate any data on the varieties concerned. Cotton bags tied with string are the best for cotton seed; plastic bags are equally durable, but paper bags need to be handled with care. They are adequate for trials on the research station itself, and it is easy to write all the necessary information on the bag itself.

Depending on the quantity to be harvested, paper bags, cloth bags or grain sacks are used. Harvested produce must be labelled in the field, and the bags should preferably be labelled before the cotton is put in them; the bag should have not only a label on the outside but a duplicate label with the same details placed inside for safety. The label should state legibly not only the treatment, plot number, replication and picking date but also, if the cotton comes from breeding plots, whether from selfed or unselfed bolls.

Normally weighing is done in the field as soon as the plots are picked, and the scale used should be appropriate to the weight of produce expected: a 50 kg scale would be unnessarily coarse where yields of the order of 5 kg are likely. The face of the scale should be easy to read, and large pans are required for cotton to hold the volume of produce for a given weight. The cotton may be weighed after bagging, provided the bags are all of the same weight, and many types of scale can be adjusted for tare, giving a zero reading with an empty bag; a spring balance hung from a home-made tripod can be used for large bags. Spring balances must be checked periodically using standard weights, as the spring is liable to weaken with use. The weighing of the produce from small plots may be affected by wind; if so, the balance can be shielded by stretching a piece of sacking to windward.

In the store, bags should always be stacked so that the labels can be read without having to turn the pile over every time a particular treatment is being sought. Plenty of storage space must be allowed for seed cotton so that the produce from different experiments can be kept separate and easily accessible; slatted shelving provides more storage space in a given floor area. The building layout for a small experimental unit is given in Fig. 15.3.

Chapter 17

Harvesting, marketing and ginning

Harvesting

Machine harvesting

Practically all the cotton produced in the USA is now harvested by
machine, but in most tropical countries machine picking has barely
progressed beyond a few experimental machines on cotton research
stations and large company estates; exceptions are Australia and to
some extent South Africa. The main reason for this situation is the
cost of labour: although this is rising steadily in the less developed
countries of the tropics, it is still only a fraction of the minimum
wage in the USA. In 1980 the cost of hand picking in the tropics
ranged between 1 and 5 US cents per kg of seed cotton, while
spindle picking in the USA (including transport to the ginnery) cost
from 5 to 10 cents (Baskin and Sistler 1980).

Harvesting machines are of two main types, the spindle picker
and the stripper. The spindle picker (Fig. 17.1(a)) depends on a
number of rotating spindles which tangle with the seed cotton in the
open bolls, pulling it away from the husk, and this type of machine
is used for all good quality cotton. The stripper (Fig. 17.1(b)) is
non-selective, and harvests everything that comes its way: bolls,
husks, leaves and some branches. Stripper harvesting is cheaper and
quicker than spindle picking, but even with elaborate cleaning
equipment in the ginnery it produces a much poorer quality of lint.

The capital outlay for machine picking is high, not only for the
picker itself but also for the equipment needed for handling cotton
in bulk and for the installation of more elaborate machinery for
cleaning in the ginnery. This machinery may only be used for two or
three months in the year, but its efficient operation and maintenance
requires a high level of skill and experience. A network of local
agents supplying spares and service plays an important part in this
maintenance. Mechanical harvesters need large fields of cotton for
efficient operation, with long straight runs and a heavy crop, as the
machine travels at the same pace whether the crop is large or small.

a

b

Fig. 17.1 Mechanical cotton harvesters (G. W. Cathey): (a) two-row spindle picker; (b) four-row stripper.

A two-row spindle picker can pick about 200 ha of cotton in a season under good conditions. Picking starts when between 60 and 80 per cent of the crop is ripe, and may be followed by a second picking; the final picking or 'scrapping' cleans up the remaining cotton.

The cotton should as far as possible be stripped of leaves before machine picking; drying out at the end of the season will cause natural leaf shedding, but where green leaves are still present on the plant at the time of harvest they can be removed by the use of chemical defoliants. Harvest aid chemicals, the term used to include defoliants, desiccants and growth regulators, are applied to more than three-quarters of the cotton acreage in the United States (Cathey 1980). If the defoliant is not fully effective, pieces of green leaf will be harvested with the seed cotton; these must be dried and removed before ginning to prevent a green stain being transferred to the lint. Cotton should be picked within two weeks of treatment with chemical defoliants, otherwise their effect is nullifed by the sprouting of a new growth of green leaf.

Harvesting by hand

Although picking is one of the most costly operations in growing cotton, it is often the least efficient. It needs a large number of people for a short time, and this often means recruiting temporary unskilled labour. Only too often these labourers are sent into the field with no instruction on how to go about the job, and both the rate of picking and the quality of the crop are far lower than they woule be if a few simple techniques had been applied. The result of this is seen in the very wide range of figures quoted for the weight of cotton picked per day, which runs from below 10 kg to over 60 kg. Parnell (1933) gives a day by day record of five outstandingly good pickers who, on a farm in the Barberton area of South Africa, averaged daily pickings for two months of 65 kg of seed cotton each; but he puts the figure for a gang average at around 18 kg per day. In Zimbabwe the best pickers in a heavy (irrigated) crop averaged up to 44 kg per day, and in a good raingrown crop up to 33 kg. The peasant grower in Uganda and Tanzania picks his cotton in a more leisurely fashion, but gets it notably free of trash, and rarely exceeds a rate of 9 kg a day (Prentice 1972).

Techniques of hand picking

The first essential is a suitable container to hold the picked cotton, leaving both hands free for picking. The simplest is a bag about the size of grain sack, tied with string or tape round the picker's waist so that the mouth hangs open in front. The tapes should be sewn to the mouth of the bag about one-third of the circumference apart, leaving two-thirds hanging loose. The bag may be made of canvas, plastic or hessian, and grain or fertilizer bags are quite suitable. A small pocket should be sewn in the mouth of the bag to hold second grade cotton, and the mouth can be held open if desired by a withy or wire ring.

Professional pickers in the USA used a cotton bag 3 m long and 70 cm in diameter, dragged along the ground between the picker's legs. It was sometimes in the form of a tube with the bottom end tied with string, which was untied to facilitate emptying; as it held a large quantity of cotton, it did not need to b'e emptied so frequently as a small bag. Jong (1965) found that up to 25 per cent more cotton could be picked in a given time using this type of bag, but it has so far gained little acceptance in other countries.

Both hands should be used for picking, and only two or three bolls picked at one time before being fed into the bag. The picker moves between two rows of cotton, picking from the near side of each row, rather than stretching over to pick from both sides of the plant. The first pickings are grade A cotton only, leaving any dirty or damaged cotton to the final picking; any grade B cotton picked by mistake goes into the small pocket sewn into the bag.

Grade A cotton is defined as 'clean selected cotton completely free of leaf and trash, white in colour without any sign of stain from any cause' (Malawi 1976). The commonest source of trash is the bract at the base of the boll; this bract withers when the boll ripens, and small brittle bits of the teeth of the bract are easily caught up in the lint, especially when a handful has to be pulled out from the heart of the plant. Other sources are shed leaves, weeds and soil: shed leaves are dry and brittle, and bits of broken leaf can stick to the lint in open bolls, especially the hairy leaves of jassid-resistant varieties; a late weeding will remove weeds, particularly grasses, which may shed their seeds on the open bolls; heavy rain splashes soil on to the lower bolls, and some soil is bound to cling to open cotton which has fallen to the ground. Trash is best picked off the seed cotton as it is picked, before it is compacted in the hand or in the sack.

Discolouration or staining of the lint is caused by pests and diseases; when they attack only one loculus in the boll, the damage is not easily seen by the picker. Pest and disease control is therefore an aid to good picking. The fungus transmitted by stainers is a direct cause of staining; damage by other insects can cause staining indirectly by the development of moulds and fungi on the damaged parts of the boll, and may hinder the full development of mature lint; bollrots caused by bacteria or fungi result in tight and discoloured loculi if they do not destroy the contents of the boll completely.

Genetic and agronomic factors can contribute to faster and hence to cheaper picking. One of these is boll size, as it is easier to pick a few large bolls than many small ones; this is a varietal character which does not affect machine picking, and the newer American varieties tend to have smaller bolls than the old ones. Another measure worth considering is skip-row planting, which is popular with pickers as easier access and more bolls per plant make for faster picking. Compact plants which are not too tall produce a concentration of bolls which makes picking easier. Storm-proof varieties are not suitable for hand picking, as the cotton does not fluff out and cannot be grasped so easily as normal cotton; the cotton clings to the husk, especially that of the seeds at the base of the loculus, which may be left behind when the rest of the cotton is pulled away.

How many pickings are necessary or desirable will depend on a variety of factors such as the weather, the size of the crop, insect damage and the rate at which the crop matures. The highest rates of picking have been achieved with about 500 kg per ha of open cotton ready to pick, and there is little advantage to be gained by allowing more to open between picks (Jong 1965, 1982). On the other hand, open cotton should not be left too long in the field. In wet weather a silvery grey fungus discolours the lint and probably weakens the fibre (Lord and Anthony 1960); in dry weather, the open bolls gradually elongate, without the seeds actually separating, and present a greater surface to trap any wind-blown debris. Such bolls are apt to tangle with the framework of the plant, and in windy weather they may be blown off the plant to the ground. Under average conditions two or three pickings should be sufficient.

Rain and dew

In the subtropics the wet and dry seasons are distinct and predictable; the crop ripens in the dry season, and rain at harvest time is a very rare phenomenon. Nearer the Equator, however, there are likely to be two wet seasons in the year (see Ch. 9) and some rain can be expected even in the drier months. Rain at harvest does little damage to a healthy crop of cotton if it is followed by dry sunny weather, although it may take some of the bloom off the cotton in open bolls. It will, however, encourage the spreead of any diseases attacking the crop directly, such as bacterial blight, and of secondary fungi present in bolls damaged by insects. There is usually some scope for adjusting the planting date so that ripening occurs in the driest months, and more frequent pickings may be needed to make the most of any dry spells that do occur. If the picking of wet cotton is unavoidable, it should be spread out to dry before it is packed into bags, under cover if necessary, but preferably exposed to sun and wind.

In most cotton-growing countries, the humidity may be high enough in the early morning to deposit dew on the open bolls, and picking should not start until this has dried off. If a heavy dew would delay picking unduly, moist cotton can be spread out on tarpaulin or plastic sheets at the side of the field until it is dry enough to pack into bags. Some of the bigger farmers in Zimbabwe use the early morning period for inspection and final grading of the seed cotton picked the previous day. At the end of each day the sacks of cotton are brought to a shed with a clean concrete floor, and each picker is credited provisionally with the weight of his sacks; the following morning, while awaiting the disappearance of the dew, the cotton of each picker is turned out on the floor for inspection; if it is up to the required grade, the picker's credit is confirmed, but if below grade it is given back to him for sorting; the produce is then heaped together at the end of the shed to await packing and transport to the ginnery.

Packing and transport

Finally the seed cotton must be packed for transport to the market or ginnery; the size of the container will depend to some extent on the form of transport – head load, pikul stick, bicycle, donkey, ox cart or lorry. The grower with a small plot of cotton may pack his cotton into jute grain sacks holding about 40 kg of seed cotton, or he may use baskets made from local materials, or a square of jute or cotton with the corners tied together (Fig. 17.2). The cotton buyer sometimes distributes suitable bags to the growers. There are two points to note in packing seed cotton for transport: the cotton should not be rammed in too tightly, and the use of pointed sticks for this purpose will damage the lint; secondly, the string used to sew or tie the mouth of the bag should be made from a soft fibre such as jute or cotton rather than sisal or plastic, hard fibres which, if mixed in with the lint during ginning, cause serious trouble in the spinning process.

Larger growers in southern Africa and Egypt use hessian packs holding about 170 kg (Fig. 17.3). These are suspended from a frame for filling, and the cotton is tramped down while they are being filled. An aid to good baling in southern Africa is the Brockhouse 'Waybaler', which produces a bale in a matter of minutes: a weighed batch of seed cotton is released into a box and compressed by winding gear into a pack which is then sealed off by sewing a flap across the open end.

Machine-picked cotton is handled in bulk. From the picker basket it is tipped into a trailer, basically a wire cage mounted on a four-wheel chassis, in which it is transported direct from field to ginnery; the picker basket holds the equivalent of about two bales of

a

b

Fig. 17.2   Bringing seed cotton to the local market: (a) Indonesia (Ciba-Geigy); (b) Nigeria (British Cotton Growing Association).

lint and trailer sizes range from 4 to 10 bales. The use of cotton modules is an alternative to the trailer system; the module is formed by loading seed cotton into a steel box, in which it is compacted by mechanical trampers; when slid out of the box it is sufficiently dense to retain its shape, and may be parked on the ground and slid on or off a flat trailer with a ramp and winch. Modules can be left in the

Fig. 17.3   Filling hessian packs with seed cotton in Zimbabwe (Prentice 1972).

field and resist weather with little economic effect on cotton quality, and can be fed directly into the opener of a gin. The module system is deemed to be too expensive for farms growing less than 200 ha of cotton, but may be owned and operated on a community basis (Baskin and Sistler 1980).

Destruction of crop residues

The main object of destroying the cotton plants remaining in the field after harvest is to reduce or prevent the carry over of pests and diseases from one cotton season to the next. It should be done as soon as possible after the harvest is completed. In many countries there is a legal close season between certain dates, during which no cotton is permitted (see Ch. 12); all cotton residues must be uprooted and burned before the date when the close season starts, and no cotton may be sown before the date on which it finishes. To obtain a satisfactory burn, the stalks are collected into heaps and allowed to dry before being set alight. While this method is suitable for small-holders and hand cultivation, it is not easy to mechanize, and the object of the close season can be achieved by chopping the stalks into small pieces, and burying them and other trash by ploughing them into the ground. Where domestic fuel is a problem, cotton stalks may be collected and stored for use in cooking; as long as they are dry and free from leaves and seed cotton, they are unlikely to support many insect pests, but can still provide a source of infection for bacterial blight.

Small, shallow-rooted plants are easily uprooted by hand pulling, but this is not so with bigger plants, especially when the soil is dry. Bigger plants may be chopped off with a hoe, leaving the roots in the ground; to prevent regrowth from the roots, the stem should be cut just below ground level. The stalks and trash are then collected into heaps or windrows for burning.

A stalk-pulling machine has been developed by the National Institute of Agricultural Engineering in Britain. It is tractor-mounted, and uproots four rows of cotton at a time by gripping them between revolving rubber tyres; tested in the Sudan it cleared 1.5 to 2.2 ha per hour, and was found to be cheaper than hand pulling (Kemp and Matthews, 1982).

Rotary slashers operating either horizontally or vertically can be used to chop up the stalks of standing cotton into small pieces; the stumps are left in the ground, and the operation must be followed by ploughing to uproot the stumps and bury the chopped material, as a substitute for burning. Probably the cheapest form of mechanization is to loosen the plant roots with a spring tine cultivator and collect the stalks into windrows with a hay rake for burning (T. Lewis pers. comm.).

Ratooning

The wild ancestors of cotton were perennials, as were the first cultivated cottons. It was only when cotton was brought to the United States, where cold stops growth in the winter, that the annual habit was developed by selection. Modern varieties, however, retain enough of their perennial tradition to survive for two or three seasons in the tropics where the winter is not too cold. If they are ratooned, or cut back after harvest, a large proportion of them will sprout again in the next season; King and Lee (1957) showed that there were differences between the local Nigerian varieties in their capacity to survive. Some varieties are still grown regularly as perennials, ratooned every year, in South America and the West Indies (see Ch. 2).

Ratooned plants are more prostrate and bushy than the original, or plant, crop, as all the branches from the mainstem are monopodial; Morris (1973) in Zimbabwe has shown that the plants can be shaped by pruning off some of the shoots without loss of yield. Some plants will not survive ratooning, but a reduction in stand down to 30,000 plants per hectare did not seriously affect yields. The ratoon crop flowers earlier than the plant crop, by as much as six weeks in Egypt (Templeton 1925), but more commonly by two or three weeks. There are conflicting reports of the effect of ratooning on lint quality (Evenson 1970).

The practice of ratooning is banned by law in most cotton-growing countries, mainly because of the danger of carry over of pests and diseases on the ratooned plants. This danger is now less than it used to be, with the advent of new insecticides and the breeding of disease-resistant varieties, and some people think that the practice of ratooning should be examined again in situations where it offers distinct advantages (Evenson 1970, Morris 1973). Conditions should be such that the ratooning is followed by a dormant period caused by cold or drought.

Any relaxation of the rules against ratooning would demand strict discipline among the cotton growers concerned to ensure that control measures were fully applied by everyone; even a few plots of cotton where pests and diseases were uncontrolled could cause widespread damage. Special situations in which ratooning might offer advantages occur in areas with a short growing season, where an early start to flowering would give time for a larger crop to be set. It might also be used to obtain a better return from hybrid seed by allowing a second crop to be taken from the F_1 hybrid plants (Evenson 1970).

Marketing of seed cotton

Crop estimates

Early forecasts of the size of the final crop are an aid to efficient marketing. Cotton packs or other containers can be ordered and distributed in the right numbers, transport at all levels from farm to railroad can be organized, the amount of cash required for the purchase of seed cotton can be arranged, traders in the producing areas can estimate the likely demand for consumer goods, forward contracts for ginned lint can be entered into, and purchasers of lint will know how much cotton they can expect to meet their requirements.

The demand for seed provides the first indication of the area to be planted. Once the cotton is established, the government field staff, usually the extension service of the Department of Agriculture, is called on to estimate the planted area, either by actual measurement of plots or by their judgement of whether the area is more or less than usual, and if so by how much. Large-scale growers are likely to have to make statistical returns of area planted and their anticipated crop.

Crop prospects are influenced at the very start of the season by the earliness or lateness of planting. Estimates of the area planted should therefore be made weekly, fortnightly or monthly so that the effect of planting date can be taken into account.

A method of predicting the cotton crop in Uganda was started by Manning (1952) based on the official estimates of sowing date and acreage, using correlations that were found to exist in previous years between these factors and cotton production. This method was developed by Kibukamusoke (1958) to include the effect of pre-sowing and post-sowing rainfall. Munro (unpublished) grouped the rainfall data into three periods: sowing and early growth; full crop cover; and flowering, ripening and harvest: he calculated regression equations separately for each district in Uganda, using data from the last 25 years (Rijks 1976). Estimates of the crop correct to within a few per cent were obtained with this method as long as there were no significant changes in the agricultural system, but a combination of new seed issues and the more effective use of insecticides caused the estimate to fall progressively short of reality from 1967 onwards (Table 17.1).

Many cotton-growing countries have a system by which the crop estimates are refined as the season progresses: India (Chadwick 1921), Egypt (Balls 1921), West Indies (Harland 1921), as well as Uganda. The amount of seed issued, area planted, date of planting, rainfall, crop growth and damage by pests and diseases, can all be incorporated into the estimate as the information becomes available. The early returns from the ginneries when buying starts are of course more reliable, but are received too late to be of much use in a forecast.

Shortly before the bolls start to split, the yield can be estimated from the number of green bolls on the plants. The average weight of seed cotton in a ripe boll is fairly constant for a given variety, and the yield will depend on the number of bolls per hectare. This can be estimated by pacing off a number of sample stretches of row at random in typical parts of the field, and counting all the bolls in these samples which are 10 or more days old, and likely to contribute to the crop. Early in the season flowers and young bolls may be

Table 17.1   *Cotton production forecast in Uganda*

Forecast and actual production for the total crop for six seasons. Production of the two new varieties, SATU and BPA, is also shown, together with the number of tins of insecticide distributed to growers.

	Production of ginned lint in thousands of bales of 400 lb (181.4 kg)					*Issues of insecticide in thousands of tins**
Season	*Forecast*	*Produced*	*Difference*	*SATU*	*BPA*	
1963–64	378	379	+ 1	1	0	477
1964–65	447	438	+ 9	3	0	25
1965–66	444	445	+ 1	19	0	54
1966–67	411	428	+ 17	82	1	97
1967–68	320	345	+ 25	148	7	210
1968–69	358	421	+ 63	225	35	250

* Each tin contains 1½ gallons (7.8 litres) of 25 per cent DDT emulsifiable concentrate, and is sufficient to enable the grower with one acre (0.405 ha) of cotton to follow the official recommendation to spray his crop four times.
Source: Arnold 1969, Table 1.

included in the count, but the resulting estimate will be less reliable.

This is best illustrated by an example. Suppose that the rows of cotton are spaced 1 m apart and the average boll weight is 5 g, including all pickable cotton, not the selected bolls used to establish the boll weight and ginning percentage of a variety. Choosing five sample lengths of 10 m each, the total number of green bolls is found to be 1000, or 20 bolls per m^2; these weigh 100 g, and the yield is 1000 kg per ha. The length of row and the number of samples may be any arbitrary figures, usually between 3 and 10 m per sample.

Once the average boll weight is known, a row length for boll counts can be selected which will dispense with any calculations in the field, because the number of bolls counted will equal the yield in kilograms per hectare, as in the example given above. The yield of a sample is given by the formula:

$$\text{kg per ha} = \frac{\text{No. of bolls} \times \text{boll weight in g} \times 10{,}000}{\text{Length of row in m} \times \text{width in m} \times 1000}$$

If the number of bolls equals the yield in kg per ha,

$$\frac{\text{Boll weight} \times 10}{\text{Length} \times \text{width}} = 1$$

$$\text{Length of row in m} = \frac{\text{Boll weight in g} \times 10}{\text{Width of row in m}}$$

The total length of row can be divided into any convenient number of samples; for example, if the boll weight is 3.6 g, and the row spacing is 0.75 m,

$$\frac{3.6 \times 10}{0.75} = 48 \text{ m}$$

$$= 6 \text{ sample rows of 8 m each}$$

Marketing systems

Much of the cotton grown in the tropics is produced by small-scale farmers with less than five hectares each; the average is often less than one hectare per farmer. Typically the seed cotton is picked by hand and bought by a state-run corporation or board, either directly or through growers' cooperatives, and the ginning and marketing of lint is organized by the same corporation. The price of seed cotton is fixed by law for each grade and grade standards are prepared and distributed for reference; there are usually two or three grades, the best grade being clean cotton free of trash, and the worst allowing the inclusion of a proportion of diseased and damaged bolls and trash, with or without an intermediate grade between them. Seed cotton is bought from the individual farmer at small local markets, and transported in bulk to the ginnery.

With larger farms and increasing sophistication the system of marketing is modified, tending towards the system operating in the USA, where the farmer picks his cotton by machine and transports it in bulk to the ginnery, where it is ginned individually and the lint is sold by the farmer to a merchant or broker. The farmer pays for the ginning, a large part of this payment being offset against the value of the seed which becomes the property of the ginner. Between these two extremes, a wide variety of marketing systems has grown up in different countries depending on local conditions.

Seed cotton prices and grades

With a large number of small farms and fixed prices for seed cotton, the price is determined by the expected value of the lint on the world market when it is sold. This will be at least six months after the price is announced, maybe up to a year, and commodity prices are notoriously unpredictable. Unexpected price changes may be guarded against by forward sales for delivery on an agreed date, but not all the crop can be sold forward because allowance has to be made for a poor cropping season. Before independence, many of the colonial governments in Africa established stabilization funds which were replenished when world prices were high, and drawn on to make up any deficit when prices were unexpectedly low; all of these were wound up after independence.

The number of grades in which seed cotton is bought and the price differential between them is largely a matter of policy, depending on the importance placed by the government and the textile industry on high-quality cotton. Modern cleaning equipment at the ginnery and the spinning mill can produce an acceptable product from low-grade seed cotton, but in the process fibre quality suffers, and there is no doubt that clean seed cotton must be the starting point if the end product is to be of the best possible quality. Clean, hand-picked cotton generally commands a premium over machine-picked cotton for this reason.

The number of grades varies from country to country. In Pakistan, for example, there is only one standard grade, and since 1976 an official price for this grade has been fixed annually and announced at the beginning of the season. Some spinners, however, find this one grade inadequate, and come to an arrangement with individual farmers to supply them with a better grade of seed cotton at a premium price. At the other extreme the system in Egypt attempts to match the grade of seed cotton with the quality of lint it will produce, and staple length as well as cleanliness is included in the criteria for assessing grade; there are five grades for each variety, and prices are directly proportional to the corresponding price of lint.

Two grades or classes of seed cotton are recognized in most tropical countries, clean cotton and stained or dirty cotton. These may be called grades A and B, grades I and II or (in Swahili) *safi* and *fifi* respectively. A wide difference in price between the two grades is intended to encourage the grower to sort his cotton, so that as much clean cotton as possible is produced to realize the maximum amount of foreign exchange in the export market. A narrower difference in price is probably a truer reflection of the market value of the lint, but may not be enough to compensate farmers for the cost of grading. The price ratio is generally of the order of 100 : 50. A commission of inquiry in Uganda (Shenfield 1962) reported that 'the differential between the prices of *safi* and *fifi* cotton is, in our opinion, greater than is required in order to keep them separate in the marketing process or in order to induce the grower to maximise his supply of *safi* and minimise that of *fifi*'. In 1960–61 the price ratio was 100 : 36; 10 years later, however, it had only risen to 100 : 44.

In Nigeria, three grades of seed cotton are recognized; usually over half the seed cotton is grade I, a third is grade II and the remainder grade III; the price paid is roughly in the ratio 100:75:60. The Zimbabwe crop as it comes from the field may be first, second or third grade, or unclassified, with prices respectively in the ratio 100 : 85 : 65 : 50. The more grades there are, the smaller the difference between them, and the greater the difficulty of classification.

Glass topped sample boxes, holding samples of the different grades of seed cotton, and available for reference at the local markets, are a great help in settling disputes; they should be prepared officially either before buying starts, or as early as possible from the first pickings of the current season's crop.

The establishment of local spinning and weaving mills may alter the quality requirements of a country, because the mills in their early stages are designed to spin relatively coarse yarns and produce coarse calico cloth; this does not demand a high quality of lint, and the spinners may find the cheaper grades are quite good enough. The author was once asked in all seriousness by a local spinner in Nigeria whether it was possible to produce more grade II cotton and less grade I. As local spinning and weaving develop, however, finer quality cloth is produced and cleaner seed cotton is required.

Organization of markets

Permanent centres such as ginneries and cooperative headquarters are supplemented during the buying season by temporary markets or buying posts. These may be open for business throughout the buying season, or may open one or two days a week when they are visited by a buying team. In northern Nigeria the markets are designed on a set pattern, with a grading shed where the seed cotton grade is determined by official examiners, an open courtyard for the growers awaiting their turn for grading, and individual stalls for the licensed buying agents who wish to purchase seed cotton at that market. The farmer receives a ticket indicating the grade allotted to the parcel, and is then free to sell to the buyer of his choice, subject to a guaranteed minimum price for that grade. Each buying stall has three sections, one for each grade, into which the cotton from the farmers' sacks or baskets is tipped; when sufficient cotton has accumulated in the buying stalls, it is bagged into sound bags containing 100 lb (45 kg) of seed cotton, and sealed with a lead seal by the examiner, whose identification number is pressed into it; the bags of each grade are sown up in a distinctive manner before the seal is affixed (Harmer 1957). These markets are erected each season with temporary materials, although some of the more important sites have permanent steel frames to support the temporary roof and walls.

The markets are usually less highly organized in other countries. Temporary buildings and fences may be erected to regulate the flow of cotton, to provide office accommodation and allow safe storage for seed cotton after it has been bought. Where buying posts are opened one day a week, there are usually a limited number of secure stores to which the seed cotton is transported on the day it is bought; no overnight storage is provided at the buying posts and no permanent staffing is therefore needed. The buying team bring with them the equipment they need – scales, bags, record books, grade samples, and cash if cotton is paid for on the spot. A representative of the local farmers is appointed to watch over their interests. Markets and buying posts are arranged so that no farmer has to transport his cotton more than five miles, but a minimum quantity is set below which the establishment of a buying post is uneconomic; on the other hand, the establishment of a post may itself be an incentive for increased cotton production.

It is common practice to buy only one grade at a time, as this makes storage easier and reduces the possibility of mixing cotton of different grades. Clean cotton is bought first, and only when the flow of cotton slackens off are the markets opened for buying the lower grades.

Ginning capacity must be sufficient to cope with the cotton crop in about four months, with sufficient storage space to act as a buffer between the cotton coming in from the markets and the capacity of the gins. The eight months between ginning seasons allows plenty of time for thorough maintenance, overhaul and replacements if required, so that breakdowns during the season are as few as possible. Storage should be provided at the ginnery itself rather than at the markets or in field stores, to reduce double handling and provide better security. If the gins are not working efficiently, or the storage space is inadequate, or the transport of seed cotton is not properly organized, the markets and field stores will become clogged with cotton, liable to pilfering and deterioration in quality; additional temporary storage arranged to cope with a crisis will be even less satisfactory.

Ginning

Ginning is the act of separating lint from seed by mechanical means, to give spinnable cotton and undamaged seed. The lint is not attached so firmly to the seed that it cannot be pulled off readily by hand, but a day's work may give less than a kilogram of lint. An early form of hand gin used in India was the *churcka*, in which the lint was drawn between a pair of wooden rollers, leaving behind the seed which could not pass between them (Fig. 17.4); but the process was still slow, and it was not until the invention of the saw gin by Eli Whitney in 1793, in the USA, that the barrier to expansion of the cotton industry was broken, allowing it to reach pre-eminence in the world trade in textile fibres.

Saw gins
The Whitney gin was the prototype of the saw gins which now

Fig. 17.4   A primitive hand gin of the Indian *churcka* type in Pakistan.

process the bulk of the world's cotton. Essentially, the saw gin consists of a series of small circular saws, between 300 and 450 mm (12–18 in.) in diameter, mounted closely on an axle and made to revolve at high speed in order to tear the lint away from a roll of seed cotton. The saws project slightly between bars or 'breasts', so spaced as to prevent the seed from going forward with the lint (Fig. 17.5). The seed cotton is fed continuously into a rounded box or trough, and the action of the saws keeps it revolving in a loosely compacted roll. The seed falls through a grid into a collecting box or seed conveyer, while the lint is whipped off the teeth of the saws by high speed brushes or an airblast. Each manufacturer has his own modifications of the basic design. The small early gins were fed by hand, but modern gins have an automatic feed and the lint, instead of awaiting removal by hand after passing through a condenser, is conveyed to the baling press mechanically or pneumatically. The average saw gin turns out about 500 kg (2 bales) of lint an hour (see p. 329), and a supervisor can handle a battery of several gins. The driving power for the gins can come from any convenient source: steam, producer gas (using cotton seed as fuel where the seed is not saleable), oil or electricity.

Roller gins

The other type of gin in use throughout the cotton world is the

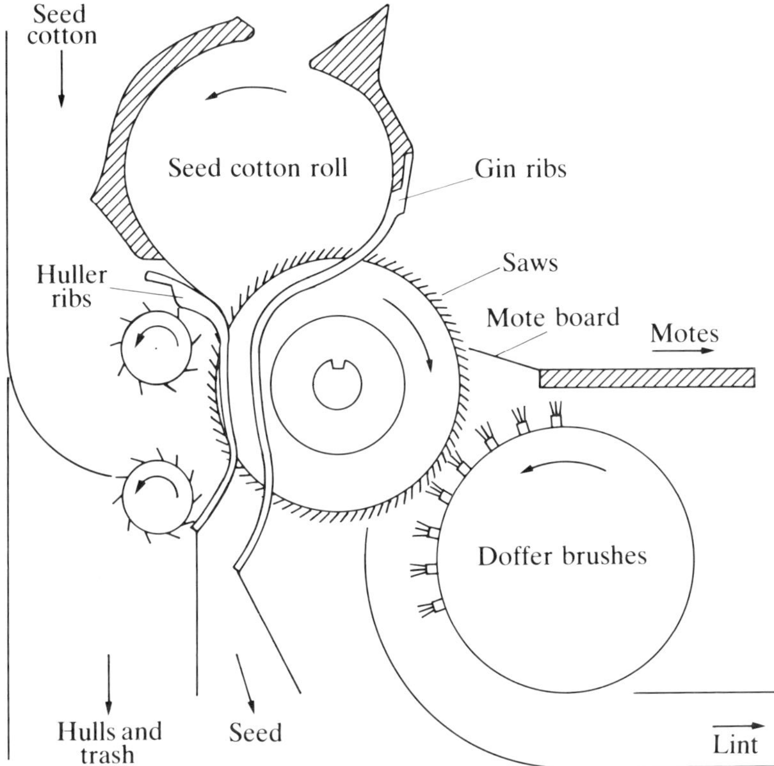

Fig. 17.5   Diagram of a saw gin.

roller gin, which works on an altogether different principle. There are many variations and refinements of the roller gin, but basically it consists of a leather-covered roller which is made to revolve in close contact with a fixed metal blade (Fig. 17.6); the lint in the seed cotton sticks to the leather roller and is pulled through the gap between the roller and blade, which is too narrow for the seed to pass; the separation is helped by a moving knife or beater bar pushing the seed away from the fixed knife. The lint is lifted off the roller and collected for onward transmission to the baling press, while the seed follows its own channel; as with saw ginning, modern ancillary equipment can be used to bring the seed cotton to the feed box of the gin, and the lint and seed to their respective collecting points, providing automatic handling to any required degree. The fixed or 'doctor' knife has to be delicately adjusted for pressure on the roller, and the scored leather of the roller, traditionally bull-hide, has to be kept in working condition, but a merit of the roller gin is its comparative simplicity. 'Double acting' roller gins have two

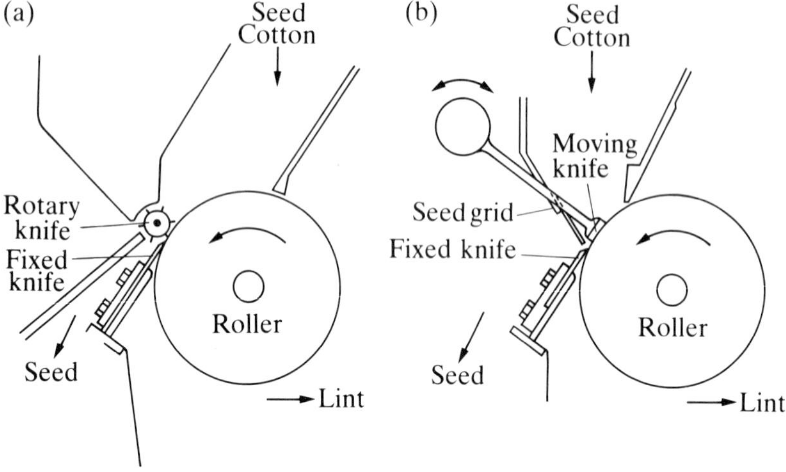

Fig. 17.6 Diagram of a roller gin: (a) Rotary knife (b) Reciprocating knife.

beater blades per roller, and rollers can be positioned in pairs to give any combination of single or double action with single or double rollers, and all of these variations have their advocates. A modern roller gin has rotating instead of oscillating beater bars, and will produce over 45 kg of lint an hour, although the output varies considerably with the kind of cotton.

Roller compared with saw gins

Broadly speaking, roller gins are used for ginning the long-staple cottons of the world, e.g. the *G. barbadense* crops of Egypt and the Sudan, the Sea Island cotton of the West Indies and the Pima cottons of New Mexico and Arizona. The action of the roller gin is gentler, and less damaging to high-quality lint, which may be broken or 'gin-cut' in a saw gin, shortening the mean staple length. Roberts (1975) considered that any cotton with a staple longer than $1\frac{3}{8}$ in. (35 mm) should be roller ginned.

For the same size of machine, the output of the roller gin is much less than that of the saw gin; a 40 in. roller gin is roughly the same size as a 40-saw gin, the former turning out about 50 kg of lint per hour and the latter about 500 kg. The saw gin requires a more powerful motor, but is more efficient in that it uses considerably less power per kilogram of lint. The roller gin produces between 1 and 2 per cent more lint than a saw gin from a given quantity of seed cotton, probably because the gentler action removes less dust and trash. The lint from a roller gin is smoother in appearance and has a sheen which is absent from the fluffy lint from a saw gin.

In spite of the greater efficiency of the saw gin, several countries prefer roller ginning even though their cotton is in the medium and short staple range. In Uganda approximately 70 per cent of the lint is between $1\frac{1}{8}$ in. and $1\frac{1}{4}$ in. (28 and 32 mm), and this provides some justification for roller ginning, but both Tanzania and Kenya also use roller gins for their shorter staple ($1\frac{1}{8}$ in., 28 mm) cotton. In Pakistan and India there is a mixture of saw and roller gins, but the roller gins tend to be antiquated units and are being replaced with saw gins. While in Africa generally roller-ginned cotton commands a premium over saw-ginned, the reverse is the case in India and Pakistan; in fact there is little or no difference in spinning quality, but blow room losses (short fibre, dirt and trash) at the spinning mill are lower from saw-ginned cotton.

A change from roller to saw ginning would be expensive in capital cost, but there are other arguments for maintaining roller ginning as a policy. The main one is that the units are small and simple, and if one unit breaks down only a small part of the ginning capacity is affected. Modern saw gins have such a fast output that one gin stand is more than enough for the average ginnery in Africa, and a breakdown not only closes the whole ginnery, but may be beyond the capacity of the local engineer to get it going again without help and spare parts from outside. Saw gins now incorporate automatic feeding, and are designed for automatic handling of lint and seed; similar auxiliary equipment can be used with roller gins, but it may be socially desirable, and economic when wages are low, to employ more people in the ginnery and carry out these operations manually. Replacements and additions to a roller ginnery can be made unit by unit as required, without incurring the heavy capital cost of a saw gin stand. The same could be said of the small 40-saw gins manufactured by Platt Bros., Oldham, but these are no longer produced. Replacing a number of small ginneries with one large one increases the problems of transporting seed cotton over bad roads with unreliable lorries; not only is the movement of lint and seed after ginning much less urgent, but lint bales occupy less volume than seed cotton and some of the seed may be retained for planting.

Ownership of ginneries

Ginneries may be owned and operated by private enterprise, as in Egypt, India, Pakistan and most of Latin America. Ginneries in East Africa were privately owned before 1952 mostly by members of the Indian community and often run in conjunction with oil mills, but some by the British Cotton Growing Association (BCGA). The BCGA (see Ch. 2) built and operated all the ginneries in Nigeria and Malawi, and some of those in Uganda, India and Pakistan. The reputation of African cotton is largely due to the high standards of ginning and baling set by the BCGA, and its influence has been felt far beyond the areas of its immediate concern.

The ownership of ginneries in East Africa has gradually been taken over either by the Marketing Board or by agricultural co-operatives (Tanzania 1953, Shenfield 1962). Ownership by quasi-government bodies is found in other countries such as Indonesia and the Republic of the Sudan; the Sudanese ginneries were built and operated by the Sudan Plantations Syndicate when the Gezira irrigation scheme was started, and are now owned by the Sudan Gezira Board.

In West Africa the Compagnie Française pour le Développement des Fibres Textiles (CFDT) was set up in 1949 by the French government to develop cotton production in Morocco, French West Africa and Madagascar (White 1957); it has continued to act as managing agent of the ginneries in most of these territories since their independence, and also organizes the primary marketing of seed cotton and the sale of lint.

Close cooperation between the ginneries, extension staff and the marketing board is important for the smooth production and pro-cessing of the crop, as well as in the marketing of lint and seed. The ginnery is the logical centre for the final gathering of crop statistics, and plays an important role in the organization of seed supplies and seed cotton buying; it may be involved in dressing and packaging of seed for distribution to growers.

The ginnery is normally located within the area of production, and should be sited with a view to economy in transport. The flow of seed cotton to the ginnery should be in roughly the same direction as the eventual movement of seed and lint to the buyer. The direction of this flow will often depend on the most convenient form of transport, which may be by road, rail, river or lake.

Ginnery capacity

The ginnery should be capable of handling the seed cotton normally produced in the zone in which it operates. In a changing situation, it is often necessary to decide whether some ginneries should be closed, whether a new ginnery should be provided, or whether additional gin stands should be installed in the existing ginneries. For this some knowledge of ginning capacity is required.

The capacity of a ginnery will depend on the output of the gins, the number of shifts worked and the length of the ginning season; this must equal the maximum production to be expected from the zone. The efficiency of the ginnery can be measured by comparing the manufacturer's rating with the actual output of lint of satisfactory quality. Stopforth (pers. comm.) estimated the efficiency of the ginneries in the Philippines at 65 per cent of the manufacturer's rating. The approximate output of different types of gin averaged over a number of seasons is roughly as follows:

Type of gin	Output in lint per hour
Single roller 40 in.	25 kg (50 lb)
Double roller 40 in.	35 kg (75 lb)
High-capacity roller	150 kg (350 lb)
40-saw	150 kg (350 lb)
80-saw	340 kg (750 lb)
120-saw	500 kg (1000 lb)
Modern 128-saw	over 1000 kg (2200 lb)

From these figures the ginnery capacity can be calculated. For example, a 40 in. single roller gin working 2 shifts of 8 hours each per day, 25 days per month, will produce (25 kg × 16 × 25 =) 10 metric tons of lint per month, equal to 55 bales of 182 kg (400 lb); 10 gins working for 6 months will produce 600 tons equal to 3300 bales. Under the same conditions an 80-saw gin will produce (350 kg × 16 × 25 =) 140 metric tons per month; in 4 months it will produce 560 tons, equal to just over 3000 bales.

It is not possible to give a figure for the optimum size of a ginnery, as this depends on so many variables. In Uganda an output of 2400 bales per season was regarded as a suitable figure for a standard ginnery (Shenfield 1962) equipped with roller gins, and the same figure has been used in Kenya (World Bank 1980). It will be obvious from the previous paragraph that a single saw gin will need more cotton than this to provide full-time working, and hence maximum return on capital. Modern saw gins have 120 saws or more and a greater output per saw, and there are various types of high-speed roller gins on the market.

The size of the ginnery depends mainly on how widely the cotton growers are dispersed; in the Sudan Gezira, agriculture is in the irrigated area, and the crop is channelled into a few large ginneries; south of the Sahara not all farmers grow cotton and farms are usually separated by grazing land and bush. The more widely dispersed the growers, the greater the difficulties of transporting seed cotton to the ginnery, and ginneries tend to be smaller; any savings in ginnery costs have to be balanced against increases in the problems and costs of transport from primary markets.

Reception of seed cotton

Cotton arriving at the ginnery is first weighed, providing a check on the quantities bought in the market. The cotton is then emptied into heaps on a concrete surface; the yards are usually uncovered, as harvesting and ginning are timed to coincide with the dry season, but if rain is likely to occur the heaps are covered with tarpaulin sheets. The main grades of seed cotton are kept separate, but the ginner himself may classify each grade into lots of apparently similar quality to give a run of homogeneous bales by the time the lot has been thoroughly mixed in the ginning. The lots may improve or deteriorate in quality as the season advances, depending on how the crop has fared in the field; the early pickings from a district are often, but not always, the best and the change may be large enough to be reflected in the price eventually received for the lint. These differences can only be realized if transport from the markets is rapid and continuous; any hold-up will result in early purchases being mixed with later ones by the time they reach the ginnery.

The lint from any one ginnery, gathered from its own zone, usually has characteristics of its own which may mask minor changes due to the advancing season; buyers of lint learn to recognize these characteristics and may purchase their requirements from specified ginneries. The high quality of the crop in the Sudan and Egypt fully justifies an elaborate pre-ginning classification of the seed cotton into six recognized grades, but this degree of refinement is unusual.

From the lots stacked in the ginnery yard the seed cotton is taken in to the ginnery either by hand using canvas sheets or through a suction tube on a flexible mounting allowing it to be moved as the cotton is cleared.

Basic equipment in the ginnery

Hand-picked and graded seed cotton does not need much cleaning, and the traditional ginnery in Africa was a very simple affair. It included an opener, a battery of roller gins, a baling press for the lint and a screw conveyer to bring the seed from the gins to the bagging point; seed cotton and lint were moved by hand.

The opener loosens up the seed cotton after having been compressed into bags for transport to the ginnery. The beating action of the opener removes a certain amount of dust and trash, and dislodges some of the insects and bollworms which have been picked with the cotton; an examination of the opener waste gives a good indication of pink bollworm infestation where this is a problem. After ginning the lint is taken to the baling press; in very dry weather it may have to be moistened by a fine spray of water before baling. It is difficult to compress very dry lint into the standard size and weight of bale, and it is economic to raise the moisture content to just above the trading limit, so that after a sea voyage it holds the maximum permitted amount of moisture when it arrives at its destination.

The ancillary equipment described in the next sections can be used in conjunction with either saw or roller gins.

Precleaning

In the ginning of machine-picked cotton in the USA, the gin itself is a very small part of the installation; a typical ginnery layout is shown in Fig. 17.7. The first process is drying, to bring the seed cotton to a moisture level which gives the most efficient cleaning and ginning, with the least possible damage to the fibre. For saw ginning, the fibre should contain between 6.5 and 8 per cent moisture, while 5 to 6 per cent moisture is desirable for roller ginning (USDA 1977). The seed cotton is fed into the top of a tower and moves down through a series of baffles against a high-volume/low-velocity airstream which is heated unless the moisture content of the cotton is already low. It may then pass through a burr and stick remover and an inclined cleaner to the extractor/cleaner/feeder which is mounted above the gin and feeds the seed cotton evenly to the gin at the required rate. The seed cotton is blown from one to the other of these cleaners in pipes fed with compressed air.

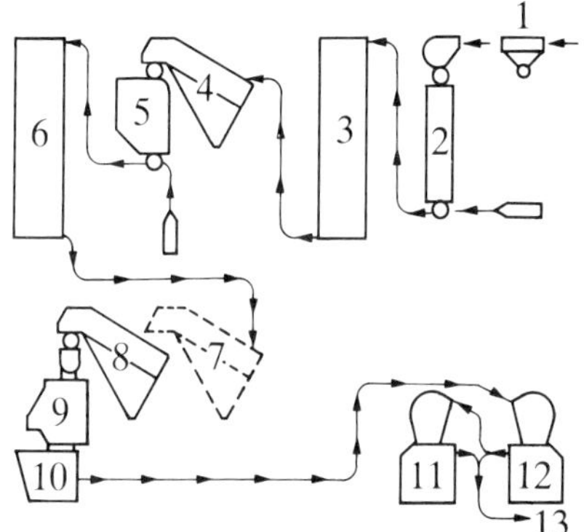

Fig. 17.7 Recommended gin machinery arrangements for machine-picked cotton (USDA 1977).

Sequence of operations:
 1. Rock and green boll trap
 2. Feed control
 3. Tower drier or equivalent
 4. Cylinder cleaner: 5–7 cylinders
 5. Stick and green leaf machine
 6. Tower drier or equivalent
 7. Cylinder: 5–7 cylinders (optional)
 8. Cylinder cleaner: 5–7 cylinders
 9. Feeder, extractor type
10. Saw gin stand
11. Saw-cylinder lint cleaner
12. Saw-cylinder lint cleaner
13. Press

Handling of ginned lint

After ginning, the lint may pass through one or two lint cleaners, which squeeze it between rollers to break up any pieces of trash or seed and loosen it with combs to allow the trash to fall away. In passing through a humidifier the moisture content is raised to about 9 per cent to facilitate baling. The lint for each bale is collected and weighed, adjusting the quantity to within the permitted limits of weight per bale; each bale may be sampled automatically at this stage for classification into its correct grade.

The baling box is lined with wrapping material, hessian, cotton or more recently plastic sheeting; the lint is fed into the box and compressed by a hydraulically operated piston. The bale is bound with hoop iron, either in separate bands or spirally, before the pressure is released. The ends of the bale may then be covered to complete the wrapping, the weight is checked, and the weight and identification marks are stencilled on one side before the bales are stacked ready for shipment (Fig. 17.8).

The unit for international trade in cotton has for many years been the USA running bale of 500 lb, containing 478 lb of lint. Most of the former colonies in Africa, however, use a smaller bale containing 400 lb of lint, and other countries use bales of different sizes (Table 17.2). The International Cotton Advisory Committee publish statistics of world trade both in bales (480 lb since 1983) and in metric tons of

Fig. 17.8 Nigerian cotton awaiting shipment (British Cotton Growing Association).

Table 17.2   *Weight of cotton bales in various countries*

Country	Average bale weight (gross) (ICAC 1977)			Official bale weight (net)
	kg	lb		lb
India	170	375		375
Pakistan	176	388		375
South Africa	180	397	(export)	500
Uganda	—	—		410
Nigeria	—	—		410
Tanzania	183	403		410
Malawi	184	406		400
Kenya	185	408		400
Upper Volta	200	441		
Mozambique	200	441		440
Argentina	206	454		
Zambia	—	—		420
Brazil	—	—		420
				425
Sudan	209	461		420
				450
Syria	209	461		
Paraguay	210	463		
China	—	—		450
Ivory Coast	215	474		
Turkey	215	474		
USSR	215	474		
Colombia	—	—		475
Mexico	218	481		485
Angola	220	485		
El Salvador	225	496		485
Nicaragua	226	498		485
Iran	227	500		450
USA	227	500		480
Australia	229	505		500
Costa Rica	230	507		
Israel	235	518		
Greece	235	518		516
Peru	240	529		
Colombia	250	551		475
				500
Morocco	250	551		
Ecuador	266	586		
Bolivia	—	—		580
Egypt	331	730	(export)	720
			(country)	800

Sources: ICAC 1977, Chokey Singh 1980, Volkart 1984, *Cotton Outlook* pers. comm. 1985.

lint (ICAC 1959 etc.), and there is an increasing tendency towards using metric tons as the international unit. A bale of 480 lb net, 55 × 20 × 27 in., has a density of 28 lb per ft^3 (USDA 1977).

Handling of seed

Cotton seed may be bagged for storage directly from the gins. Bagging is preferable to bulk storage, unless the seed is to receive further treatment at the ginnery; without special equipment the seed is much easier to handle in bags. Seed is usually packed in hessian grain sacks holding about 50 kg, but plastic and paper sacks have been greatly improved in recent years and are now quite satisfactory and often cheaper than hessian bags.

Seed of American Upland varieties is usually fuzzy, and the short hairs or fuzz remaining on the seed after ginning are recoverable as linters. Mechanical delinters are often installed in the ginnery, and the seed is passed to them from the gins before it is bagged. The value of the linters varies, but it can be expected to cover the cost of delinting and usually shows a profit (see Ch. 5). Acid and flame delinting destroy the linters and are not part of the commercial ginning process, being mainly confined to treatment of planting seed.

Gin settings

The correct setting of the gins, to give the maximum output consistent with undamaged lint, is a skilled job, and an experienced gin fitter can tell by the appearance of the lint and seed when the gins are improperly set and what adjustments to make.

In a roller gin, the roller must be truly cylindrical, with spiral saw cuts to help grip the lint and draw it past the fixed knife. The fixed knife must be sharp and pressing evenly on the roller at the correct point on its circumference; pressure and location can both be adjusted. The moving knife must overlap the fixed knife by the correct amount, and in general longer-staple cotton requires more overlap. Bearings must be firm and well lubricated, with no excess of oil to contaminate the lint. The manufacturer recommends the speed at which the gin should be operated; a faster speed may give a greater output, but will risk damaging the lint and seed.

There are fewer adjustments possible on a saw gin; the breast is adjustable to allow greater or less penetration of the saws into the roll-box, and the setting of the seed grid below the roll and of the shutter controlling removal of dirt and trash can both be varied. The rate of feed must be enough to maintain the roll of seed cotton in the roll-box at the correct density; a tighter roll gives more output, but tends to break the fibres and lower quality. The saws must be sharp and the teeth undamaged; saw sharpening machines are available, but many ginneries replace the saws with new ones every season rather than attempt to sharpen the old saws.

Ginning percentage

The percentage weight of lint in seed cotton is variously known as 'ginning percentage', 'lint percentage' or 'ginning outturn' (GOT), and is usually around one-third. It is fair to use this figure, $33\frac{1}{3}$ per cent, for a quick conversion of seed cotton figures into lint, but the actual percentage may range from 25 per cent to over 40 per cent. The exact figure is important to the ginner as his return, whether he buys the seed cotton or is paid a commission on ginning, is based on the weight of lint produced. To a ginner who processes 100 tons of seed cotton, the difference between 34 and 35 per cent ginning outturn is one ton of lint, worth $1,760 at 80 cents per lb; put another way, a ginnery producing 2400 bales of lint at 35 instead of 34 per cent will gain 70 bales of lint over the season. The percentage of seed rises as that of lint falls, but the lint is worth about 12 times as much as the same weight of seed (see Table 17.8).

A badly set gin can lead to an apparent drop in ginning percentage by failing to remove all the lint from the seed. Drying and cleaning machinery in the ginnery reduces the ginning percentage by removing moisture, dust, trash, motes and broken particles of lint during the various operations. In a well-run ginnery using a good class of cotton, such loss or 'waste' is unlikely to exceed 4 per cent in a normal season, but may range from 2 to over 7 per cent. Ginning percentage also varies from season to season according to the conditions the crop has encountered. Over nine seasons, one of the ginneries in Zimbabwe using the same variety of cotton and the same machinery, showed a range in ginning percentage over eight years from 30.7 to 33.2 per cent, with a mean of 32.25 per cent.

The marketing of lint

Samples of lint are taken from the pressed bales at the ginnery, and the lint is sold on the basis of these samples. Ideally, every bale should be sampled, but this is rarely achieved outside the USA. For example, in Pakistan only 3 per cent of all bales are sampled; in Kenya two bales are sampled from each lot of 50 bales (4 per cent); in Zimbabwe and Syria every 10th bale is sampled (10 per cent); every bale has been sampled in Malawi since 1970. The usual sample is rolled in strong paper and weighs about 0.25 kg (8 oz), and further samples may be taken as the bales pass through the hands of brokers to the eventual user; the appearance of the bale when it reaches the spinning mill may be very different from the neat bales produced at the ginnery (Fig. 17.8).

Cotton lint is usually sold by the bale through brokers operating in one of the cotton exchanges. When a country produces a significant quantity of cotton for export, brokers from the exchange may have their own representatives stationed there, even though the exchange is in a different country. Alternatively, the producers or their cotton board may send a representative to the cotton exchange.

The futures market quotes prices for delivery on an agreed date up to a year or more ahead. Cotton producers can use this market as a hedge if they wish to publish fixed or minimum prices for seed cotton before the season starts, or at any time before the markets open for buying seed cotton. The government or marketing board guaranteeing the local price will want to ensure that this can be met without any financial loss, and will sell some of the crop forward for future delivery. How much it sells in this way will depend on how prices are expected to move, and how much cotton it can be certain of delivering on the agreed date. If prices are likely to fall, it will want to sell as much of the crop forward as possible; but the seller must be sure that he can deliver the quantity contracted for, even if growing conditions are unfavourable. Early and reliable estimates of the season's crop, discussed earlier in this chapter, are therefore of the greatest importance.

In the nineteenth century Lancashire was not only the largest consumer of cotton, but also the centre for international trading through the Liverpool Cotton Exchange. A highly sophisticated system was built up, depending on the services of individual brokers, buyers, agents, merchants, classifiers and graders, and providing spot and futures markets along with storage and warehousing facilities for cotton from all over the world.

The Liverpool Cotton Association set up internationally recognized standards of grade and staple, backed by an arbitration service to settle any disputes. During 1913 and 1914 representatives of the American and European cotton exchanges and spinners' associations discussed the adoption of international standards for grading American Upland cotton and certain modifications were agreed, to reconcile the existing Liverpool and proposed American standards. The Liverpool Cotton Association deferred final action on the matter, but the standards agreed upon were adopted as the official cotton standards of the United States in December 1914 (Meadows 1921).

With the decline of the Lancashire cotton trade, the United States have taken over the lead in setting standards: cotton prices even in Liverpool are now quoted in US cents per pound. The Liverpool standards have been replaced by the slightly different Universal standards for grade (Table 17.3) and American standards

Table 17.3   *Grade standards and code numbers for American Upland cotton*

Grade	Colour class							
	0	*1*	*2*	*3*	*4*	*5*	*6*	*7*
No. and Description	White		Light spotted	Spotted	Tinged	Yellow stained	Light grey	Grey
	Plus grade	Full grade						
	Code numbers for grade and colour							
0 Above grade		01						
1 Good middling		11*	12	13	14	15	16	17
2 Strict middling		21*	22	23*	24	25	26	27
3 Middling	30	31*	32	33*	34*	35	36	37
4 Strict low middling	40	41*	42	43*	44*		46	47
5 Low middling	50	51*	52	53*	54*			
6 Strict good ordinary	60	61*						
7 Good ordinary	70	71*						
8 Below grade		81	82	83	84	85	86	87

* Denotes standards for which sample boxes are available (USDA, Agricultural Marketing Service, Form CN-383, 1979).

Table 17.4 *Classification of staple length*

Description	Code		Staple length	
			inches	*millimetres*
Short	24		Below 13/16	Below 21
Medium	26	*	13/16	21
	28	*	7/8	22
	29	*	29/32	23
	30	*	15/16	24
	31	*	31/32	24
	32	*	1 inch	25
Medium-long	33	*	$1\frac{1}{32}$	26
	34	*	$1\frac{1}{16}$	27
	35	*	$1\frac{3}{32}$	28
Long	36	*	$1\frac{1}{8}$	28
	37	*	$1\frac{5}{32}$	29
	38	*	$1\frac{3}{16}$	30
	39	*	$1\frac{7}{32}$	31
	40	*	$1\frac{1}{4}$	32
	41		$1\frac{9}{32}$	32
	42	**	$1\frac{5}{16}$	33
	43		$1\frac{11}{32}$	34
Extra-long	44	**	$1\frac{3}{8}$	35
	45		$1\frac{13}{32}$	36
	46	**	$1\frac{7}{16}$	36
	47		$1\frac{15}{32}$	37
	48	**	$1\frac{1}{2}$	38
	49		$1\frac{17}{32}$	39
	50		$1\frac{9}{16}$	40

and upward in steps of $\frac{1}{32}$ in.

* American Upland } denote standards for which sample rolls are available
** American Pima } (USDA, Agricultural Marketing Service, Form CN-383, 1979).

for staple length (Table 17.4). Cotton fibre testing methods have been largely standardized (see next chapter), and USDA calibration samples can be supplied to other laboratories; each 1 lb sample is accompanied by its fibre test data for all routine items except micronaire and Pressley strength at zero gauge, for which International Cotton Calibration Standards have been agreed (USDA 1982). While the American system of grading is used by other producers of Upland style cotton, many countries, including India and Russia, have their own systems of lint grades.

Cotton prices

After a period of comparatively stable prices after the Second

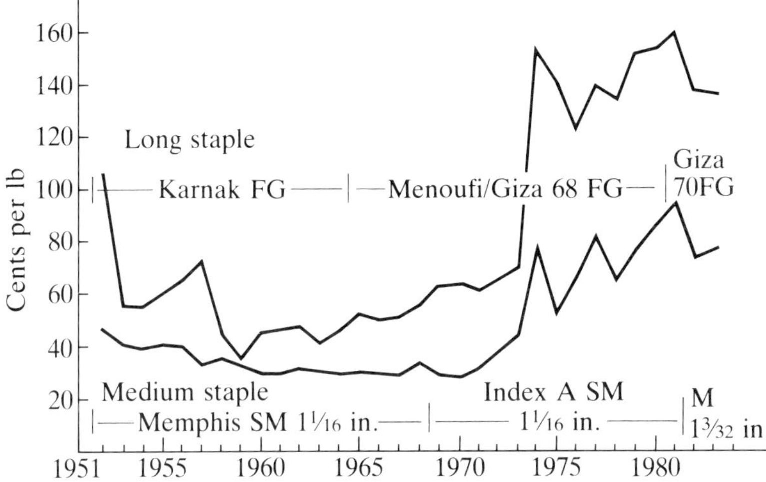

Fig. 17.9   Cotton prices, 1951–83, c.i.f. Liverpool, (*Cotton Outlook; ICAC*).

Table 17.5   *Cotton prices – c.i.f. North Europe*

Country	Type	Price* (US cents per lb)	
United States	Orleans/Texas M 1 in.	66.76	
	Memphis Terr. M 1³⁄₃₂ in.	76.30	
	California SM 1⅛ in.	79.79	
Mexico	M 1³⁄₃₂ in.	75.28	A
Nicaragua	M 1³⁄₃₂ in.	72.17	A
Guatemala	M 1³⁄₃₂ in.	72.87	A
Brazil	São Paulo Type 5 1¹⁄₁₆ in.	72.90	N
Argentina	Grade C 1³⁄₃₂ in.	63.75	NA
Syria	M 1³⁄₃₂ in.	79.82	A
Turkey	Izmir M 1³⁄₃₂ in.	77.55	A
Greece	M 1³⁄₃₂ in.	81.00	A
USSR	M 1³⁄₃₂ in. Utoroi	73.02	
Tanzania	AR Mwanza No. 1	88.08	A
Pakistan	NT Sind S.G.	65.65	
Egypt	Giza 67 FG	115.73	
	Giza 70 FG	137.34	
Sudan	G 5 VS	93.79	
	G 5 B	90.76	
Peru	Tanguis Grade 3	83.42	
	Pima G.1 1⁹⁄₁₆ in.	121.68	
United Kingdom	Outlook Index A M 1³⁄₃₂ in.	73.76	
	Outlook Index B Coarse count	64.38	

* Average August 1981–July 1982.
N = nominal.
A = average for less than 12 months.
Source: ICAC 1983, courtesy of *Cotton Outlook*, Liverpool.

Table 17.6   *Premiums and discounts for grade and staple length of 1981-crop American Upland cotton*

Grade	Code	Staple length (inches)								
		$^{13}/_{16}$ through $^{29}/_{32}$ (26–29)	$^{15}/_{16}$ (30)	$^{31}/_{32}$ (31)	1 (32)	$1^{1}/_{32}$ (33)	$1^{1}/_{16}$ (34)	$1^{3}/_{32}$ (35)	$1^{1}/_{8}$ (36)	$1^{5}/_{32}$ & longer (37 & longer)
					Points per pound					
White										
SM & Better	(11 & 21)	−1010	− 840	− 585	− 420	− 20	+ 175	+ 210	+ 240	+ 345
MID PLUS	(30)	−1025	− 855	− 605	− 440	− 40	+ 150	+ 190	+ 225	+ 325
MID	(31)	−1035	− 870	− 610	− 450	− 55	+ 135	+ 170	+ 205	+ 300
SLM PLUS	(40)	−1075	− 900	− 655	− 515	− 135	+ 55	+ 90	+ 125	+ 220
SLM	(41)	−1100	− 930	− 685	− 560	− 185	B	+ 40	+ 70	+ 160
LM PLUS	(50)	−1185	−1025	− 785	− 685	− 360	− 205	− 175	− 160	− 85
LM	(51)	−1235	−1075	− 840	− 745	− 470	− 320	− 295	− 255	− 195
SGO PLUS	(60)	−1500	−1390	−1270	−1205	− 925	− 850	− 835	− 810	− 810
SGO	(61)	−1545	−1445	−1320	−1255	−1000	− 930	− 920	− 895	− 895
GO PLUS	(70)	−1810	−1720	−1620	−1570	−1320	−1260	−1250	−1225	−1225
GO	(71)	−1855	−1765	−1665	−1620	−1385	−1335	−1320	−1300	−1300
Light spotted										
SM & Better	(12 & 22)	−1060	− 895	− 650	− 515	− 100	+ 75	+ 110	+ 140	+ 235
MID	(32)	−1095	− 935	− 680	− 560	− 185	− 10	+ 25	+ 60	+ 155
SLM	(42)	−1190	−1060	− 830	− 720	− 460	− 315	− 295	− 250	− 195
LM	(52)	−1425	−1295	−1160	−1130	− 925	− 860	− 845	− 825	− 825
Spotted										
SM & Better	(13 & 23)	−1295	−1210	−1130	−1015	− 630	− 515	− 505	− 480	− 480
MID	(33)	−1375	−1295	−1210	−1110	− 805	− 705	− 700	− 675	− 675

Table 17.6   *(contd)*

Grade	Code	Staple length (inches)								
		$^{13}/_{16}$ through $^{29}/_{32}$ (26–29)	$^{15}/_{16}$ (30)	$^{31}/_{32}$ (31)	1 (32)	$1^{1}/_{32}$ (33)	$1^{1}/_{16}$ (34)	$1^{3}/_{32}$ (35)	$1^{1}/_{8}$ (36)	$1^{5}/_{32}$ & longer (37 & longer)
					Points per pound					
SLM	(43)	−1515	−1435	−1355	−1305	−1080	−1030	−1020	−1000	−1000
LM	(53)	−1705	−1600	−1540	−1510	−1300	−1265	−1260	−1240	−1240
Tinged										
SM	(24)	−1565	−1490	−1460	−1435	−1410	−1395	−1395	−1155	−1155
MID	(34)	−1610	−1530	−1500	−1475	−1450	−1440	−1440	−1200	−1200
SLM	(44)	−1685	−1615	−1595	−1580	−1555	−1550	−1545	−1300	−1300
LM	(54)	−1810	−1740	−1715	−1695	−1675	−1665	−1660	−1415	−1415
Light gray										
SM & Better	(16 & 26)	−1265	−1080	− 805	− 670	− 235	+ 5	+ 55	+ 110	+ 200
MID	(36)	−1415	−1230	−1000	− 900	− 540	− 310	− 275	− 200	− 145
SLM	(46)	−1745	−1615	−1450	−1330	−1035	− 935	− 900	− 775	− 775
Gray										
SM & Better	(17 & 27)	−1415	−1230	−1010	− 910	− 565	− 355	− 325	− 255	− 195
MID	(37)	−1750	−1620	−1455	−1330	−1105	− 990	− 955	− 840	− 840
SLM	(47)	−2105	−1975	−1855	−1765	−1520	−1420	−1390	−1260	−1260

Grade symbols: SM – Strict Middling; MID – Middling; SLM – Strict Low Middling; LM – Low Middling; SGO – Strict Good Ordinary; GO – Good Ordinary.

B = National loan rate of 52.46 cents per pound for basic grade, strict low middling $1^{1}/_{16}$ inches, net weight.

Source: Federal Register 1981.

Table 17.7   *Value differences for grade, staple and colour*

COTTON based on Universal standards for grade and on US staple standards

MEMPHIS TERRITORY
Basis Middling 1¹/₁₆ in.

Staple (in.)	White						
	GM	SM	MID	SLM	LM	SGO	GO
1¹/₃₂	—	—	—	—	—	—	—
1¹/₁₆	—	150	B	–250	–450	—	—
1³/₃₂	—	225	75	–175	–375	—	—
1⅛	—	275	125	–125	–325	—	—

SPOTTED — Middling: −250 points off White
TINGED — Middling: −450 points off White
GRAY — Middling: −350 points off White

OTHER AMERICAN RAINGROWN
Basis Middling Inch.

Staple (in.)	White						
	GM	SM	MID	SLM	LM	SGO	GO
²⁹/₃₂	—	–125	–300	–500	–750	—	—
¹⁵/₁₆	—	– 50	–225	–425	–675	—	—
³¹/₃₂	—	75	–100	–300	–550	—	—
INCH	—	175	B	–200	–450	—	—
1¹/₃₂	—	325	160	– 50	–300	—	—
1¹/₁₆	—	475	300	100	–150	—	—
1³/₃₂	—	—	—	—	—	—	—

SPOTTED — Middling: −250 points off White
TINGED — Middling: −450 points off White
GRAY — Middling: −350 points off White

CALIFORNIA Acala SJV Basis Middling 1⅛ in.

Staple (in.)	GM	SM	MID	SLM	LM
INCH	—	—	—	—	—
1¹/₃₂	—	—	—	—	—
1¹/₁₆	—	—	—	—	—
1³/₃₂	150	50	– 25	–275	—
1⅛	175	75	B	–250	—
1⁵/₃₂	275	175	100	–150	—

CALIFORNIA/ARIZONA DPL
Basis Middling 1¹/₁₆ in.

	GM	SM	MID	SLM	LM
	—	—	—	—	—
	175	75	–25	–175	–250
	200	100	B	–150	–225
	225	125	25	–125	–200
	275	175	75	– 75	–150
	—	—	—	—	—

MEXICAN Basis Middling 1¹/₁₆ in.

Staple (in.)	GM	SM	MID	SLM
INCH	– 75	–175	–375	–575
1¹/₃₂	225	125	– 75	–275
1¹/₁₆	300	200	B	–200
1³/₃₂	350	250	50	–150
1⅛	430	330	130	– 70

EL PASO Basis Middling 1⅛ in.

Staple (in.)	GM	SM	MID	SLM	LM
1¹/₁₆	– 40	–150	–265	–460	–600
1³/₃₂	10	–100	–215	–410	–550
1⅛	110	B	–115	–310	–450
1⁵/₃₂	185	75	– 40	–235	–375
1³/₁₆	260	150	35	–160	–300
1⁷/₃₂	435	325	210	15	–125
1¼	585	475	360	165	25

Source: Weekly circular from the Liverpool Cotton Association.

World War, both medium- and long-staple cotton doubled in price in 1973. Since then prices have fluctuated widely: the 'A' index for medium-staple cotton has ranged from a low of 37.22 cents per pound in August 1986 to a high of 100.62 cents in September 1980. Monthly and yearly averages tend to smooth out some of these

Table 17.8   *Cotton and cotton seed products*

Components of	% by weight	Value		
		cents per lb (1)	*$*	*%*
Seed cotton:				
Lint	35	70.3	24.60	88
Seed	60	5.6	3.36	12
Waste	5	—	—	—
Total	100		27.96	100
Cotton seed: (2)				
Oil (3)	16.6	29.1	4.83	52
Meal (4)	46.2	7.1	3.28	35
Hulls	23.6	1.9	0.44	5
Linters (5)	8.9	8.0	0.71	8
Waste (6)	4.7	—	—	—
Total	100.0		9.26	100

(1) Prices of cotton products in the USA, 1974–78.
(2) Weight of cotton seed products in the USA, 1962–76.
(3) Refined oil.
(4) 41% protein, less than 1% oil.
(5) First cut 3.0%; second cut 5.9%.
(6) Including acid oil.

Sources: Hise and Ethridge 1980, ICAC 1983.

Table 17.9   *Composition of vegetable oil seeds*

Crop	% oil in seed	% protein in cake
Copra	60	30
Groundnut	45	45
Sunflower	25–35	37
Kapok	20–25	26
Soya	18–20	45*
Cotton	18–20	40*
Rice bran	16–20	18*

* solvent extraction; others expeller.

fluctuations, but they are still to be seen in the yearly averages shown in Fig. 17.9. The 'A' index published by *Cotton Outlook* is derived from a group of medium-staple cottons, giving an average price for middling grade $1\frac{3}{32}$ in. cotton, c.i.f. Northern Europe; prior to August 1981 quotations were for strict middling $1\frac{1}{16}$ in. Long-staple cotton is represented in Fig. 17.9 by the best quality Egyptian, latterly Giza 68 and Giza 70, with a staple of about $1\frac{1}{2}$ in.

    Each class of cotton has its own standard, the quoted price of the standard providing a reference point or base for any individual lot in that class (Table 17.5), with so many points on or off according to the classer's estimate of grade, colour and length; a point is one hundredth of a cent. These differentials are not constant, being affected by variation in the base price and in the supply and demand for each individual type; there is room for bargaining in any individual contract. The values shown in Table 17.6 give a general indication of the value of a sample in October 1981 in relation to the price of SLM $1\frac{1}{16}$ in. white, but smaller tables for each class of cotton are published regularly by the cotton exchanges. An example of the tables for American raingrown cotton from the weekly bulletin of the Liverpool Cotton Association is shown in Table 17.7, when the 'A' index was 76.75 cents per pound.

Cotton seed

Every ton of cotton picked in the field yields about 600 kg of seed and 350 kg of cotton lint, but the value of the lint is about seven times that of the seed. Seed is therefore treated as a by-product of the industry, but it can provide significant quantities of edible oil and a protein-rich food for livestock.

The four major components of cotton seed in order of extraction are linters, hulls, oil and cake or meal (Table 17.8). Delinting has already been discussed in Chapter 15, and may be carried out at the ginnery or at the oil mill. It is possible to separate the hulls from the kernels or 'meats' without delinting, but the linters will nearly always show a small profit even if they have to be exported (Hise and Ethridge 1980). The seed is cut or cracked and the hulls separated mechanically from the meats. The hulls are bulky and of little value, and are often used as fuel in the factory boilers; but they can be reduced to a fine powder known as 'hull bran' and used as a filler in animal feed formulations. The cracked seeds can be put through a screw press without separating out the hulls, giving undecorticated cotton cake with a protein content of about 25 per cent, but there is no longer an export market for this type of cake.

The oil can be extracted from cotton seed either mechanically in a screw press or chemically with a solvent. Most of the older mills use the screw press, yielding oil and cotton seed cake which still contains at least 5 per cent of oil. When the oil content of the seed is below 20 per cent, as in cotton, soya and rice bran (Table 17.9), this means that less than 75 per cent of the available oil is being extracted, and this is now considered to be uneconomic; solvent extraction, leaving less than 1 per cent of oil in the meal, is being used in new plant for treating seed of this kind. Whichever method of extraction is used, the meats are heated or cooked to enable the oil to flow freely and to rupture the oil-bearing cells. This also reduces the free gossypol in the cake or meal, as most of it is fixed by chemical combination with other constituents of the seed.

In solvent extraction, the hulls must be removed so that the meats can be rolled into thin flakes to expose a maximum surface to the solvent. The solvent is a petroleum product, N. Hexane, with a very low boiling point (65 °C); it is either flooded over the prepared flakes, or percolated through them. The solvent is recovered from the extracted flakes by heating and condensation, after which the flakes, now almost free of oil, are cooled and humidified. The resulting product is called an 'extraction', to distinguish it from the cotton seed 'cake' produced by pressing. The solvent containing the oil, known as 'miscella', is then distilled; the solvent is condensed for re-use, leaving the oil free from solvent for further refining.

Cotton seed oil has to be refined to get rid of the inherent impurities – free fatty acid (FFA), colour, gums, odour and gossypol. The refined oil has low FFA, is light in colour and has no odour or taste, and has been accepted all over the world as a first-class edible oil comparable with vegetable oils from other sources.

Cotton seed extractions, with about 40 per cent protein, is one of the standard components of animal feeding stuffs. Compared with extractions from other oil-seeds, cotton seed protein is deficient in certain amino acids and contains some gossypol; these do not detract from its value for feeding ruminant livestock, which can synthesize the required proteins in the rumen and are not affected by gossypol. In fact, untreated cotton seed can be fed directly to cattle with no ill effects. Non-ruminants like pigs and poultry are susceptible to gossypol poisoning, and the cake or meal must be free of gossypol or treated to deactivate it. The cooking of the meats in the extraction process reduces the amount of free gossypol by combination with some of the amino acids, which thereby lose their food value; treatment with iron sulphate will neutralize the gossypol, and glandless (gossypol-free) varieties have been produced (see Ch. 14).

Two other defects are directly linked with poor storage conditions: high FFA and aflotoxin. To prevent them, cotton seed for oil extraction needs the same storage conditions as cotton seed for planting (see Ch. 15), principally the avoidance of high moisture content and high temperatures.

Chapter 18

Lint quality

The textile industry

Some of the cotton produced in parts of Asia and Africa is still ginned, spun and woven into cloth by hand in the home. This is notably so in Nigeria, where hand-woven cloth was very fashionable at the time of independence, but even there thread from the local spinning mills is replacing hand-spun thread for domestic weaving. Ghandi's name is associated with the promotion of spinning and weaving in the villages of India, but the great bulk of the world's cotton is destined for industrial spinning mills, which may be located on a different continent from the cotton field.

After leaving the ginnery, cotton lint enters a different world – consumption as opposed to production – where it is variously known as 'raw cotton', 'cotton fibre', or simply as 'cotton'. The cotton trade, with its buyers, merchants, brokers, exchanges and futures markets (see Ch. 17), evolved as the intermediary by which the requirements of the textile industry are matched to the supply of different types of cotton, ranging from the short, harsh Comilla cotton produced in Bangladesh to the long, fine and silky Sea Island cotton from the West Indies (Fig. 18.1). As the trade is international, it needs internationally recognized standards of quality, which in turn has led to the definition of the components of quality, and accepted procedures for measuring them.

Lint quality is characteristic of the variety of cotton which is being grown, and the cotton belts of the world grow the varieties most generally suitable to their circumstances: climatic, biological, social and economic. The varieties have been selected by trial and error over the years, and proposals for innovation should be treated with caution. The textile industry of a country may require to import a style of cotton which is not produced locally, and the cotton breeder is often asked to provide a variety with the quality needed to substitute for these imports. This is not simply a matter of obtaining seed from the country of origin of the imports. It will require several seasons of testing to see if the imported variety

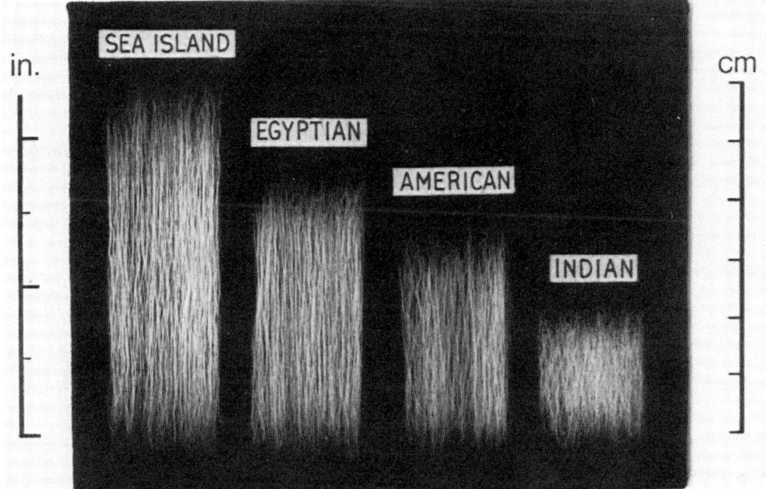

Fig. 18.1 Range of cotton fibre length (International Institute for Cotton).

retains the same qualities in its new environment, and the yield will most probably be less than that of the indigenous variety; a complete new programme of crossing and selection (see Ch. 14) may be necessary. In the field, arrangements will have to be made to keep the new and the indigenous varieties separate at all stages from seed distribution to marketing and ginning (see Chs. 15 and 17). All these factors must be taken into consideration before deciding to go ahead with any import substitution. In extreme cases, it may be best to export the entire crop and import cotton for use by the local textile industry.

Whatever the local circumstances, all those concerned in cotton production should have some knowledge of how their product is disposed of and of the requirements of their customers; for this they must understand the various components of quality. The main criteria for judging quality are grade and staple length, traditionally determined by skilled cotton classers using a standard procedure described in *The Classification of Cotton* (USDA 1980). The skill of the classer can only be learned through a long apprenticeship, but his verdict is widely accepted in the world's cotton markets, and is used in the spinning mill to decide the mixture of bales required to produce a given quality of yarn. The acceptability of laboratory testing devices depends on their agreement with the classer's assessment, but they also measure qualities not determined directly by the classer. Laboratory tests are valuable guides to lint quality, but they all have their limitations, some of which are mentioned in the following paragraphs. The difficulties and pitfalls in their interpretation are discussed in a paper by Lord (1962).

Grade

'Grade is composed of three factors – color, leaf and preparation' (USDA 1980). American Upland cotton is graded in steps from 'ordinary' through 'middling' to 'fair' (see Table 17.3), while American Pima grades are numbered from 1 to 9, grade 1 being the best. Colour and trash content can be measured in the laboratory, but there is no standard test for preparation, a general term covering faults introduced during storage and handling, and by incorrect setting and maintenance of the ginning machinery.

Colour tests involve two measurements, reflectance and yellowness, obtained on the Nickerson-Hunter Cotton Colorimeter; a special diagram is used to relate the two measurements to the official grade standards. The amount of trash is measured by the Shirley Analyser; typical trash or non-lint content of different grades is shown in Table 18.1, and a similar relationship is found between grade and the amount of waste extracted during scutching (picking) and carding in the spinning mill.

Lint length

Seed cotton

There is no absolute value for the lint length of a variety or sample; even on a single seed the lint hairs are not all of the same length, being generally longest at the chalazal or blunt end and shortest at the micropilar end. Add this to the variation from boll to boll, plant to plant and field to field, and the resulting lint will present the range in lengths seen in a Baer diagram (Fig. 18.4).

Table 18.1   *Trash and manufacturing waste*

Grade	American Upland		Grade	American Pima	
	Non-lint content (%)*	*Picker & card waste (%)*		*Non-lint content* (%)*	*Picker & card waste (%)*
Strict middling	1.9	5.2	2	1.9	6.4
Middling	2.3	5.5	3	2.3	6.7
Strict low middling	3.1	6.0	4	3.0	7.4
Low middling	4.4	6.9	5	3.7	8.0
Strict good orinary	5.6	7.7	6	4.7	8.9
Good ordinary	7.2	8.8	7	6.0	10.1
			8	8.4	12.3
			9	9.9	12.9

* Shirley Analyser.
Source: USDA 1982.

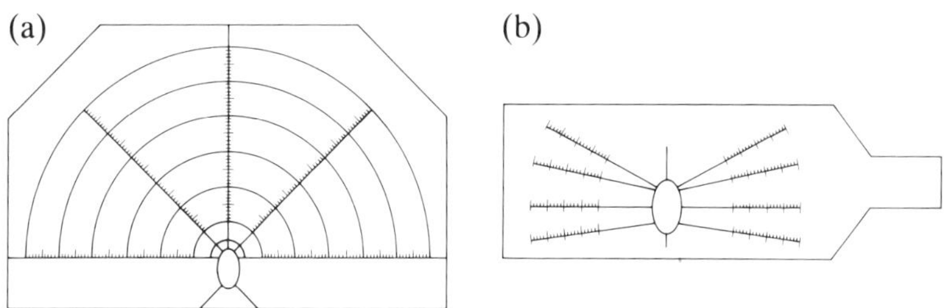

Fig. 18.2   Measurement of lint length in seed cotton (Fielding 1948): (a) lint combed as a halo and measured with a Bailey protractor; (b) lint combed as a butterfly and measured with a Barberton scale. The protractors are made from clear plastic and graduated in mm (half actual size).

Lint length is generally measured on a sample of lint after ginning, but the plant breeder may want an early measurement from a few seeds of a single plant. This can be done by combing out the lint in the form of a halo or butterfly (Fig. 18.2), and measuring the length with a transparent plastic protractor which has a hole at the centre to fit over the seed; an ordinary ruler can be used, but is less convenient. A steel comb is used to tease out and straighten the fibres, which are then pressed flat on a black velvet covered board. Readings are taken at each of the radii calibrated on the protractor. The average of four or five readings on each of five seeds gives a good working measurement for comparing the lint length of single plant selections. It is not easy to decide the exact point to measure, as a number of extra-long hairs must be disregarded, but this does not matter as long as the same standard is used in all measurements. Another source of error is the loss of hairs in combing, which may amount to over half the lint, and almost certainly removes a proportion of the longer fibres.

Hand stapling

Returning to the sample of ginned lint, the cotton classer will first examine it to determine the grade; he will handle the cotton to assess qualities such as 'feel' and 'body', and then proceed to measure the staple length. For this he takes a handful of lint and breaks it in two; from the exposed surface of one half he draws fibres in small pinches between his finger and thumb until he has a tuft of the size he wants; from one end of this tuft he draws out the projecting fibres a few at a time (Fig. 18.3), collecting them on top of each other, straight and parallel, with the ends more or less coinciding; this process is repeated as often as necessary, and the

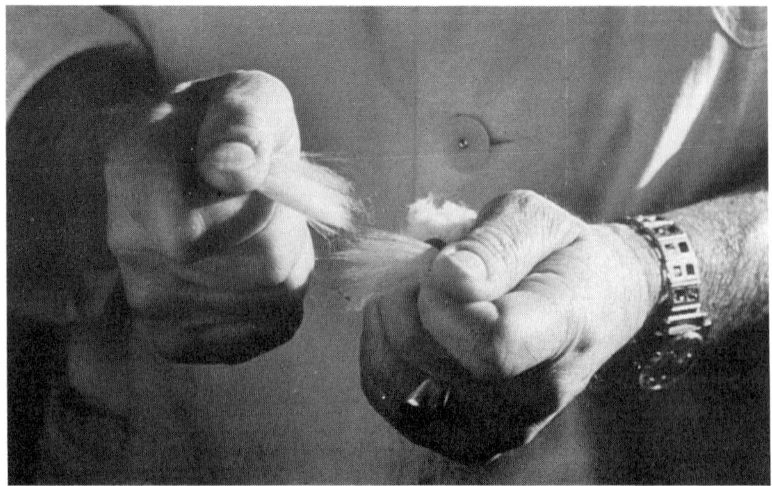

Fig. 18.3   Hand stapling of ginned lint (The Textile Institute).

finished tuft is placed on a dark background, on the coat sleeve or on the back of the hand. The length is judged by eye or measured with a ruler. A classer often will specialize in a certain class of cotton in which he becomes an expert.

Staple lengths are measured to the nearest $\frac{1}{32}$ in. or to the nearest millimetre, but are divided for convenience into five classes as shown in Table 17.4.

Baer diagram

One of the earliest machines used to provide a more precise and objective measurement of length was the Baer Sorter. The apparatus has nine lower combs 5 mm apart and projecting upwards, and three upper combs facing downwards which mesh between the first four lower combs. A sample weighing 15 mg and containing about 3000 hairs, similar to the first tuft drawn by the grader, is drawn and doubled several times until the hairs are mixed, straightened and laid as parallel as possible. This sample is then pressed on to the lower combs at the right-hand end of the sorter with about half the length projecting; a series of tufts are now drawn from the sample with a clamp and placed on the combs at the left-hand end so that the ends of the fibres are all flush with the nearest comb. The sorter is now turned round and the top combs placed in position so as to grip the sample; starting with the longest fibres, tufts of successive lengths are pulled down and placed in order against the baseline marked on a black velvet pad, giving an array of the type shown in Fig. 18.4 (Clegg 1931). The Suter-Webb Fiber Sorter operates in the same way, separating the fibres into different length groups at $\frac{1}{8}$ in. intervals; each length group is accurately weighed, to obtain the

(a)

(b)

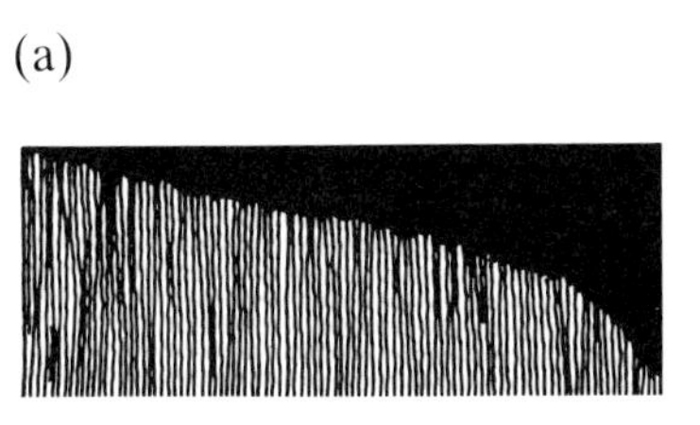
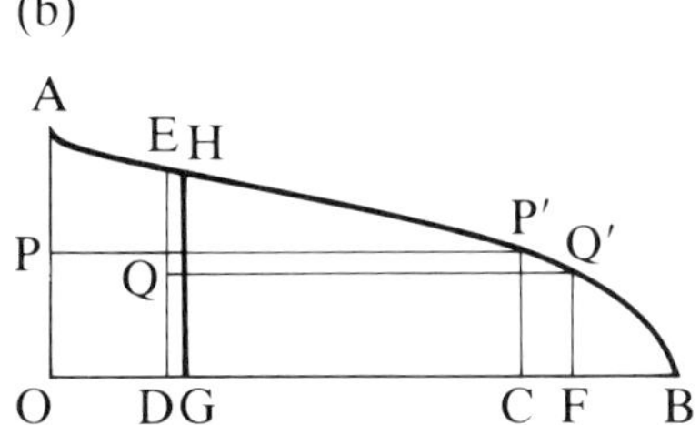

Fig. 18.4   Baer diagram and effective length (Clegg 1931).

(a) Drawing of typical Baer diagram.
(b) Treatment of trace from Baer diagram: OP = ½ OA,
    OD = ¼ OC, DQ = ½ DE, OG = ¼ OF.
    Effective length = GH
    Percentage of short hairs = 100 × FB/OB

length–weight distribution used in calculating various fibre length values.

The most interesting value derived from the Baer diagram is the 'effective length', in that it approximates more closely than other values to the staple length as determined by the classer. For long-staple cotton the values are about the same, but for Upland cotton of $\frac{15}{16}$ to $1\frac{3}{8}$ in. (24 to 35 mm) the effective length is about 10 per cent more than the staple length. The effective length is derived from the Baer diagram as shown in Fig. 18.4. The diagram also provides a measure of the percentage of short hairs (by number), but this may be inaccurate because of the difficulty in sorting the shortest fibres, and may be a measure of breakage rather than inherently short fibre. Short fibres are removed by carding and combing in the spinning mill, and the proportion of short fibre is an indication of the amount of waste in these processes (Clegg 1931).

Unfortunately the production of a Baer diagram is a slow and laborious process, taking about an hour for each sample, less for short-staple cotton but up to $1\frac{1}{2}$ hours for extra-long staple. Modern methods of measuring staple length speed up the process by eliminating the necessity for rearranging the fibres so that they all start from the same baseline.

The tuft method

A compromise between the Baer diagram and hand stapling is provided by the tuft method used in Pakistan and described by Brown (1975b). Tufts are drawn from the exposed surface of a broken handful of lint as in hand stapling, but they are then laid out with the ends overlapping on a piece of paper and rolled into a sliver; this is passed once or twice through a hand-operated draw frame to straighten and parallelize the fibres. A sample from the sliver is placed on a rack of three parallel combs, and the fibres withdrawn a few at a time with a hand clamp; they are laid on a black velvet covered board with the gripped ends of the fibres superimposed. The length of the resulting tuft is measured from this line.

Photoelectric staplers

In the Shirley photoelectric stapler[1] the sample is formed into a flat tuft of straight and parallel fibres laid out on a black velvet covered board (Fig. 18.1). This is passed under two parallel bars of light at right angles to the fibres, and the amount of light reflected from each bar is measured. The reflection is greatest at the centre of the sample and tapers off at each end; the difference in reflection between the two bars of light shows two well-defined peaks where the fibres thin out at each end of the tuft, and the distance between these two peaks measures the fibre length.

The best known photoelectric stapler is the Fibrograph.[2] Test specimens are prepared by the Fibrosampler, which collects the lint on a series of fine hooks and clamps it in a holder. The protruding fibres are combed and brushed to straighten them and to remove loose fibres, and the holder is placed in position in the Fibrograph where it is scanned by a bar of light, the fibres being held straight by a blast of air. The manual and Servo models register the length-frequency distribution of the fibres in the sample, from which the mean length and upper half mean length are obtained graphically; the digital model registers the 2.5 per cent span length, the 50 per cent span length and the 50/2.5 uniformity ratio directly on an electronic display panel. The 2.5 per cent span length is the length equalled or exceeded by 2.5 per cent of the fibres. The upper half mean length and the 2.5 per cent span length are theoretically different, but are essentially the same in their practical application, and are usually closely related to the classer's staple length (USDA 1982).

Fibre fineness

It is possible to measure fineness directly as the mean diameter of a number of fibres in cross section under a microscope, but the irregular shape of a collapsed lint hair (see Fig. 4.14) presents several problems. A more common measurement is the mean hair weight per unit length, which is found by cutting a standard length from a sample bundle of hairs, weighing the cut lengths and counting the fibres in the bundle. The bundle of hairs is held in a clamp of known width, say 1 cm, and the protruding ends are trimmed off; this not only gives a standard length, but eliminates the ends of the fibres, which tend to be tapered with less secondary thickening than the centre portion. In the USA bundles of uncut fibres are taken from each length of the Suter-Webb array, omitting the two shortest classes, and each bundle is weighed and counted separately.

The traditional units in the USA are micrograms per inch, and in Europe micrograms per centimetre; these are being replaced by 'tex' units of linear density, or mass per unit length. Tex is measured in grams per kilometre of fibre or yarn, and fibre fineness is usually expressed as millitex, or micrograms per metre (Ramey 1982).

These methods of measuring fineness are rarely used, as a much quicker and reliable alternative is available. This depends on the principle that the resistance to the flow of air through a bundle of fibres varies in proportion to the surface area of the fibres; a standard weight of cotton is compressed into a plug of standard

Table 18.2   *Micronaire and causticaire readings*

Description	Micronaire reading	Causticaire maturity index (%)
Very low	Below 3.5	Below 72
Low	3.5–3.9	72–75
Average	4.0–4.4	76–79
High	4.5–5.0	80–83
Very high	Above 5.0	Above 83

Source: USDA 1982.

volume, and the pressure required to force air through this plug is measured. Several instruments have been designed to operate on this principle, of which the most important is the Micronaire; the original calibration was designed to measure gravimetric fineness in μg/per in.; although it is now known to measure surface area, the original units have been retained as 'micronaire values' or 'micronaire scale units', and have been generally accepted in the cotton trade (Table 18.2).

The samples used for the micronaire test must be clean and well opened, and it is advisable to put the sample through a specially designed blender, but the test is not especially sensitive to changes in preparation techniques and atmospheric conditions (Ramey 1982).

Fibre maturity

Immature fibres weigh less and offer less resistance to the flow of air than do mature fibres with well-developed secondary thickening, and measures of fineness do not discriminate between intrinsic fineness and immaturity. Only when dealing with a single variety can it be assumed that local and seasonal changes in the micronaire values are mainly an indication of changes in maturity. When different varieties are involved, it is quite possible for an immature sample of a coarse cotton to appear to be as fine as a mature sample of a fine variety. Tests of fineness must therefore be supplemented by tests of maturity, not only in breeding work but also in the spinning mill, where immature cotton will produce neps, weak thread and irregular dying.

The most reliable method of measuring maturity is that of counting mature, immature and dead fibres in a sample treated with 18 per cent caustic soda. This is similar to the mercerizing process, and causes the cellulose in the fibres to swell; in fully mature hairs (N) the lumen disappears and the fibre looks like a solid rod when viewed through a microscope; immature (I) and dead hairs (D) remain thin-walled and convoluted (Fig. 18.5). As a measure of maturity, the Shirley Institute introduced the concept of 'maturity ratio' (M), which is calculated from the formula:

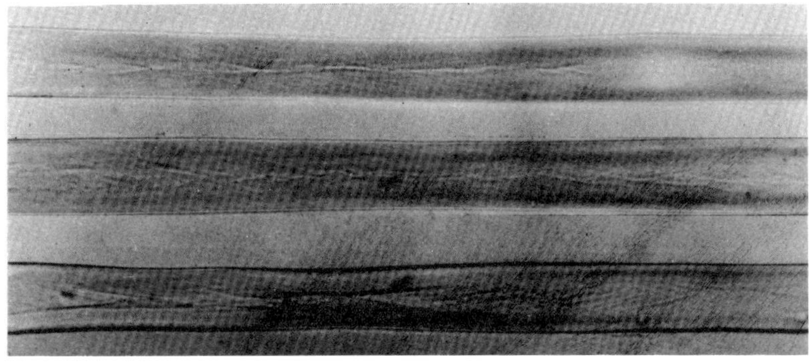

a

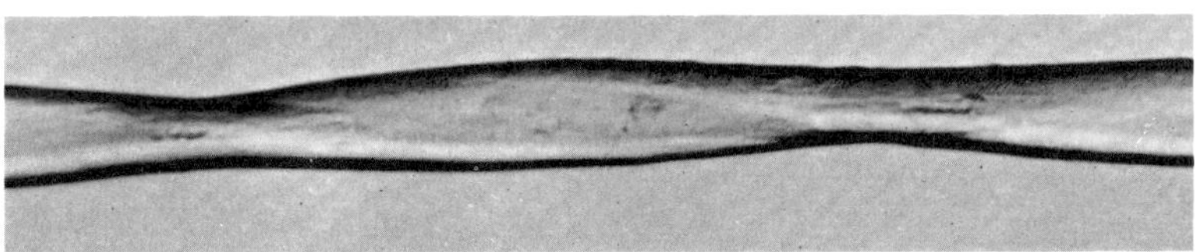

b

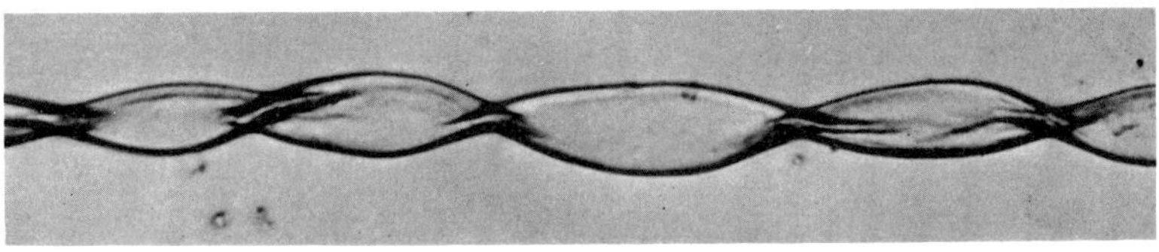

c

Fig. 18.5 Microphotograph of: (a) mature; (b) immature; and (c) dead hairs, all treated with caustic soda (British Standards Institution: extracts from BS 3085: 1981).

$$M = (\%N - \%D)/200 + 0.70$$

An estimate can now be made of what the fineness of a sample would be if the cotton were fully mature by adjusting the measured fineness (H) to allow for the maturity of the sample. A maturity ratio of unity is taken as standard, and the 'standard hair weight' (H_S) is calculated from the formula $H_S = H/M$; this is a measure of the intrinsic fineness of the cotton. Standard fibre weight and maturity are related to the micronaire value by the formula:

$$MH = M^2H_S = 3.86\,X^2 + 18.16\,X + 13$$

where M = maturity ratio, H_S = standard fibre weight and X = micronaire value (Lord 1981). Figure 18.6 illustrates the relationship, and how it can be used to compare the lint of different varieties; the difference between the three main Uganda varieties of that time can be clearly seen (Munro 1966).

Maturity ratio	Standard fibre weight																		
	130	135	140	145	150	155	160	165	170	175	180	185	190	195	200	205	210	215	220
0.70	2.1	2.2	2.2	2.3	2.3	2.4	2.4	2.5	2.5	2.6	2.6	2.7	2.8	2.8	2.9	3.0	3.0	3.1	3.1
0.72	2.2	2.2	2.3	2.3	2.4	2.5	2.5	2.6	2.6	2.7	2.8	2.8	2.9	2.9	3.0	3.1	3.2	3.2	3.3
0.74	2.3	2.3	2.4	2.4	2.5	2.6	2.6	2.7	2.8	2.8	2.9	3.0	3.0	3.1	3.2	3.2	3.3	3.3	3.4
0.76	2.4	2.4	2.5	2.6	2.6	2.7	2.8	2.8	2.9	3.0	3.0	3.1	3.2	3.3	3.3	3.4	3.5	3.5	3.6
0.78	2.5	2.5	2.6	2.6	2.7	2.8	2.9	2.9	3.0	3.1	3.2	3.2	3.3	3.4	3.4	3.5	3.6	3.6	3.7
0.80	2.6	2.6	2.7	2.8	2.8	2.9	3.0	3.1	3.1	3.2	3.3	3.3	3.4	3.5	3.6	3.6	3.7	3.8	3.8
0.82	2.7	2.7	2.8	2.9	3.0	3.0	3.1	3.2	3.3	3.3	3.4	3.5	3.6	3.7	3.7	3.8	3.9	3.9	4.0
0.84	2.8	2.8	2.9	3.0	3.1	3.1	3.2	3.3	3.4	3.5	3.5	3.6	3.7	3.8	3.9	3.9	4.0	4.1	4.1
0.86	2.9	2.9	3.0	3.1	3.2	3.3	3.4	3.4	3.5	3.6	3.7	3.8	3.9	3.9	4.0	4.1	4.2	4.2	4.3
0.88	3.0	3.1	3.1	3.2	3.3	3.4	3.5	3.6	3.7	3.7	3.8	3.9	4.0	4.1	4.1	4.2	4.3	4.4	4.5
0.90	3.1	3.2	3.2	3.3	3.4	3.5	3.6	3.7	3.8	3.9	4.0	4.1	4.1	4.2	4.3	4.4	4.5	4.5	4.6
0.92	3.2	3.3	3.4	3.5	3.5	3.6	3.7	3.8	3.9	4.0	4.1	4.2	4.3	4.4	4.4	4.5	4.6	4.7	4.8
0.94	3.3	3.4	3.5	3.6	3.7	3.8	3.9	4.0	4.1	4.1	4.2	4.3	4.4	4.5	4.6	4.7	4.7	4.8	4.9
0.96	3.4	3.5	3.6	3.7	3.8	3.9	4.0	4.1	4.2	4.3	4.4	4.5	4.5	4.6	4.7	4.8	4.9	5.0	5.1
0.98	3.5	3.6	3.7	3.8	3.9	4.0	4.1	4.2	4.3	4.3	4.4	4.5	4.6	4.6	4.7	4.8	4.9	5.1	5.2
1.00	3.6	3.7	3.8	3.9	4.0	4.1	4.2	4.3	4.4	4.5	4.6	4.7	4.8	4.9	5.0	5.1	5.2	5.3	5.4

(Boxed regions within the table are labelled UPA, BP52 and BPA.)

Fig. 18.6 Micronaire values (in body of table) associated with maturity ratio and standard fibre weight (Munro 1966).

For a much quicker but less accurate test, the micronaire principle can be used to measure maturity. This is done by measuring the air pressure required to maintain a constant flow through the same standard sample when compressed to two different volumes, the difference between the two readings varying with the maturity. The Arealometer, invented in 1951, was the first machine to use this principle, but it is no longer manufactured; more recently, a project sponsored by the International Institute for Cotton (IIC) and the Shirley Institute has developed the IIC-Shirley Fineness/Maturity Tester[1], which has been recognized in Japan and the USA for maturity tests (Lord 1981).

The Causticaire test is used by the cotton testing service in the USA (USDA 1982). This measures the airflow through specimens of lint before and after mercerizing with caustic soda. The readings are on a special scale proportional to the flow of air, not in micronaire units; a maturity index and the fibre fineness in micrograms per inch are derived from the two readings (Table 18.2).

Fibre strength

The inherent strength of individual cotton fibres is an important factor in the strength of the thread spun from them, and strong fibres usually give less trouble in processing. There are two instru-

Table 18.3   *Fibre strength*

| Description | Medium staple (³²/₃₂–³⁵/₃₂ in.) | | Extra-long staple (⁴¹/₃₂ in. and longer) | |
	Zero gauge '000 psi	¹/₈ in. g per tex	Zero gauge '000 psi	¹/₈ in. g per tex
Very low	70–76	16–18	93–96	27–29
Low	77–83	19–21	97–100	30–32
Average	84–90	22–24	101–104	33–35
High	91–97	25–27	105–108	36–38
Very high	98–104	28–30	109–112	39–41

Source: USDA 1982.

ments commonly used to measure fibre strength – the Pressley and the Stelometer. In both of these the strength is measured by spreading a bundle of parallel fibres across two clamps, so that each clamp grips one end of the bundle. Force is applied to separate the clamps, and gradually increased until the bundle breaks; the broken halves are collected and weighed.

In the Pressley machine the clamps are usually set at 'zero gauge', that is, the clamps are touching each other at the start of the test, and the results are reported in thousands of pounds per square inch (Table 18.3). In the Stelometer, there is a gap of one-eighth of an inch between the clamps, '$\frac{1}{8}$in. gauge', and the results are reported in grams per tex. Both instruments can, however, be set for zero and $\frac{1}{8}$in. gauge tests, and the units are convertible; at zero gauge, the reading in thousand pounds per square inch (S_1) is roughly twice the grams per tex (S_3), the exact formula given in USDA (1982) being:

$$S_3 = S_1 \times 0.496$$

The zero gauge Pressley test is more commonly used in the cotton trade in America, while the $\frac{1}{8}$in. gauge Stelometer test is more widely used in Europe. The results of the $\frac{1}{8}$in. gauge tests are highly correlated with fibre length (Table 18.3), and more closely related to the results of spinning tests than are those of the zero gauge. The Stelometer at $\frac{1}{8}$in. gauge measures fibre elongation as well as fibre strength.

Fibre test reports

The report from a fibre testing laboratory takes the form shown in Table 18.4, listing the values found for each sample of the measured fibre characters which have been described briefly in the preceding pages. The table shows a range of cotton types from Sea Island to the Old World diploid cottons of India.

Table 18.4 *Fibre characters of selected cottons*

Description of sample	Effective length (¹/₃₂ in.)	Maturity ratio	Micronaire value	Fibre weight per cm Mean	Fibre weight per cm Standard	Bundle strength (g per tex) at ¹/₈ in.
Sea Island						
St Vincent V 135	62	0.92	2.9	108	117	30
Egyptian						
Karnak EX	49	0.97	3.3	116	120	30
Sakel GS	50.5	1.01	3.9	141	141	29
Ashmouni FG	38	0.97	4.6	187	193	23
American Upland						
USA 1⅛ SM/GM	41	0.90	4.4	185	205	20
Uganda BP 52	41	0.84	3.3	139	166	21
Tanzania CLA	38	0.82	3.5	153	186	20
Tanzania MZA	38	0.83	3.8	166	201	19
Nigeria NA 1	36	0.84	4.0	177	211	18
USA ¹⁵/₁₆ M	33	0.90	5.0	219	243	19
Indian						
Bengals	25	1.05	7.3	336	320	17

Source: Lord 1961.

Fibre tests are carried out at 70 °F (21 °C) and 65 per cent relative humidity, and the samples are conditioned in this standard atmosphere for about 24 hours before testing. The sample is thoroughly blended before sub-samples are taken for each test. In order to ensure that different testing laboratories all work to the same standards, the USDA Agricultural Marketing Service can supply calibration samples[3] of the official cotton standards for grade and staple (see Tables 17.3 and 17.4), the international standards for micronaire, Pressley 'O' gauge and fibrograph length, and calibration cottons for other fibre tests.

Spinning tests

Spinning

Anyone can produce a short length of thread by drawing a few fibres from a handful of cotton lint with the finger and thumb, twisting the ends of the fibres together and pulling at the same time. These are the actions common to all spinning, from the hand-held spindle through the spinning wheel to the modern spinning mill with its complex machinery.

The most skilful spinner cannot produce fine thread from a short, coarse cotton, and the qualities of the raw cotton entering the mill must match the fineness of the finished yarn, which is usually measured by the 'count'. The finer the thread the greater the length produced from a pound of cotton, and the count is the number of

Table 18.5   *Classification of yarn counts*

Class	Counts of
Thick	1–10
Coarse	11–22
Medium	23–46
Medium fine	47–80
Fine	81 and over

Source: Ruston 1954.

hanks of 840 yards which go to make up a pound of yarn. A particular mill might be described as producing forties counts (40s) from medium-staple cotton; yarns may be divided into five classes according to the count as shown in Table 18.5. The metric yarn count is the number of kilometers making up 1 kilogram of yarn; this metric count is 1.693 times the English count (Volkart 1984).

The main processes in a spinning mill are opening, carding, drawing, doubling and spinning. A number of bales are chosen to provide a suitable mix, and cotton is taken from each of them in turn by the picker or scutcher; this is fed into the opener, which loosens up the compressed lint and mixes the cotton thoroughly, producing a blanket or lap of a suitable width to feed into the carding machine. In this the fibres are teased out between two mats of fine combs, leaving them more or less parallel and separating out some of the short fibre and impurities as card waste. Carding may be followed by combing, an extra process used only in the production of very fine yarns to remove nearly all the short fibres which are unsuitable for fine spinning; combing increases the loss of fibre as waste, but this is balanced by the improvement in the quality of the finished yarn.

From the carding and combing machines, the fibres are collected into a sliver, which is a rope of loosely packed fibres; the fibres in the sliver are progressively straightened and laid parallel by alternately drawing and doubling. In drawing, the sliver passes through two pairs of rollers, the second pair moving faster than the first; this increases the length and reduces the diameter of the sliver. The original diameter is restored by laying two or more slivers together (doubling) before feeding them again into a drawframe; drawing and doubling can be repeated until the sliver is ready for the final drawing; this leaves a fine sliver or roving which is wound on to a bobbin ready for transporting and fitting to the spinning frame, where the necessary twist is imparted to the thread. The machines originally invented for this process, such as the spinning jenny and the spinning mule, are now obsolete, and most mills are equipped for ring spinning, in which the twist is imparted by passing the thread through a metal ring which travels round the bobbin on a

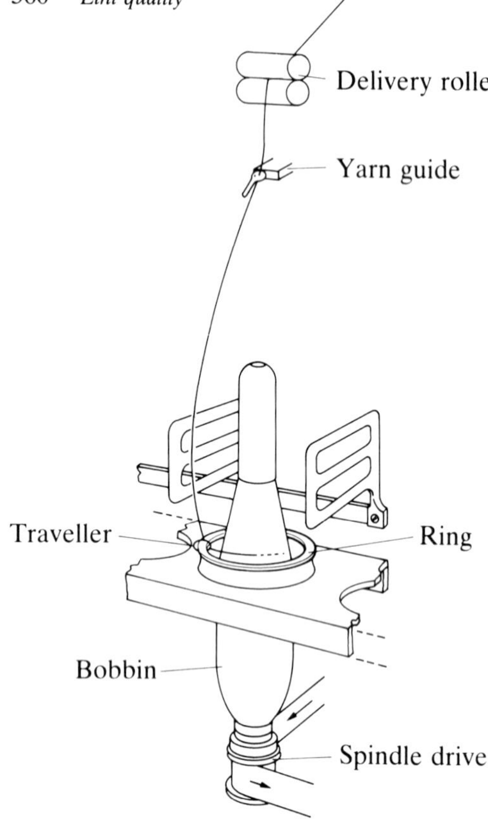

Fig. 18.7   The ring frame spindle.

circular track (Fig. 18.7). Spindle speeds have been progressively increased until the modern ring spindle has reached a stage of development where little further advance can be expected.

Recently a completely new method of yarn production, known as open-end, rotor or break spinning, has been introduced; the first commercial machines were produced in Czechoslovakia in 1965. In this system the spindle is replaced by a high-speed rotor (Fig. 18.8) revolving at some 45,000 rpm – over three times the speed achieved in ring spinning. Clean fibres (usually in the form of a sliver) are fed to an opener where they are further cleaned, separated and conveyed individually in an airstream to the rotor. A fibre ring forms on the inner peripheral surface of the rotor; to start spinning a seed yarn is introduced into the rotor to pick up the fibres in the ring. As the fibres are withdrawn, they are twisted into yarn by the rotation of the rotor, and wound on to a suitable package (IIC 1974).

The present range of equipment is designed to spin coarse and medium yarns from 4s to 28s count; in this range there is a

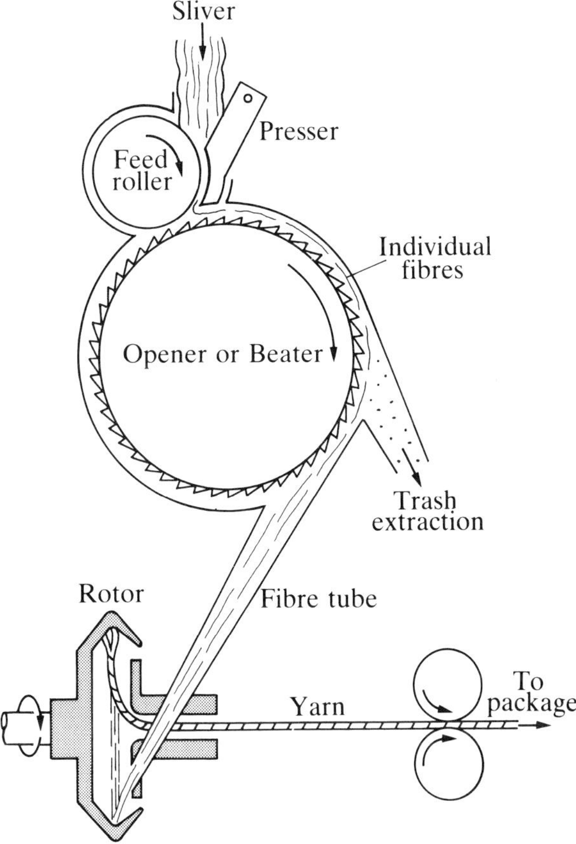

Fig. 18.8 Principles of open-end spinning (International Institute for Cotton).

considerable saving on power costs as well as a threefold increase in the rate of yarn production. The yarn is on average about 15 to 20 per cent weaker than comparable ring-spun yarns, but it is more regular with fewer weak places, and has better resistance to abrasion; it also differs in its behaviour in the processing of yarn and fabrics. The fibre properties required in raw cotton for break spinning are different from those for ring spinning, but have not yet been clearly defined: they are probably less exacting, except in regard to trash and dirt. The cotton presented to the rotor must be as clean as possible, and is subjected to very thorough precleaning treatment, some systems having a trash extractor fitted to each rotor.

More detailed descriptions of the various processes in the spinning mill can be found in specialized publications, such as the textbook by Lord (1961).

Small-scale spinning tests

Testing a new variety of cotton in a commercial spinning mill needs a sample measured in bales rather than pounds or kilograms; highly integrated modern mills require a very large sample to give an adequate test, although four or five bales will be sufficient in a mill of traditional design. It also requires the cooperation of the spinner, who has to put up with the disruption of the normal flow of cotton through the mill. It is only in the final stages of the development of a new variety that such a test is practicable.

The Shirley Institute has brought the spinning of small samples of lint to a high pitch of perfection in its small-scale and large-scale spinning tests, requiring only 50 g and 5 kg respectively (Lord 1961); it tests thousands of samples every year on behalf of spinners and plant breeders, and many overseas laboratories have been equipped with the machinery designed by the Institute. The characters measured in fibre tests give a good indication of the quality of the lint, but do not always predict accurately its behaviour in the spinning mill. The spinning test uses machinery similar to that of a commercial mill, and the yarn produced is very similar in character to what will be produced on a commercial scale.

The results of the fibre tests are used to indicate the count to which the cotton should be spun, the twist factor, and whether it should be combed before spinning. After the appropriate settings have been made, the sample is passed through the small-scale mill and the spun yarn is tested for strength and appearance. The strength is measured on a 'lea tester', in which a lea (120 yards or 110 m) of yarn is wrapped round a special reel and the loop of yarn so formed is placed over two hooks; the hooks move apart at a

Table 18.6 *Range of highest standard counts on various cotton samples*

Cotton	Highest standard counts
Sea Island	
St Vincent V 135	148–176 combed
Egyptian	
Sudan GS	108–122 combed
Ashmouni	34–41 carded
American Upland type	
Uganda BP 52	59–65 carded
Tanzania CLA	56–62 carded
Tanzania MZA	37–47 carded
Nigeria NA 1	34–41 carded
USA $^{15}/_{16}$ M	15–21 carded
USA $1^{1}/_{16}$ M	33–60 carded

Source: Ruston 1954

Fig. 18.9 Wrap board for judging yarn appearance. (Courtaulds Central Testing Department).

constant speed, gradually increasing the load until the loop of yarn finally breaks. The breaking load adjusted for the count of the sample is reported as the lea count strength product. By spinning the same cotton at two different counts, it is possible to estimate the count at which it would give a product of 2000 carded or 2250 combed; this is called the highest standard count. The highest standard counts of various cotton types are shown in Table 18.6 and a comparison with Table 18.4 shows how this count is related to the fibre characters of these types, especially the length, strength and fineness.

Yarn appearance is judged by winding the thread round a black board at regularly spaced intervals (Fig. 18.9); it can then be compared with the ASTM international standards[4], photographs of yarn of standard grades wound on boards in the same way. The most important features to look for are the regularity of the yarn and the frequency of 'neps'. Neps are specks on the yarn consisting of a tangle of fibres; they may be 'tangled neps' consisting of immature fibres which have not been straightened out, or 'seed coat neps' which are bundles of fibres, usually immature, attached to small fragments of seed coat. Tangled neps are usually associated with a low maturity ratio, that is, a high proportion of immature fibres. Seed coat neps may occur even in a mature sample of cotton, and arise from a structural weakness in the seed coat at the chalazal end of the seed, allowing a fragment to break off during ginning. This defect is heritable, and can be reduced or eliminated by selection and breeding.

Motes are immature undeveloped seeds which have survived the ginnery cleaning processes and the carding, where they should have been removed. They appear to be caused by environmental factors, and there is little point in trying to reduce mote number by selection (Pearson 1949, Hughes 1968).

Neps and motes do not affect the strength of the yarn, but show up as defects in the finished cloth; in dyed fabrics they appear as spots of a different colour, since mature and immature fibres take up different amounts of dye. Unevenness in the thickness of the yarn usually means weakness, as the yarn breaks more easily at the thinner segments.

---

1. Shirley Developments Ltd, Didsbury, Manchester, England.
2. Spinlab, Knoxville, Tennessee, USA.
3. Standards Section, Cotton Division, AMS, USDA, Box 17723, Memphis, TN 38122, USA.
4. ASTM (American Society for Testing and Materials), 1916 Race Street, Philadelphia, PA 19103, USA; or in Europe, American Technical Publishers Ltd, 68a Wilbury Way, Hitchin, Herts SG4 OTP, England.

Chapter 19

Cotton and man-made fibres

Textile fibres

In the industrial world the main competitor of cotton is the new range of man-made fibres, which has increased spectacularly since the end of the second world war (Fig. 19.1). Wool production has grown comparatively slowly, and is little more than one-tenth of that of cotton; the production of silk and linen are comparatively unimportant statistically (Table 19.1).

The first man-made fibres were produced as a substitute for silk rather than cotton, and were known as 'artificial silk'. The silkworm extrudes a protein gum through two holes in its lower lip; this gum dries and hardens on contact with air, producing a continuous double filament 500 metres or more in length, which is used to form the cocoon in which the silk worm pupates. Man-made fibres are produced in much the same way, by forcing a viscous liquid through

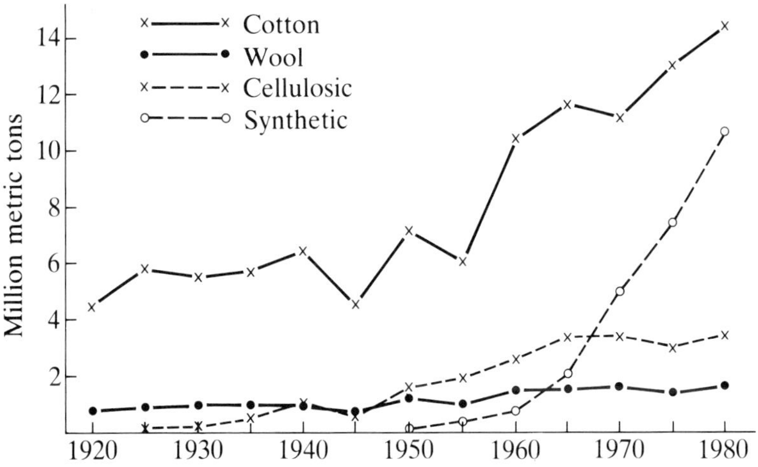

Fig. 19.1 Production of textile fibres in the world, 1920–80 (various sources).

Table 19.1　*World fibre production*

Type of fibre	5-year average 1976–80	
	thousand metric tons	*% of world total*
Cellulosic fibres		
Yarn and monofilaments	1,173	4
Staple and tow	2,112	7
Synthetic fibres (except olefin)		
Yarn and monofilaments	4,533	16
Staple and tow	5,239	19
	13.057	46
Natural fibres		
Raw cotton	13,582	48
Raw wool	1,531	5
Raw silk	52	0.2
Total natural	15,165	54
World total	28,222	100

Source: Textile Organon, June 1983.

holes in the head of a 'spinneret' (Fig. 19.2). The extruded filament is solidified in one of three ways, depending on how the liquid is prepared: if the polymer (see below) is in an alkaline solution, it is hardened by extruding it into an acid bath (Fig. 19.3), and the process is called 'wet spinning'; if it is dissolved in a volatile solvent, it is extruded into warm air where the solvent evaporates, called 'dry-spinning'; if the polymer is liquefied by melting and extruded into cold air to solidify it, the process is called 'melt spinning'.

Two or more filaments may be spun or 'thrown' with a slight twist into continuous lengths, using a process similar to the throwing of silk, to produce 'filament yarns'; or the filaments may be chopped up into designated staple lengths, called 'staple fibre', and then spun into yarn using conventional cotton or wool machinery; 'spun silk' is made in the same way from waste filaments from all stages in the production of silk. The length of the staple can be varied as required to blend conveniently with natural or other staple fibres.

The advantages common to all man-made fibres are their uniformity, freedom from waste, and relative stability in supply and price. The first two are inherent in the controlled conditions of manufacture, but regularity of supply and price is not always in favour of man-made fibres. The wider fluctuations in the price of cotton means that it will at times be cheaper than man-made fibres;

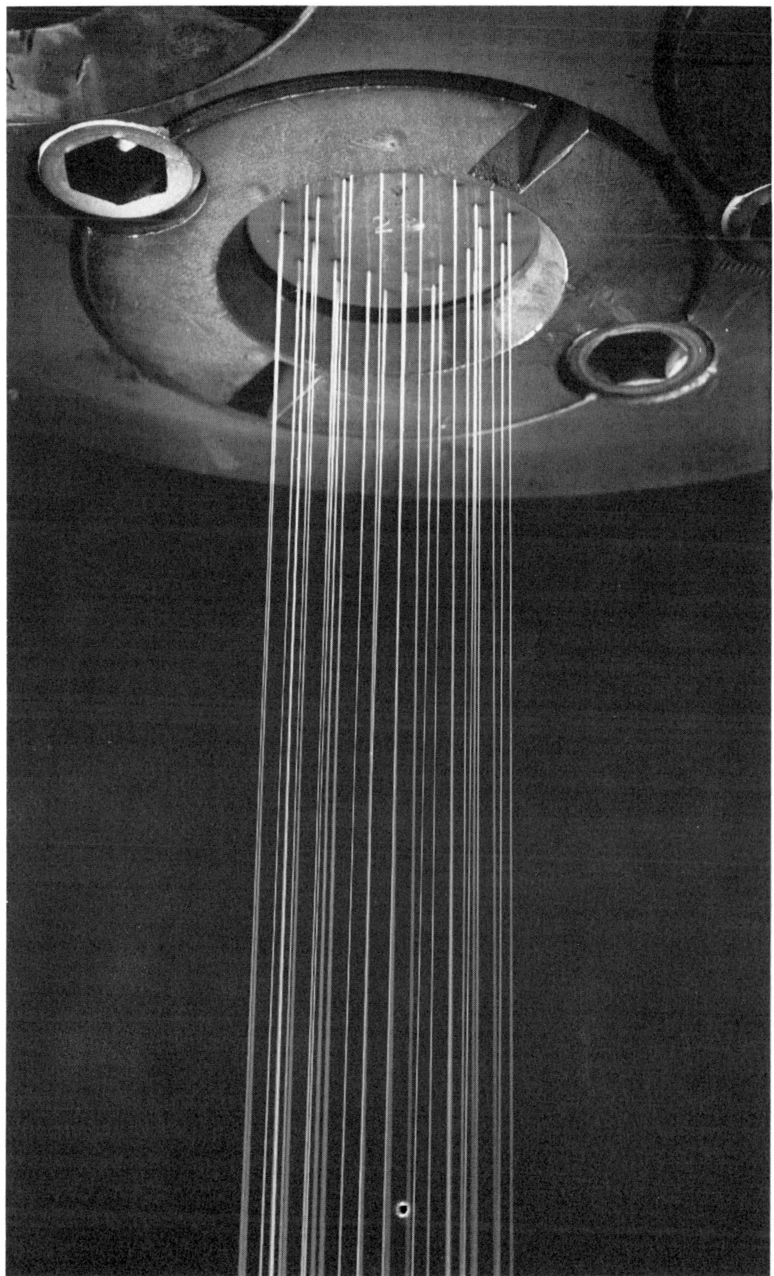

Fig. 19.2 Spinneret extruding 28 filaments of 100 denier nylon (enlarged) (Courtaulds PLC).

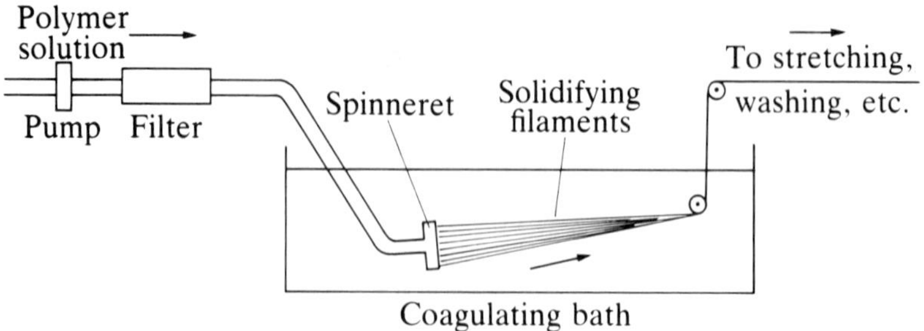

Fig. 19.3   Wet spinning of viscose rayon: (a) pump; (b) filter; (c) acid bath; (d) spinneret; (e) filaments; (f) pulley; (g) to stretching, washing and other processes.

the raw materials of the latter, mainly wood pulp, oil, coal and natural gas, are also subject to unforeseen price changes. The price of the fibre is important, but is not a major selling point, as it is only a small percentage of the cost of the finished article; the material in a cotton shirt weighs less than half a pound (140–200 g), and costs about 25 pence or 35 cents as raw cotton, and about twice as much in the form or spun yarn.

The demand for textiles varies from year to year, and changing fashions and consumer preferences influence the demand for different types of fibre, but the supply of man-made fibres cannot be adapted quickly to changing circumstances. It requires capital investment and industrial plant, and it may be several years before the necessary finance and construction are completed and production can start. If the demand falls off, a company is loath to abandon its investment, and will do all it can to keep the plant in operation. The cotton grower, on the other hand, can substantially decrease or increase the cotton area within a year by switching to or from alternative annual crops; both large-scale and peasant growers react quickly to changes in price. In the long term, shortage of suitable land may limit the production of cotton; there is no such limitation on the production of man-made fibres (Burkitt 1981).

Polymers

Both natural and man-made fibres are made up of large molecules called 'polymers', in which smaller molecules called 'monomers' are repeatedly linked together in a regular pattern. The arrangement of the monomers in the polymer affects its physical properties, fibrous properties being the result of a thread-like structure in which the basic groups are arranged in long chains, possibly with some branching and cross-linkage.

The single groups or monomers can be built up into polymers of any required size (n). An example of polymerization is the formation of polythene from ethylene, which can be represented as follows:

$$(CH_2=Ch_2) + (CH_2=CH_2) + \ldots \rightarrow$$
$$\text{ethylene}$$

$$[\,-CH_2-Ch_2-CH_2-Ch_2-\,]_n$$
$$\text{polythene}$$

Cellulose is formed when glucose molecules are polymerized with the elimination of water:

$$n\ C_6H_{12}O_6 \rightarrow \qquad [\,-C_6H_{10}O_5-\,]_n + n\ H_2O$$
$$\text{glucose} \qquad\qquad \text{cellulose} \qquad \text{water}$$

The cellulose in cotton is a polymer formed from about 10,000 glucose units; wool is a long chain polymer built up from amide groups into a protein (Cook 1980).

Early man-made fibres

The first artificial silk was produced by Count Chardonnet in France in 1870, by imitating the silkworm; he prepared a viscous liquid from mulberry leaves and extruded an artificial fibre through an early form of spinneret. Later attempts to produce artificial silk were based on pure cellulose, but although lustrous filaments with a superficial resemblance to silk were produced, they had little of the warmth, resilience and beauty of natural silk. Gradually it was realized that the future of these materials lay in the development of new products in their own right, and not as inferior copies of silk.

Towards the end of the nineteenth century various artificial fibres were produced from cellulose: cellulose nitrate by Count Chardonnet in 1884; the Bemberg or cuprammonium process in 1897; cellulose xanthate patented in 1892 and produced commercially by Courtaulds in 1904; cellulose acetate patented by Dr Dreyfus in 1911 and produced commercially from 1921. During the 1920s work was started on the use of staple fibres, as opposed to the filament yarns which had been produced hitherto. The term 'rayon' replaced artificial silk to give a separate identity to the cellulose fibres, and included all the regenerated cellulose fibres but not the cellulose acetates; rayon has now been superseded by the term 'viscose' (Thomson 1978). A recent classification of the man-made textile fibres is given in Table 19.2, and the relative importance of man-made and natural textile fibres in recent years is shown in Table 19.1.

Regenerated fibres

Most of the regenerated fibres are made from natural cellulose, although a few protein fibres are still produced. The pure cellulose

Table 19.2   *Classification of man-made fibres*

General name	Chemical composition	Some trade names
Cellulosic fibres		
Viscose	Cellulose	Evlan, Durafil, Fibro, Sarille, Viloft, S.I. Fibre (Super inflated viscose)
Acetate	Cellulose acetate	Dicel
Triacetate	Cellulose acetate	Tricel
	Cellulose acetate and nylon	Tricelon
	Cellulose acetate and polyester	Tricelesta
Synthetic fibres		
Acrylic	Polyacrylonitrile	Courtelle, Learcyl
Elastane	Polyurethane	Lycra
Modacrylic	Polyacrylonitrile and vinylidene chloride	Teklan
Nylon	Polyamide	Bri-nylon, Buclon, Counterstat, Enkalon, Limbaki, Pavanne, Tactel, Timbrelle, Qazul, Quintesse
Polyester	Polyester	Limbaki, Mitrelle, Superloft, Terinda, Terylene, Trevira
Nylon/polyester	Polyamide and polyester	Parafil, Paraweb
Polyolefin	Polyethylene Polypropylene	

Source: British Man-Made Fibres Federation pers. comm. 1985.

from cotton linters was used originally, but the main source is now wood pulp, which contains 50 per cent cellulose and is a cheaper source. The most widely used and cheapest form of man-made fibre is viscose rayon, the modern form of one of the earliest types of artificial silk. The spinning solution is prepared by the reaction of cellulose with caustic soda and carbon disuphide, which produces cellulose xanthate; this in turn is dissolved in caustic soda to give a viscous solution, or 'viscose', which is extruded into an aqueous solution of sulphuric acid and sodium sulphate (Fig. 19.3).

Viscose rayon is reasonably strong, between wool and cotton, when dry, but loses half its strength when wet and can easily be overstressed; the use of heat, moisture and chemical treatments has to be carefully controlled during the processing of fabrics. It is not so resilient as cotton, but staple fibres can be crimped during manufacture; it is highly absorbent of moisture, and has a similar affinity for dye to cotton. In general it is a useful low-priced alternative to cotton and wool, and the many types of staple which have been developed are widely used in blends with other fibres.

In the 'cuprammonium' process, the cellulose is mixed with

aqueous ammonia, copper sulphate and caustic soda. The spinning solution is extruded into water, which produces a very slow coagulation during which the filaments are greatly extended before being hardened in an acid bath; this produces a very fine filament although the holes in the spinneret are larger than those used for viscose. Cuprammonium rayon is similar in its properties to viscose, but the recovery of copper and ammonia is expensive and the fineness of the filaments is its main advantage. Its popularity as a substitute for silk in stockings and quality rayon goods has now been lost to the synthetic nylons.

Various modified forms of rayon have special properties, such as chemical crimp and extra strength; the strength is achieved at the expense of elasticity and comfort in clothing, but is suitable for industrial uses such as reinforcing fabrics for tyres and conveyor belts. 'Polynosic' or 'Modal' rayons have a wet strength about equal to that of cotton, and have properties more akin to cotton than any other man-made fibres; in handle and appearance the fibre is similar to a good quality mercerized cotton, and it can be processed, dyed and finished alone or in mixture with cotton without the modifications necessary for some cellulosic fibres (Miller 1973).

In the 'acetate' process, the cellulose is converted into cellulose acetate, which remains chemically unchanged when extruded; acetate fibre is thus chemically different from cellulose and has quite different properties. The original triacetate is usually reduced to diacetate by adding small quantities of water, so that it can be dissolved in acetone and extruded into warm air for dry spinning. Triacetate can be dissolved in methylene chloride and dry spun, but the solvent and its recovery are more expensive than acetone. Cellulose acetate is not as strong as rayon nor as resistant to abrasion, but it is more resilient and handles like silk; it is naturally crease-resistant, but has a low melting point and must be handled at temperatures well below the boiling point of water. Its dye affinity is quite different from that of viscose or cotton, and special 'disperse' dyes are needed.

Regenerated protein fibres have been made from milk casein, groundnuts, soya beans and maize. They have some of the properties of wool, but when wet their strength is negligible, and very little of this type of fibre is now produced.

Synthetic fibres

In the regenerated fibres discussed above, the regeneration tends to degrade or shorten the polymer chains of the original cellulose or other primary material. In contrast, the production of synthetic fibres consists of building up long chain polymers from a new range of unit molecules, many of which are themselves synthesized from simpler chemicals. The discovery of the first synthetic fibre, nylon, was made by a research team of the American firm of Du Pont in 1934 after six years of work on polymers; commercial manufacture began in 1939, and a similar material was developed by I.G. Farben in Germany during the Second World War.

Polyamides or nylons

The polymers used for making nylon are known chemically as polyamides. Benzine or phenol, derived from bituminous coal, are combined with an amine to produce a nylon salt or amide which is polymerized by heating under the right conditions. The polymer melts at 250 °C and filaments are produced by melt spinning. After cooling, the filament is stretched to several times its original length ('cold drawing'), which greatly increases its strength and elasticity.

Nylon exhibits some unusual properties which distinguished it from all other fibres which existed at the time of its discovery. It combines strength, elasticity and toughness in a much higher degree, exceeding silk in all these properties; the fineness of filament and degree of lustre could be controlled during manufacture. Secondly, it can be given a permanent set or crease by heating to temperatures between 100 and 180 °C, a treatment which alters the internal structure of the fibres; this set is retained as long as the setting temperature is not exceeded in normal use, and is one of the qualities contributing to 'easy care'. Thirdly, nylon fibres absorb much less water than the cellulose fibres, and therefore dry more quickly after washing; this 'drip-dry' quality is another contribution to easy-care fabrics.

At the same time, poor water absorption rendered such clothing as shirts and underwear less comfortable to wear, static electricity tended to attract particles of dirt and made the fabric cling to the skin, and the translucence of fine fabrics was not always appropriate. Research and development has minimized some of these disadvantages, and nylon has established itself as one of the major fibres for both clothing and industrial uses. Nylon staple can be blended with wool, cotton and rayon, and often quite small percentages give a large increase in resistance to wear. Blended fabrics with a high proportion of nylon cannot, however, be shaped and manipulated by shrinking, and traditional methods of cutting and tailoring have to be modified.

Polyesters

The polyesters are produced by the reaction of organic acids with alcohols instead of amines, and form long chain molecules similar to those of the polyamides. Patents for the production of polyester fibres were taken out by the Calico Printers Association in England in 1941, but they were not developed commercially until the early 1950s, when Imperial Chemical Industries (ICI) started producing Terylene and Du Pont the similar fibre Dacron in the USA.

The polymer is prepared by reaction of the ingredients in a vacuum at high temperatures, and the filaments by melt spinning. The melting point is only slightly higher than that of nylon, but the drawing is done before the filaments are allowed to cool. Terylene is a difficult fibre to dye, as the dye molecules do not penetrate nor attach themselves to the chemical groups. Disperse dyestuffs developed for cellulose acetate fibres are used, and the best results are obtained by dyeing under pressure at temperatures higher than 100 °C.

Terylene can be heat set like nylon; its moisture retention is extremely low, giving it excellent drip-dry qualities. The uses of nylon and Terylene in the production of fabrics are very similar, but Terylene has certain advantages when blended as a staple with wool and cotton, while the stretch resistance of nylon renders it more suitable for ladies stockings.

Acrylics

Acrylonitrile is a liquid which polymerizes when emulsified with water containing ammonium persulphate; most of the commercial acrylic fibres are 'copolymers', in which the acrylic monomers are combined with other monomers in forming the chain. The properties of these copolymers depend on their precise chemical constitution, which is not usually disclosed by the manufacturer. The acrylics cannot be melt spun, and their development was delayed by the search for suitable solvents for dry spinning; commercial production started about 1950.

Most acrylics are used as staple fibre processed to resemble wool in warm, lightweight sweaters and blankets; they can be drip-dried and are very resistant to sunlight and weathering. The filament is not as strong as other synthetics, but similar in strength and extension to wool. Fabrics cannot be completely heat set, and tend to stretch in warm, moist conditions. Different acrylics vary in filament thickness, staple length, handling and dyeing capabilities, providing a range from which the manufacturer can select the type best suited to the texture, blend and colour effects that he requires.

Modacrylic fibres are 'modified' acrylics in that they contain less than 85 per cent acrylonitrile; Dynel, for example, combines 60 per cent vinyl chloride with 40 per cent acrylonitrile in a copolymer. Their outstanding property is flame resistance. In general their other properties are similar to those of the acrylics, but they are rather more sensitive to heat in washing and ironing, and some can be heat set like nylon and the polyesters.

Elastane or spandex fibres

Very few fibres can stretch more than 50 per cent without breaking, and elastic fibres in the past have depended on natural rubber for their elasticity; rubber filaments are formed by extruding treated latex through a spinneret. But rubber deteriorates with age and use, and the search for more durable synthetic substitutes has been going on for a long time. The most successful have been the polyurethanes, given the generic name 'spandex', which can be stretched to five or six times their length before they break, and have a high rate of recovery from stretching up to 200 per cent. They may be used as filament yarns *per se*, covered by twisting other filament yarns around them, or form the core for a sheath of staple fibres; when covered in this way the amount of stretch is much less than that of the core, but sufficient to improve the fit and comfort of garments made from them.

Other fibres and plastics

The fibres described above are those which may replace or be combined with cotton in textile manufacture, as this is the main field of competition. Many other types of natural and synthetic fibres are used in industry, but their impact on the market for cotton is relatively small.

Mixing and blending

The term 'mixture' is used where each of the different fibres is spun into a separate yarn, and these yarns are then woven together into a fabric, sometimes called 'union cloth'. A common type of mixture has a strong warp and a bulky weft, giving a stronger material than would be possible using only the bulkier, and weaker yarn. A 'blend' is an intimate mixture of fibres of different kinds, blended at some point in the spinning process so that the individual yarn contains a number of different types of fibre. The blending of cotton with man-made fibres is usually made after carding, as the card settings for the two fibres are so different. It is difficult to say which are the most popular mixtures and blends, and what proportion of the textile market they represent, as very few statistics are available.

Mixtures and blends may be used for three reasons. The first is economy, where an expensive fibre is diluted with a cheaper one, retaining as far as possible the quality of the expensive fibre. The cheapest fibres are usually cotton and viscose rayon, the acetates slightly more expensive. Synthetic fibres cost about 12 per cent more than cotton (July, 1985), and wool is about three times as expensive. Prices of the natural fibres fluctuate more quickly and widely than do those of the man-made fibres.

The second reason is to combine the desirable properties of different fibres where no one fibre is ideal. One of the most popular blends is that of cotton and polyester: the cotton gives absorbency and comfort and a good surface texture: the polyester makes the fabric hard-wearing, crease-resistant and drip-dry. The ratio of cotton to polyester is commonly 35:65, but there is a growing tendency for cotton's share of the blend to increase (Miles 1980); Hardingham (1978) says the ideal blend ranges from 60:40 to 70:30. The blend of wool and cotton in 'viyella' fabrics combines the warmth and comfort of wool with the strength and good washing properties of cotton.

The third reason is to obtain a decorative or colour effect. Mixtures of different coloured yarns have long been used to produce distinctive colour patterns. Different fibres vary in their affinity for dye, and this may be used to produce a colour effect when the blend or mixture is dyed after weaving. Some fibres have a distinctive appearance, lustre or texture, which can be used to give a decorative effect to the finished fabric.

Generally speaking a blend containing less than 10 per cent of a different fibre will not differ appreciably from the pure fibre; above 20 per cent the different fibre will begin to be noticeable; as it increases above 50 per cent it will increasingly dominate the characteristics of the fabric until near 90 per cent it takes over completely. If the heat setting properties of a synthetic fibre are required, the blend must contain at least 50 per cent.

Competition in textiles

Cotton quality

The increase in the market share of man-made fibres, particularly synthetics, is a challenge to the competitiveness of cotton. Cotton has unique properties which have not yet been matched by synthetic substitutes. It now accounts for less than half the world market for textile fibres, compared with over 70 per cent in 1950, but the absolute consumption of cotton is still rising in the less-developed countries. To maintain its position in the industrialized countries, it will be necessary to increase the attractiveness of cotton as a raw material, and to sell it at a competive price.

A major disadvantage of cotton compared with man-made staple fibre is the amount of waste in the manufacturing process (see Table 18.1). This consists mainly of trash, dirt and short fibre. The removal of trash and dirt is not only a direct loss, but involves additional processes and has a deleterious effect on lint quality. In order to deliver a clean product, cleanliness should start in the field and be maintained throughout marketing, ginning, baling and transport to the mill. Hand picking combined with strict grading standards results in cotton which commands a premium over machine-picked or poorly graded cotton of the same staple length. All cotton contains a proportion of short fibre, but cleaning machinery, badly set gins and overloading of the gins increase this proportion by breakage of the longer fibres. Some varieties have a greater percentage of short fibre than others, and this can sometimes be reduced by breeding.

Other complaints about cotton relate to stickiness caused by whitefly and aphids in the field, and the presence of hard fibres in the lint arising from the use of unsuitable packing materials and string.

Cotton processing

Cotton yarns and fabrics can be treated to improve their quality in certain respects. The earliest treatment, still widely used, was discovered by John Mercer in 1844, and has been called 'mercerizing' ever since; the yarn is treated with caustic soda and stretched during rinsing, and results in a more lustrous appearance with increased strength and absorbency. To prevent woven fabrics from shrinking after washing, controlled shrinking can be carried out during finishing by the Sanforized and other processes. The use of synthetic resins to impart crease resistance to cotton has been used commercially since the 1930s; the resins combine with the fibre molecules to form cross-links between them, but the process reduces the strength of the material and its resistance to abrasion. More recently, the use of liquid ammonia has been found to relax the internal stresses in the fabric, in the same way as mercerizing but to a lesser extent. In heavyweight fabrics like denim, the ammonia treatment alone confers some easy-care properties and improves its handling qualities and its stability in laundering. Lighter fabrics require some resin treatment to produce satisfactory easy-care finishes, but the ammonia treatment allows better penetration of the resins, so that less is required and the harmful effects of the resin treatment are reduced or eliminated (IIC 1982). Durable and non-durable water repellant finishes can be applied to give shower-proof garments, while fabrics treated with rubber, oils and silicones can be rendered completely waterproof. Flame resistance and some stretch properties can be imparted to pure cotton fabrics by special treatments.

Many of these properties are achieved more easily by blending cotton with man-made fibres which already possess them, and will impart them to the mixture. The type of fibre and the proportions may be varied, but some blends are more successful than others; staple fibres are produced in a wide range of length and fineness, and the best combination of these can be selected to suit any particular blend, while other characters such as dye absorption and elasticity may be important in achieving the best results.

Promotion of cotton

Much of the success of man-made fibres can be attributed to advertising and promotion, not only of the fibres themselves but also of the textiles made from them; the cost of advertising textiles may be subsidized by the fibre manufacturers. The dispersed nature of the cotton industry, and the former predominance of cotton in the market for textile fibres, has meant that cotton has not been similarly promoted in the past. An exception has been Egyptian cotton, which to many of the general public in Europe is synonymous with quality; in the USA cotton growers have combined to establish an integrated fibre company called Cotton International Incorporated (Cotton Inc.), to promote the use of cotton in that country, and there has been the independent promotion of the Pima variety as 'quality cotton'. In 1966 the International Institute for Cotton (IIC) was founded by a group of cotton-producing countries to increase the consumption of raw cotton and cotton products, starting with Western Europe and Japan; its headquarters are in Brussels, and it has divisions for cotton promotion, technical research and market research. It cooperates with a number of international organizations and textile research institutes in various countries. Cotton Inc. and the IIC have established symbols (Fig. 19.4) to distinguish fabrics made from pure cotton, on the same lines as the woolmark.

Fig. 19.4 Symbol used by the International Institute for Cotton to denote pure cotton (I.I.C.).

Table 19.3   *Production of fabrics by region*

Region	5-year average production 1976–80			
	Cotton	*Man-made*	*Total*	*% cotton*
	'000 metric tons			
North America	1,003	1,124	2,127	47
South America	95	48	143	66
Western Europe	943	703	1,646	57
Eastern Europe	397	26	423	94
Asia and Oceania	1,964	757	2,721	72
Africa	209	2	211	99
USSR	1,011	134	1,145	88
China PR	1,158	*	1,158	100
Total fabrics	6,780	2,794	9,574	70.8
Fibre production	12,755	11,049	23,804	53.6

* Not reported.

Source: ICAC 1983.

The penetration of man-made fibres (Table 19.3) has been greatest in the more industrialized countries – USA, Western Europe and Japan – and their consumption of cotton has decreased over the last 20 years; in the less-developed countries, particularly those which produce cotton themselves, cotton consumption has increased. This is partly the result of a shift to more local spinning and weaving, but in some countries there is direct government intervention to safeguard the cotton industry: India applies heavy excise duties on synthetic fibres, several South American countries impose high import duties, and in other countries the production and importation of synthetic fibres is discouraged in various ways (Symonds 1980). In the Far East, South Korea, Taiwan and Thailand have based their textile production on locally manufactured synthetic fibres and imported cotton.

The increase in world demand for textile fibres is expected to continue; not only is the world population increasing, but the *per capita* consumption is likely to increase as the standard of living improves in the less-developed countries. In the developed countries, consumption has stabilized around 10 kg per head, while in some of the less-developed countries it is as low as 2 kg. To maintain its share in this increase, cotton promotion must be intensified in the developed countries; many of the less-developed countries are themselves cotton producers, and any increase in consumption is likely to favour cotton. It is a useful cash crop providing a high level of employment; in countries which have not yet developed their own textile industry, it is a major source of foreign exchange, while in others it saves foreign exchange by import substitution. Some of the production of synthetic fibres will be diverted from textiles to new and specialized uses where it does not compete with cotton.

Industrial fibres

Cotton, wool and man-made fibres are classed as 'textile fibres', and competition in the field of clothing and apparel receives the most attention from the press and public; yet only half of the cotton produced world-wide ends up in what are classed as cotton fabrics, and only a quarter of the man-made fibres. In Western Europe, the production of cotton yarn is less than half the production of staple and filament yarns, yet more fabrics are produced from cotton than from man-made fibres (Fig. 19.5). It is not easy to trace where much of this fibre goes. While it is well known that man-made fibres have largely replaced cotton for tyre cords because of their extra strength, and plastic insulation has replaced cotton in electrical wiring, it is almost impossible to find out the actual quantities and trends in consumption in many countries. Where statistics are available, they are usually given as undifferentiated totals, or exclude one or other type of fibre. Another problem is the units of measurement, as in the case of fibre production. This is reported as yards or square yards of fabric, and must be converted into bales or tons to trace the movement and distribution of the different fibres. The conversion

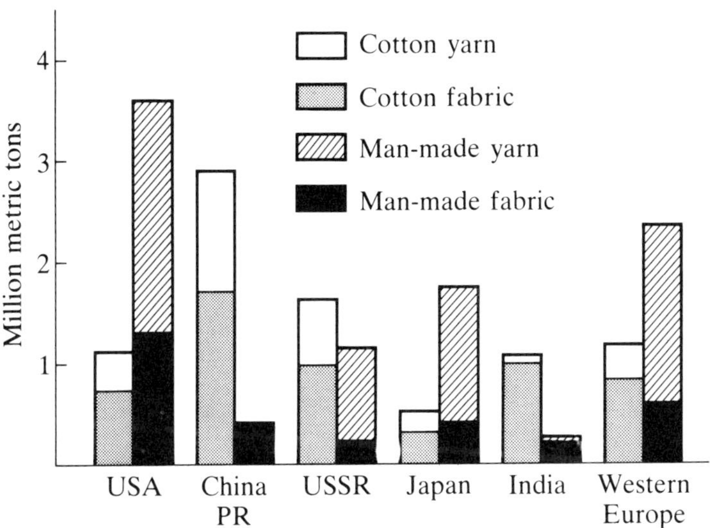

Fig. 19.5 Production of yarn and fabrics, 1980 (ICAC 1983).

factors vary widely, as shown by the equivalent in square yards of one pound of raw fibre:

	Textile Organon	*ICAC*	
		Hong Kong	*USA*
Cotton fabrics	7.0	4.08	2.31
Man-made fabrics: filament	4.4 ⎱	4.50	2.29
staple	3.3 ⎰		

Each factor presumably represents an average of the range of products included in the statistics. Another example is the conversion of raw cotton consumed to yarn produced; the yarn is consistently less than the cotton consumed, but how much of the difference represents mill waste in cleaning, carding and combing (see Table 18.1) and how much goes into the manufacture of products other than cotton yarn? The study of trends over the years is made difficult by changes in the basis of classification of industrial products, with new classes being added and some of the older ones disappearing.

The National Cotton Council (1965, 1973, 1983) of America avoids the question of units by recording the number of bales of fibre used in producing each class of product. They divide the products into three main groups: apparel, household uses and industrial uses (Fig. 19.6); Textile Organon (1983a, 1983b) uses a similar grouping: apparel, home furnishings, industrial and other uses, and exports. Since the early 1960s the consumption of cotton in the USA has fallen by about one-third, from about 1.9 to around 1.2 million tons (Fig. 19.6), with the greatest reduction in industrial use. Man-made fibres show a different pattern. There has been a fourfold increase in household uses, apparel has doubled, but industrial use has actually fallen, both proportionally and in absolute terms. Household uses include carpets, sheets and bedding, furniture covering and curtains. Among industrial uses, medical supplies provide the largest outlet for cotton, rising from 114,000 to 165,000 bales (480 lb; 218 kg), but cotton as a percentage of total consumption has fallen from 84 to 61 per cent. Next in importance is industrial thread, but here the cotton consumption has fallen both in total and in percentage. Abrasives are the only category which show an increase in cotton consumption both in bales and in percentage; wall-coverings and wiping and polishing cloths have both shown an increase in cotton consumption. Other important uses are in machinery belting and various protective uses such as tarpaulins, tents, trailer covers and boat covers, but in all these categories cotton

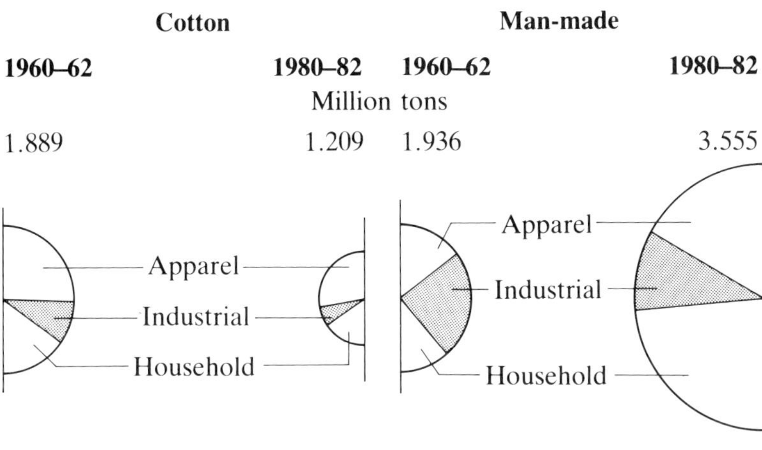

Fig. 19.6 End use of cotton textile fibres in the USA (National Cotton Council 1965, 1983).

consumption has fallen (National Cotton Council 1983). What has happened in the USA gives some idea of the trends which can be expected as the production of man-made fibres develops in other countries.

Key to the principal disorders of cotton

The key that follows is intended primarily for use in the field, and its object is to enable users who may not have a technical knowledge of entomology or plant pathology to diagnose, with the unaided eye, any of the principal disorders from which cultivated cotton suffers.

The symptoms shown by the plant form the basis of the key; where insect attack is responsible, it should be possible with the key to diagnose the group of insects involved, even though none can be found when the examination is made. The characteristics of the insect agents are also given, however, and may supply confirmatory evidence; the final confirmation of the identity is a matter for the specialist.

The key is constructed of couplets or pairs of alternatives that are mutually exclusive where only a single disorder is concerned. Cotton plants, however, often exhibit several disorders simultaneously. If, therefore, use of the key indicates a disorder that involves only some of the symptoms observed, one should see what alternatives might have been followed in the absence of the symptoms characterizing the first disorder.

Adapted from Pearson (1958).

Abbreviations

Bact.	Bacteria
Col.	Coleoptera (beetles)
Hem.	Hemiptera (bugs)
Isop.	Isoptera (termites)
Lep.	Lepidoptera (moths)
Orth.	Orthoptera (locusts, grasshoppers, etc.)

Seedlings

(1) Seedlings collapsed or withering, but not eaten above ground level   2
— Seedlings eaten above ground level   4

(2) Seedlings showing blackened lesions above ground level 3
— No such symptoms visible, but root gnawed at or below ground level; shiny-black, many-legged millepede found amongst remains of the sown seed – Millepedes (Myriapoda).
(3) Cotyledons showing dark, glassy-green lesions that turn dark brown or black and extend down the stem – Seedling form of bacterial blight, *Xanthomonas malvacearum* (Bact.).
— Stem blackened from slightly above ground level down to root, large patches of plants often affected – Sore-shin, *Rhizoctonia solani* (Fungus).
(4) Stems clipped through just above ground level, but leaves not much eaten; pale brown, wingless insects, up to 1 cm in length, moving in columns exposed on ground surface – Harvester termites, *Hodotermes mossambicus* (Isop.).
— Holes chewed in stems and leaves, or seedlings completely eaten 5
(5) Hopping insects, often gregarious; small faecal pellets (1–2 mm) on ground – Locusts, grasshoppers, crickets (Orth.).
— Caterpillars, often feeding at night and gregarious – Cutworms: *Agrotis ypsilon, Euxoa* spp., *Spodoptera exigua, S. littoralis*, etc. (Noctuidae, Lep.).

Whole plants

(1) Leaves wilting, lower part of stem, or roots, affected 2
— Leaves not wilting, but otherwise abnormal; lower part of stem and roots not affected 5
(2) Leaves wilt suddenly and dry up from the edge inwards between the main veins; easily detached. Vessels visible as brown dots at leaf scar and as brown streaks if mainstem is split open. Roots unaffected externally. Usually sporadic, but susceptible strains may be affected in patches – Wilt, *Fusarium vasinfectum* or *Verticillium dahliae* (Fungi).
— Leaves wilt slowly, base of stem, or roots, affected externally 3
(3) Upper part of mainstem, and woody core of root, not affected, but bark becomes spongy and wet from root tip to slightly above ground level, easily rubbed off, disclosing minute, black pustules on surface of wood, which is stained yellowish-grey and possesses characteristic smell. Usually in patches – *Rhizoctonia* spp. (Fungi).
— Upper part of stem not affected, woody parts of root channelled; usually sporadic 4
(4) Root damage mainly superficial 5
— Woody parts of root penetrated by discoloured channels at

or below ground level. The mainstem may eventually break at this point – Root-boring larvae: *Eutinobothrus (Gasterocercodes)* spp. (Curculionidae, Col.) in South America; *Sphenoptera* spp. (Buprestidae, Col.) in India, Africa.

(5) Large numbers of small, white, wingless insects round roots, often constructing earthen runs covering basal part of stem – Termites (Isop.)

— Small numbers of small white grubs decorticating root – Cotton beetle, *Syagrus* spp. (Eumolpidae, Col.)

(6) Leaves and stem mechanically damaged   7

— Leaves and stem otherwise affected   9

(7) Patches of plants with leaves and soft part of stem and branches eaten   8

— Whole fields of plants showing irregular tears in leaves and pale brown, corky scars confined to one side of branches and mainstem – Hail damage.

(8) Irregular pieces excised from leaves; small faecal pellets (1–2 mm) on ground – Locusts, grasshoppers (Acrididae, Orth.).

— Whole plants or branches browsed, spoor recognizable – Goats, antelope, elephants.

(9) Leaves not crinkled and without mosaic pattern, but these and fruiting branches progressively dwarfed and flower buds shed   10

— Leaves and bracteoles becoming crinkled, leathery and curling up at edges. Leaf veins swollen and appear clogged when viewed against light, sometimes with foliar outgrowths. In extreme cases, stems whip-like and spiralling, all foliar parts much reduced. Sometimes associated with fine mosaic pattern of dark and light green on leaves. Symptoms only commonly developed on *Gossypium barbadense*. Clouds of minute, white insects take wing when plant disturbed; young forms on lower surface of leaf, scale-like sedentary; associated with sticky secretion covering leaves and tips of stems – Leaf curl virus, transmitted by cotton whitefly, *Bemisia* spp. (Aleyrodidae, Hem.).

(10) Main veins of leaves clogged; occurs in West Africa – 'Bunchy top', thought to be a virus disease.

— Affected leaves and upper surface of branches reddened, finally much secondary development of small, abortive foliage. Only known from Congo and Malawi. Causative insect rarely seen after symptoms develop: adults 2 mm long, yellowish body, transparent wings decorated with brown spots on margin; immature forms resemble those of leaf hoppers, but sluggish – 'Psyllose', caused by *Paurocephala gossypii* (Psyllidae, Hem.). (Symptoms could be confused with those due to severe boron deficiency – see p. 119).

Stems and Branches

(1) Symptoms restricted to upper part of plant, lower part remains healthy   2
— Symptoms not restricted to upper part of plant   6
(2) Mainstem appears to fork at a somewhat swollen node; at an earlier stage, terminal bud killed and stem bored for a short distance down from tip by a young bollworm (see Green Bolls, 15) – *Earias* spp. or, more rarely, *Diparopsis* spp. (Noctuidae, Lep.).
— Upper part of plant wilts   3
(3) Extreme tip of mainstem becomes flaccid and collapses. Large black or brown, strong-smelling bug sometimes found sucking mainstem near tip – *Anoplocnemis curvipes* (Coreidae, Hem.).
— Damage not restricted to tip of stem   4
(4) Girdle of frayed fibres round mainstem, above which plant wilted or broken off; legless grub tunnelling in the stem   5
— Stem not girdled, but hollowed out; yellowish white, club-shaped grub about 2.5 cm long present in tunnel; or conspicuous exit-hole – Stem borer, *Sphenoptera* spp. (Buprestidae, Col.).
(5) Tunnel in centre of stem, grub slightly curved, adult a large weevil – Stem-girdling weevil, *Alcidodes* spp. (Curculionidae, Col.).
— Tunnels superficial, circular frass-ejection holes to exterior of stem, grub elongate, cylindrical, head withdrawn into body, adult beetle with long antennae – Stem girdler, *Tragiscoschema* spp. (Lamiidae, Col.).
(6) Stems and branches tend to dry out and become brittle; circular holes, under 1 mm diam., near nodes, which are often swollen and contain very small, white curved grubs; small, dull-black weevils common on the plant – Stem weevil, *Apion* spp. (Curculionidae, Col.).
— Not as above   7
(7) Lesions on main stem and branches   8
— Stems without lesions, but covered in colonies of small, sedentary insects, scale-like, or secreting a whitish waxy substance, usually confined to unhealthy plants – Mealybugs, scale insects (Coccoidea, Hem.).
(8) Dark glassy green lesions on petioles, pedicels, branches and mainstem (often descending the plant in that order),

becoming black and sunken; branches withered and blackened at tips, breaking or shedding – Blackarm form of bacterial blight, *Xanthomonas malvacearium* (Bact.).

— Dark green, sunken lesions, splitting longitudinally and forming brown, corky callus; growth distorted and dwarfed, leaves bunched dark green, crinkled. Insect agents not always seen; red or yellow-bodied, fragile bugs about 1 cm long with spine on back of thorax – *Helopeltis* spp. (Miridae, Hem.).

Leaves

(1) General shape and colour of leaf normal — 2
— General shape or colour of leaf abnormal — 11
(2) Holes in leaves — 3
— No holes in leaves — 5
(3) Holes relatively large, irregular, with sharp edges – Leaf-eating caterpillars (many species), *Syagrus* spp. (small black beetles), locusts and grasshoppers.
— Holes relatively small — 4
(4) Small circular holes in either surface of leaf, often not carried right through; very small, jumping beetles common – Flea beetle, *Podagrica* spp. (Halticidae, Col.).
— Small irregularly shaped holes, accompanied by narrow serpentine channels below transparent leaf cuticle; America only – Leaf perforator, *Bucculatrix thurberiella* (Lyonetiidae, Lep.).
(5) Spots on leaves — 7
— No spots on leaves — 6
(6) Narrow, serpentine channels, containing minute, reddish larvae, in green tissue below transparent cuticle on upper surface of leaf – Leaf miner, *Acrocercops bifasciata* (Gracillariidae, Lep.).
— Dark, often swollen, bands across the petioles, with corresponding internal necrosis of the pith – Boron deficiency.
(7) Spots rounded — 8
— Spots angular — 9
(8) Circular spots, with dark red margin and greyish centre, sometimes developing into holes – Leaf spot, *Cercospora gossypina* (Fungus).
— Larger, irregularly rounded, brown spots showing concentric zonation, tending to coalesce and cause leaf fall – *Alternaria* spp. (Fungus).
(9) Minute, angular spots with orange pustules in centre, on lower surface only – Rust, *Phakopsora desmium* (Fungus).
— Angular spots, not as above, on both surfaces — 10

(10) Angular spots, bounded by small veins, at first semi-transparent, later covered in white down – Powdery mildew, *Ramularia areola* (Fungus).

— Angular spots, bounded by small veins, at first dark green, water soaked, then dark brown to black. Spots frequently coalesce forming large areas which become tattered. Elongate lesions similarly formed along main veins, often running down petiole and causing leaf fall – Angular leaf spot form of bacterial blight, *Xanthomonas malvacearum* (Bact.).

(11) Holes in leaves   12

— No holes, but sometimes cracks in leaves   13

(12) Holes irregular, leaf-edge rolled into cylinder secured by webbing and often containing pale green, semi-translucent, black-headed caterpillar – Leaf roller., *Sylepta derogata* (Pyralidae. Lep).

— Minute holes on very young leaves which, as the latter unfold, develop into larger, irregular holes with healed, rounded edges to them; in extreme cases, leaves completely fretted, venation irregular and size much reduced. Insect agents not always seen; green or pale brown bugs, winged adults somewhat squat and humped, 4 mm long, or barely 2 mm long – *Lygus* spp. or *Campylomma* spp. (Miridae, Hem.).

(13) Small, angular, light brown spots, not coalescing, close to main veins; lamina dark green, crumpled; in extreme cases, leaves bunched together and distorted, with scorched appearance. Insect agents as in Stems and Branches (8), *Helopeltis* spp. (Miridae, Hem.).

— No spots on leaves   14

(14) Leaves with lobes abnormally pointed   15

— Leaves not as above   16

(15) Lower surface a dull, smoky green; lobes appear abnormally pointed due to their edges curling under; subsequently, small, straight cracks all over the leaf, which in extreme cases become tattered. Numerous, exceedingly minute transparent, glistening mites on under surface – 'Acariose', caused by *Hemitarsonemus latus* (Acarina).

— Leaves normal colour, but young leaves narrow with long, pointed lobes, and petioles sometimes twisted – 2,4–D herbicide, perhaps applied to adjacent crops.

(16) Leaves reddened or yellowed in zones or patches   17

— Leaves not as above   20

(17) Centre of leaf discoloured   18

— Edges of leaf turning pale green, then yellow and finally red, in successive zones more or less bounded by the transverse veins, rolling under and finally withering. Numbers of small, active, leaf hopper adults, of which the wingless immature forms are characterized by rapid, crab-like movements on lower leaf surface – Cotton jassid, *Empoasca* spp. (Cicadellidae, Hem.).

(18) Leaves showing yellow mottling near convergence of veins towards petiole; on under-surface are minute red spider mites, *Tetranychus* spp. (Tetranychidae, Acarina).

— Mottling between veins   19

(19) Poorly defined yellow patches between veins, turning brown and whole leaf wilting – Wilt, *Verticillium dahliae* (Fungi).

— Leaf shape normal, colour mottled, initially yellow, then bright red, but a narrow strip on either side of the veins remaining bright green – Physiological leaf reddening.

(20) Leaves puckering, rolling under at edges, stunted; under-side with numerous, soft-bodied, sedentary, green or black plant lice; leaf surface sticky, shiny, often covered by black sooty mould – Cotton Aphid, *Aphis gossypii* (Aphidae, Hem.)

— Lamina normal shape but sometimes reduced in size, surface silvery, eventually pale yellow, with papery texture and numerous minute, black droplets on surface. Numbers of very small, elongate insects with cone-shaped-mouth parts and greatly reduced wings – Thrips (Thysanoptera).

Flower Buds

(1) Flower bud shed when very small, before reaching the size of the subtending stipule. Often associated with leaf fretting; in such cases, fruiting branches truncated, plant tall and straggly with much secondary branching. Small green or brown bugs haunting terminal buds, shy, quick fliers – *Lygus* spp., *Campylomma* spp. or *Adelphocoris* (Miridae, Hem.).

— Flower bud larger when shed   2

(2) No distinct hole in flower bud   3

— Circular hole upward of 1 mm diam. in base, bracteoles often flaring back – Bollworm, usually spp. of *Diparopsis, Earias* or *Heliothis* (Noctuidae, Lep.), see Green Bolls (15) for key.

(3) Lesion or puncture mark on flower bud   4

— No visible cause for shedding, bracteoles not flared – Physiological shedding, which may be in response to the crop of bolls already being carried, drought, waterlogging, or attack by leaf-feeding insects or mites.

(4) Distinct lesion at base of calyx or on pedicel   5
— Lesions not only at base   6
(5) Dark glassy-green area, turning black, at base of calyx or on pedicel, often with angular leaf spot (Leaves, 10) on the bracteole, or blackarm (Stems and Branches, 8) on the branches – Bacterial blight, *Xanthomonas malvacearum* (Bact.).
— Sunken lesion on pedicel, with associated symptoms of *Helopeltis* attack (Stems and Branches, 8) – *Helopeltis* spp. (Miridae, Hem.).
(6) Minute, black punctures on the unopened corolla, usually marked by a small drop or smear of yellow fluid. Medium-small, elongate, delicate, opalescent bug, sheltering below bracteoles – *Creontiades* spp. (Miridae, Hem.).
— Eggs are laid in small punctures half way up the sides of the bud, and capped with yellow wax. Adult weevils cause larger punctures when feeding on the walls of buds and young bolls, causing the squares to flare and shed – Boll weevil, *Anthonomus* spp. (Curculionidae, Col.).

Flowers

(1) Petals eaten, medium or large beetles in evidence, boldly marked in red or yellow and black – *Coryna* spp. or *Mylabris* spp. (Meloidae, Col.).
— Petals not eaten   2
(2) Stamens eaten, frequently whole staminal column gone and hole into apex of the ovary – Bollworm (Pink bollworm often found in this position, see Green Bolls, 10).
— Petals shortened and folded in towards the centre; extra-floral nectaries at the base of the flower become discoloured – Boron deficiency.

Green Bolls

(1) Boll shed, or aborted when very small   2
— Boll not shed, nor aborted when very small   7
(2) No hole in boll wall   3
— Hole in boll wall – Bollworm, probably early stage of either *Earias, Heliothis* or *Diparopsis* (Noctuidae, Lep.).
(3) Distinct lesion at base of boll or on pedicel   4
— No distinct lesions on boll or pedicel   5
(4) Dark glassy-green area, later turning black, at base of calyx, often spreading on to the bracteole, or below calyx, at base of boll, or on pedicel; boll contents stained olive-

green to brown – Bacterial boll rot, a form of bacterial blight, *Xanthomonas malvacearum* (Bact.).

— Sunken lesion on pedicel, with associated symptoms of *Helopeltis* attack – *Helopeltis* spp. (Miridae, Hem.).

(5) Minute punctures through boll wall, visible externally as black dots or very small, angular, sunken areas – *Adelphocoris*, or *Creontiades* spp. (Miridae, Hem.).

— No punctures visible externally   6

(6) Punctures visible, if boll is opened, as small dots on inner surface of boll wall, often surrounded by small, water-soaked area or small button of spongy, white, proliferated tissue. Boll contents may be reduced to spongy, slimy, greyish mass – Stainer bugs (Hem.), either Pentatomidae or more often *Dysdercus* spp. (Pyrrhocoridae). If accompanied by yellowish-green staining, fungi of internal boll disease also present.

— No visible causes – Physiological shedding (Flower Buds, 3).

(7) No hole in boll wall   8
— Circular hole present in boll wall   12
(8) Lesions on boll surface   9
— No lesions on boll surface   10

(9) (a) Irregular rounded, dark glassy-green areas, up to 1 cm diam., sometimes slightly raised, subsequently sunken and turning dark brown from centre outwards, distributed generally over surface or, more usually, at extreme base of boll. Olive-green or brown staining and rotting of boll contents below lesion, often affecting extreme base of loculus and passing up central axis of boll – Bacterial boll disease, a form of bacterial blight, *Xanthomonas malvacearum* (Bact.).

(b) Irregularly rounded, sunken, blackened areas, with reddish-purplish margin and salmon-pink powdery appearance developing in the centre; lint below stained pink; often secondary – Anthracnose, *Colletotrichum gossypii* (Fungus).

(c) Circular spots, up to 3 mm diam., distributed over apical part of boll, where that is not protected by the bracteoles; at first dark green, but not glassy, later forming shallow brown craters through which, in young bolls, a slimy, pallid or corky proliferated tissue is extruded. Lesions restricted to boll wall and do not pass through it – *Helopeltis* spp. (Miridae, Hem.).

(10) Boll superficially healthy, but on cutting open, caterpillar found channelling among seeds; not exceeding 2 cm length, pearly white when small, but later stages with two transverse

pink bands on each segment, the anterior one being the broader – Pink bollworm, *Pectinophora gossypiella* (Gelechiidae, Lep.).

— Boll wall superficially healthy, but when cut open showing symptoms on inner surface as in (6) above. Embryo of seed often killed, interior of seed filled by floury tissue representing proliferated inner seed coat; lint sodden and matted, sometimes reduced to papery membrane   11

(11) No staining, injury caused solely by feeding of uninfected sucking bugs, adults of which are about 1.5 cm long:

(a) Pale green, shield-shaped adults, not usually very numerous – *Nezara* spp. (Pentatomidae, Hem.).

(b) Iridescent green-, blue- and bronze-patterned adults, obovate and strongly convex in form, sometimes numerous – *Calidea* spp. (Pentatomidae, Hem.).

(c) Somewhat elongate, flattened adults, yellow or red with black markings, frequently found pairing; young forms bright red, feeding gregariously in open bolls – Cotton stainers, *Dysdercus* spp. (Pyrrhocoridae, Hem.).

— Lint stained pale yellowish or greenish yellow – Stigmato-mycosis or internal boll disease, caused by *Nematospora* spp. (Fungi), transmitted by sucking bugs as above.

(12) Caterpillar present inside boll   13

— No caterpillar present   16

(13) Hole small, closed by thin membrane consisting of the boll wall cuticle; larva within as in (10) above – Pink bollworm, *Pectinophora gossypiella* (Gelechiidae, Lep.).

— Hole not as above, larva different   14

(14) Hole small, tunnel often not passing straight through boll wall, larva not exceeding 2 cm length, at first smoky grey with slightly darker spots, later uniform reddish-pink colour, wriggling backward actively when head is touched – False codling moth, *Cryptophlebia leucotreta* (Olethreutidae, Lep.).

— Hole usually large, passing straight through boll wall, larva not as above   15

(15) (a) Larva not exceeding 2.5 cm length; thick-set, with brownish markings and numerous fleshy protuberances; Old World only – Spiny bollworm, *Earias biplaga*, or Egyptian bollworm, *E. insulana* (Noctuidae, Lep.).

(b) Larva may attain 3 cm length; pale, duck-egg green ground-colour, with reddish, trident-shaped markings on each segment; Africa and Yemen – Red or Sudan boll-worm, *Diparopsis* spp. (Noctuidae, Lep.).

(c) Larva may attain 4 cm length; white to pale pink ground-colour, with pink trident- or M-shaped marks flanked by a pink spot on each segment; South America only – South American red bollworm, False pink bollworm, Colombian pink bollworm, *Sacadodes pyralis* (Noctuidae, Lep.).

(d) Larva large, perhaps exceeding 4 cm length; cream or tan with black spots when young, older larvae with light and dark longitudinal bands, general colour variable depending on diet – Bollworm, American bollworm, Tobacco budworm, *Heliothis* spp. (Noctuidae, Lep.).

(16) Exit-hole small, neatly rounded, boll contents only partly eaten 17

— Exit-hole large, sometimes irregular, boll contents largely eaten 18

(17) Irregular brown lines on inside of boll wall where larva has tunnelled before entering boll cavity; extensive rotting – False codling moth, *Cryptophlebia lencotreta* (Olethreutidae, Lep.).

— Clean, well-defined tunnels in boll contents, inside of seed eaten out leaving seed coat – Pink bollworm, *Pectinophora gossypiella* (Gelechiidae, Lep.).

(18) Boll more or less full of larval excrement – Spiny bollworm, *Earias* spp., or red or Sudan bollworm, *Diparopsis* spp. (Noctuidae, Lep.).

— Boll contents cleanly eaten out, mass of usually pale, straw-coloured or greenish larval excrement between the outside of the boll and the bracteoles – Bollworm, American bollworm, Tobacco budworm, *Heliothis* spp. (Noctuidae, Lep.).

Mature bolls

(1) Boll opening more or less normally 2
— Boll not opening normally, carpels not fully reflexed 6
(2) Bugs present, sucking the seeds; lint may not be stained 3
— No insects present, but boll contents damaged 4
(3) Small, black or dark brown, strongly smelling bugs, up to 4 mm long, both nymphs and adults present – Cotton seed bug, *Oxycarenus* spp. (Lygaeidae, Hem.).

— Usually only nymphs present; bright red when small, and red or yellow, banded with white, when larger – Cotton stainers, *Dysdercus* spp. (Pyrrhocoridae, Hem.).

(4) Lint stained 5
— Lint not stained, but seed cotton pulled from boll and seed kernels bitten out – Field mice, usually *Mastomys coucha*.
(5) Pale yellowish stain on lint, general or patchy; locules sometimes partially aborted, showing silvery patches of undeveloped lint. Puncture marks discernible on inner

surface of boll wall – Stigmatomycosis or internal boll disease caused by *Nematospora* spp. (fungi), transmitted by stainer bugs (Pentatomidae, or Pyrrhocoridae, Hem.).
— Bright, sulphur yellow, pink or reddish stain on lint, usually not associated with breakdown of boll contents or damage to the boll wall – *Epicoccum*, *Fusarium*, or other fungi, or bacteria entering after boll has split.
(6) Hole in carpel wall – Bollworm (the species can sometimes be decided from the nature of the damage, see Green Bolls).
— No hole in carpel wall   7
(7) Small to medium-sized caterpillar channelling in the boll contents   8
— No caterpillars present among the boll contents   10
(8) Very small caterpillar scavenging among debris; dirty white, with two equally narrow, transverse, pink stripes on each segment, often with a general suffusion of pinkish purple when still small – Pink scavenger worm, *Pyroderces simplex* (Cosmopterygidae, Lep.).
— Medium-sized caterpillar, not as above   9
(9) Appearance of caterpillar as in Green Bolls (10), about 12 mm long when fully grown; often spinning cocoon within a seed or pair of seeds, or loosely among the lint – Pink bollworm, *Pectinophora gossypiella* (Gelechiidae, Lep.).
— Caterpillar similar to that of *P. gossypiella*, but distinguished by the larger plates at the bases of the body setae, giving a spotted appearance, and the smaller size (8–9 mm long when fully grown); pupa usually without a cocoon – *Mometa zemiodes* (Gelechiidae, Lep.).
(10) Boll wall covered in sooty powder, boll contents completely rotted – Secondary infection, often of *Diplodia gossypina* (Fungus).
— Not as above   11
(11) Strands of lint webbing together the partially reflexed carpels; locules more or less aborted, stained greenish-yellow, or in extreme cases lint reduced to brown, papery membrane closely pressed to seeds, giving kidney-like appearance – Stigmatomycosis or internal boll disease, caused by *Nematospora* spp. (Fungi), transmitted by stainer bugs (Pentatomidae or Pyrrhocoridae, Hem.).
— Similar to preceding, but without webbing; damage usually starting from the extreme base of the loculus; colour usually darker, yellow to olive brown; associated with blackened external lesion at base of boll – Bacterial bollrot, a form of bacterial blight, *Xanthomonas malvacearum* (Bact.).

Production and yield of cotton lint by countries (5-year average, 1977/78 to 1981/82)

Country	Production (1000 metric tons)	Yield (kg per ha)
North America		
Costa Rica	3	428
Cuba	1	268
Dominican Republic	1	207
El Salvador	60	776
Guatemala	132	1187
Haiti	1	268
Honduras	8	739
Jamaica	—	268
Mexico	338	934
Nicaragua	79	659
United States	2902	545
South America		
Argentina	155	290
Bolivia	10	424
Brazil	569	288
Colombia	110	506
Ecuador	12	673
Paraguay	88	289
Peru	91	641
Venezuela	14	407
Western Europe		
Greece	129	853
Italy	—	189
Spain	50	838
Yugoslavia	1	261
West Africa		
Angola	1	170
Benin	7	277
Cameroon	26	460
Central African Republic	10	109
Chad	37	186
Ghana	2	222

Country	Production (1000 metric tons)	Yield (kg per ha)
Guinea Bissau	—	342
Ivory Coast	52	458
Mali	45	430
Morocco	6	467
Niger	1	233
Nigeria	31	61
Senegal	12	328
Togo	6	360
Upper Volta	22	298
Zaïre	8	87
East Africa		
Burundi	3	263
Egypt	470	903
Ethiopia	25	503
Kenya	9	74
Madagascar	11	584
Malawi	8	226
Mozambique	18	132
South Africa	53	504
Sudan	140	337
Tanzania	52	120
Uganda	9	33
Zambia	6	189
Zimbabwe	62	506
Socialist		
Albania	5	277
Bulgaria	9	404
China Peoples Republic	2421	505
Korea Democratic Republic	1	134
USSR	2737	886
Vietnam	1	91
Asia		
Afghanistan	34	363
Bangladesh	3	294
Burma	17	82
China (Taiwan)	—	537
India	1336	167
Iran	109	459
Iraq	7	367
Israel	77	1358
Korea Republic	2	272
Pakistan	648	319
Sri Lanka	—	258
Syria	134	852
Thailand	47	387
Turkey	503	748
Yemen Arab Republic	9	359
Yemen Peoples Republic	4	281

Country	Production (1000 metric tons)	Yield (kg per ha)
Oceania		
Australia	83	1198
Indonesia	2	322
Philippines	3	281

Production of less than 500 tons indicated by — .
Source: ICAC 1983.

References

The following conventions have been used for frequently occurring references:

Cott. Res. Corp.	Empire Cotton Growing Corporation (1921–66)
	Cotton Research Corporation (1966–75)
Cott. Gr. Rev.	Empire Cotton Growing Review (1924–66)
	Cotton Growing Review (1967–75)
Progress Rep. Exp. Stas.	Progress Reports from Experiment Stations. Empire Cotton Growing Corporation, London.
	Published in one volume per annum (1921–51)
Cott. Res. Rep.	Progress Reports from Experiment Stations (1951–71)
	Cotton Research Reports. (1971–75)
	Published separately for each country by the Cotton Research Corporation, London.
Res. Mem.	Research Memoirs. A series of reprints published by the Cotton Research Corporation, London.
USDA	United States Department of Agriculture
Beltwide Conf.	Beltwide Cotton Production Research Conferences. Annual proceedings, published by the National Cotton Council of America, Memphis, Tennesse.

Abernathy, J. R. *et al.* (1979) Cotton and sorghum pests in Texas. In *Pest Management Strategies*, US Govt. Printing Office, pp 1–75. (Cited by Walker, J. K. 1980).

Adkisson, P. L., Niles, G. A., Walker, J. K., Bird, L. S. and **Scott, H. B.** (1982) Controlling cotton's insect pests: a new system. *Science* **216**:19–22.

Alefeld, F. G. C. (1861) Ueber die Stellung der Gattung *Gossypium* und mehrer Anderer. *Bot. Zeit.* **19**:229–301.

Allard, R. W. (1960) *Principles of Plant Breeding*. Wiley: New York.

Amin, J. V. and **Powell, R. D.** (1966) Low temperature damage in cotton. *Beltwide Conf.*: 260–1.

Andrews, F. W. (1938) Investigations on blackarm disease of cotton III. The mode of infection on the newly planted crop. *Emp. J. Exp. Agric.* **6**:207–18.

Andrews, G. L. (1975) Evaluation of *Heliothis* nuclear polyhedrosis virus in a cottonseed oil bait for control of *H. virescens* and *H. zea* on cotton. *J. Econ. Ent.* **68**:87–90.

Andries, J. A. *et al.* (1968) Further studies on the effect of leaf shape on the incidence of boll rot and economic characters of cotton. *Beltwide Conf.*: 147–58.

Anthony, K. R. M. and **Bravo, R.** (1970) Cotton production in Colombia. *Cott. Gr. Rev.* **47**:81–92.

Ark, P. A. (1958) Longevity of *X. malvacearum* in dried cotton plant. *Plant Dis. Reptr.* **42**:293.

Armstrong, G. M. and **Armstrong, J. K.** (1960) *American, Egyptian and Indian cotton wilt Fusaria: their pathenogenicity and relation to other wilt Fusaria*, USDA Tech. Bull. No 1219.

Arnold, M. H. (1969) General review. Cott. Res. Rep., Uganda, 1968–69, p 4.

Arnold, M. H. (1970) *Cotton improvement in East Africa*. In C. L. A. Leakey (ed.) *Crop Improvement in East Africa*. Comm. Agric. Bureaux: Farnham Royal. Ch. 7 pp 178–208. (Res. Mem. No 80).

Arnold, M. H. (1972) Modal selection in BP52. *Cott. Gr. Rev.* **49**:107–25. (Res. Mem. No 85).

Arnold, M. H. (ed.) (1976) *Agricultural Research for Development*. Camb. Univ. Press: London.

Arnold, M. H. and **Arnold, K. M.** (1959) Diseases. Cott. Res. Rep., Tanganyika, 1957–58, p 17.

Arnold, M. H. and **Brown, S. J.** (1968) Variation in the host-parasite relationship of a crop disease. *J. Agric. Sci. Camb.* **71**:19–36. (Res. Mem. No 71).

Arnold, M. H., Costelloe, B. E. and **Church, J. M. F.** (1968) BPA and SATU, Uganda's two new cotton varieties. *Cott. Gr. Rev.* **45**:162–74.

Arnold, M. H. and **Innes, N. L.** (1976) Plant breeding. In Arnold (1976) Ch. 8 pp 197–246.

Arnold, M. H., Innes, N. L. and **Brown, S. J.** (1976) Resistance breeding. In Arnold (1976) Ch. 7 pp 175–96.

Arnold, M. H., Passmore, R. G., Jones, E. and **Tollervey, F. E.** (1976) The Namulonge farm. In Arnold (1976) Ch. 9 pp 247–302.

Association of Official Seed Analysts (1978) Rules for testing seed. *J. Seed Tech.* **3**:1–126.

Bachini, J. E. G. (1980) Cotton pests in South America. *Outlook on Agriculture* **10**:198–201.

Bailey, A. E. (ed.) (1948) *Cottonseed and Cottonseed Products*. Interscience Publishers: New York.

Bailey, M. A. and **Trought, T.** (1926) *The development of the Egyptian cotton plant.* Tech. Sci. Service Bull. No 60, Min. of Agric., Egypt.

Balls, W. L. (1912) *The Cotton Plant in Egypt*. Macmillan: London.

Balls, W. L. (1914) Science and the supply of fine cotton. *Science Progress* No 34. (Quoted by Harland 1917).

Balls, W. L. (1915) *The Development and Properties of Raw Cotton*. Black: London.

Balls, W. L. (1921) The possibility with cotton crops for exact reporting and fore-casting. World Cotton Conference, Liverpool and Manchester, pp 191–4.

Balls, W. L. (1924) The determiners of cellulose structure as seen in the cell walls of cotton hairs. *Proc. Roy. Soc. B.* **95**:72. (Quoted in Hutchinson *et al.* 1947).

Balls, W. L. (1953) *The Yields of a Crop*. Spon: London.

Balsinhas, A. (1963) (Alphabetical list of botanical names of spontaneous plants observed in fallows and cotton fields). *Publ. B. Divulg. Serv. Agric. Mozambique* **27**:41. (Portuguese, quoted by Kasasian 1969).

Barlow, F. D. Jr (1952) *Cotton in South America*. National Cotton Council: Memphis, Tenn.

Basinski, J. J. (1960) Cotton in China. *Cott. Gr. Rev.* **37**:141–2.

Baskin, C. C. (1981a) Storage of bulk cottonseed. *Beltwide Conf.*: 308–10.

Baskin, C. C. (1981b) Storage of seed cotton. *Beltwide Conf.*: 310–11.

Baskin, C. C. (1981c) Tetrazolium evaluation of cottonseed. *Beltwide Conf.*: 312.

Baskin, G. R. and **Sistler, F. E.** (1980) A computer model for trailer optimization of a cotton harvesting system. *Beltwide Conf.*: 108–11.

Batra, G. R. and **Gupta, D. S.** (1970) Screening of varieties of cotton for resistance to jassid. *Cott. Gr. Rev.* **47**:285–91.

Beasley, J. O. (1940). The origin of the American tetraploid *Gossypium* species. *Am. Nat.* **74**:285–6.

Beasley, J. O. (1942) Meiotic chromosome behaviour in species, species hybrids, haploids and induced polyploids of *Gossypium. Genetics* **27**:25–54.

Bederker, V. K. (1957) A short note on weeds in Hyderabad Gaorani cotton. *Ind. Cott. Gr. Rev.* **11**:148–51. (Quoted by Kasasian 1969).

Bedford, H. W. (1937) Entomological section, 1936. Ann. Rep. of Agric. Res. Service, Sudan.

Bentham, G. and **Hooker, J. D.** (1867) *Genera plantarum* 3 vols. London.

Birch, H. F. (1952) The relationship between phosphate response and base saturation in acid soils. *J. Agric. Sci. Camb.* **42**:276–85.

Birch, H. F. (1964) Mineralization of plant nitrogen following alternate wet and dry conditions. *Plant and Soil* **20**:43–9.

Bird, L. S. (1966) Report of the bacterial blight committee. *Beltwide Conf.*: 21.

Bird, L. S. (1975) Genetic improvement of cotton for multi-adversity resistance. *Beltwide Conf.*: 150–2.

Bird, L. S. (1981) Cottonseed and germination – stand establishment. *Beltwide Conf.*: 318–21.

Bird, L. S. *et al.* (1981) Performance and characteristics of new cultivars indicate further genetic gains in developing multi-adversity resistance cottons. *Beltwide Conf.*: 17–20.

Blaney, H. F. and **Criddle, W. D.** (1945) A method of estimating water requirements in irrigated areas from climatic data. USDA Soil Conservation Service. (Quoted by Roe 1950).

Blank, L. M. (1953) The rot that attacks 2,000 species. *Plant Diseases, the Yearbook of Agriculture 1953.* USDA: Washington DC. pp 289–301.

BM-MFF (1978) *Guide to man-made fibres.* British Man-Made Fibres Federation: London.

Bowden, J. (1970) Cotton pests. In Jameson (1970) Ch. 12B pp 178–85.

Bowden, J. and **Ingram, W. R.** (1958) Spraying trials on peasant cotton in Uganda in 1957. *Cott. Gr. Rev.* **35**:239–43.

Bradley, T. (1968) Cott. Res. Rep., Kenya, 1965–66.

Braud, M. (1974) (The control of mineral nutrition of cotton by foliar analysis). *Coton Fibr. Trop.* **29**:215–25 (French).

British Agrochemicals (1985) *Annual report and handbook 1984–85.* British Agrochemical Association: London.

Brown, C. H. (1948) Egyptian cotton breeding technique. *Cott. Gr. Rev.* **25**:35–7.

Brown, C. H. (1953) *Egyptian Cotton.* Leonard Hill: London.

Brown, K. J. (1962) Effect of early rainfall on the response of cotton to nitrogen. *Cott. Gr. Rev.* **39**:177–80.

Brown, K. J. (1963) Cotton growing industry in the Syrian Arab Republic. *Cott. Gr. Rev.* **40**:98–103.

Brown, K. J. (1965) Response of three strains of cotton to flower removal. *Cott. Gr. Rev.* **42**:279–86.

Brown, K. J. (1971) Plant density and yield of cotton in northern Nigeria. *Cott. Gr. Rev.* **48**:255–66.

Brown, K. J. (1975a) Rainfall reliability in central Kenya. *Cott. Gr. Rev.* **52**:38–45.

Brown, K. J. (1975b) Staple length determined by the tuft method. *Cott. Gr. Rev.* **52**:149–51.

Brown, S. J. (1969) Bacterial blight of cotton: a note on trash-borne infection in

Uganda. *Cott. Gr. Rev.* **46**:197–201.

Brown, S. J. (1976) Plant pathology. In Arnold (1976) Ch. 6 pp 151–74.

BSI (1969) *Recommended common names for pesticides*. British Standards Institution: London. 1969 onwards.

Buchanan, G. A., Wells, L. W. and **McWhorter, C. G.** (1980) Report of 1979 cotton weed loss committee. *Beltwide Conf.*: 189–200.

Bugbee, W. M. and **Presley, J. T.** (1967) A rapid inoculation technique to evaluate the resistance of cotton to *Verticillium albo-atrum*. *Phytopath.* **57**:1264.

Bull, D. L. (1981) Status of insect growth regulators. *Beltwide Conf.*: 143–7.

Burhan, H. O. (1969) Rotation responses of cotton in the Gezira. In Siddiq and Hughes (1969) pp 51–7.

Burhan, H. O. and **Babiker, I. A.** (1968) Investigation of nitrogen fertilization of cotton by tissue analysis. I. *Expl. Agric.* **4**:311–23.

Burkitt, F. H. (1981) Cotton in the '80s. *Textile Asia* Dec 1981.

Burt, B. C. (1913) American cotton in the central circle, United Provinces. *Agric. J. India* **8**:339. (Quoted by Hutchinson *et al.* 1947).

Burton, J. M. (1972) Cotton growing on Triangle sugar estate, Rhodesia. *Cott. Gr. Rev.* **49**:236–41.

Carbon, J. de (1957) La culture de coton en milieu Africain. W. African Cott. Res. Conf. (App. III). Min. of Agric.: Samaru, Nigeria.

Carruthers, I. (1985) Irrigation, drainage and food supplies. *Ceres*, in press.

Cathey, G. W. (1980) Harvest-aid chemicals and practices for cotton. *Outlook on Agriculture* **10**:191–7.

Cauquil, J. (1977) Etudes sur un maladie d'origine virale du cotonnier: la maladie bleue. *Coton Fibr. Trop.* **32**:259–78.

CEC (1948) *Industrial fibres*. Commonwealth Economic Committee, HMSO: London.

Central African Archives (1952) *The Zambezi Journal of James Stewart*, 1862–63. Oppenheimer Series No 6. Chatto and Windus: London.

Chadwick, D. I. (1921) In discussion of paper by Balls (1921). World Cotton Conference, Liverpool and Manchester, p 196.

Chandler, J. M. (1981) Postemergence control of bermudagrass biotypes. *Beltwide Conf.*: 174.

Chao, S. C. and **Li, G. H.** (1965) (Life history and habits of the spiny bollworm (*Earias cupreoviridis* Walker) in Szechuan). *Acta phytophyl. sin.* **4**:301–14. (Chinese, quoted by Reed 1974).

Chokey Singh (ed.) (1980) *Cotton at a glance*. Bull. No 3. Central Institute for Cotton Research: Nagpur.

Ciba-Geigy (1972) *Cotton*. Technical Monograph No 3. Ciba-Geigy Agrochemicals: Basle, Switzerland.

Clegg, G. G. (1931) The stapling of cottons. Laboratory methods in use at the Shirley Institute, Spring 1931. *Shirley Institute Memoirs – Ser. A* **3**:33–62.

Clower, D. F. (1980) Changes in *Heliothis* spp. attacking cotton in recent years, and how they have affected control. *Beltwide Conf.*: 139–41.

Coaker, T. H. (1958) Experiments with a virus disease of the cotton bollworm, *Heliothis armigera* (Hbn.). *Ann. Appl. Biol.* **46**:536–41.

Cochran, W. G. and **Cox, G. M.** (1957) *Experimental Designs* (2nd edn). John Wiley: New York.

Colwick, R. F. and **Barker, G. L.** (1980) Effects of subsoiling in a controlled-traffic system in a silty clay loam soil – Fourth year results. *Beltwide Conf.*: 130–1.

Cook, C. (ed.) (1980) The world of science VI Special topics. In *Pears Cyclopaedia 1980–81*. Pelham Books: London.

Costa, A. S. (1956) Anthocyanosis, a virus disease of cotton in Brazil. *Phytopath. Z.* **28**:167–86. (Quoted by Halliwell 1966).

Costelloe, B. E. (1968) Routine method of acid-delinting cotton seed for experimental purposes. *Cott. Gr. Rev.* **45**:219–22.

Costelloe, B. E. (1970) Inheritance and effect on lint of seedcoat fuzz in Upland cotton. *Cott. Gr. Rev.* **47**:8–19.

Cotton Disease Council (1981) 1980 cotton disease loss estimate committee report. *Beltwide Conf.*: 3.

Cowley, E. J. (1966) Development of the cotton growing industry in Nigeria, with special reference to the work of the British Cotton Growing Association. *Cott. Gr. Rev.* **43**:169–95.

Cox, D. R. (1958) *Planning of Experiments.* John Wiley: New York.

Cross, J. E. (1962) Gossypol content of cotton seed. *Cott. Gr. Rev.* **39**:136–9.

Cross, J. E. (1964) Field differences in pathenogenicity between Tanganyika populations of *X. malvacearum. Cott. Gr. Rev.* **41**:38–43.

Crowther, F. (1941) Studies in growth analysis of the cotton plant under irrigation in the Sudan. II. Seasonal variation in development and yield. *Ann. Bot., N.S.* **5**:509–33.

Crowther, F. (1943) Influence of weeds on cotton in the Sudan Gezira. *Emp. J. Exp. Agric.* **11**:1–14.

Dale, J. E. (1962) Fruit shedding and yield in cotton. *Cott. Gr. Rev.* **39**:170–6.

Daniels, J. (1965) Abyan root rot of cotton in Aden. *Cott. Gr. Rev.* **42**:104–22.

Davis, D. D. (1979) Hybrid cotton: specific problems and potentials. *Adv. Agron.* **30**:129–57.

Davis, D. D. and **Palomo, A.** (1980) Yield stability of interspecific hybrid NX-1. *Beltwide Conf.*: 81.

Delouche, J. C. (1981) Harvest and post-harvest factors affecting the quality of cotton planting seed and seed quality evaluation. *Beltwide Conf.*: 289–305.

Demetriadi, M. A. (1963) Long staple cotton in North-East Brazil. *Cott. Gr. Rev.* **40**:179–83.

Demol, J. and **Bannink, L.** (1957) Evolution des méthodes culturales dans la région cotonnière nord du Congo Belge. West African Cott. Res. Conf. (App. V). Min. of Agric.: Samaru, Nigeria.

Dineur, P. (1957) La psyllose. West African Cott. Res. Conf. (App. XVIII). Min. of Agric.: Samaru, Nigeria.

Donald, C. M. (1963) Competition among crop and pasture plants. *Adv. Agron.* **15**:1–118.

Doorenbos, J. and **Pruitt, W. O.** (1977) *Crop water requirements.* FAO Irrigation and Drainage paper No 24 (revised). FAO: Rome.

Douglas, A. G., Brooks, O. L. and **Perry, C. E.** (1967) Influence of mechanical harvester damage on cottonseed germination and seedling vigour. *Beltwide Conf.*: 129–34.

Dowson, W. J. (1957) *Plant Diseases due to Bacteria.* Camb. Univ. Press: Cambridge.

Dransfield, M. (1968) *Seed dressing trials on cotton in Northern Nigeria, 1953–67.* Samaru Misc. Paper No 26. Inst. for Agric. Res., Ahmadu Bello Univ.: Nigeria.

Dransfield, M. (1972) Pathology. Cott. Res. Rep., Nigeria, 1970–71.

Dransfield, M. and **Beeden, P.** (1974) *A new cotton seed dressing for northern Nigeria.* Samaru Misc. Paper No 46. Inst. for Agric. Res., Ahmadu Bello Univ.: Nigeria.

Dunlap, A. A. (1945) *Fruiting and shedding of cotton in relation to light and other limiting factors.* Bull. No 677. Texas Agric. Expt. Sta.: College Station, Texas.

Dunlap, A. A. (1948) 2,4-D injury to cotton from airplane dusting of rice. *Phytopathology* **38**:638–44.

Eaton, F. M. (1931) Cotton roots. *J. Agric. Res.* **43**:875–83.

Eaton, F. M. (1955) Physiology of the cotton plant. *Ann. Rev. Pl. Physiol.* **6**:299–328.

Ebbels, D. L. (1975) Fusarium wilt of cotton: a review with special reference to Tanzania. *Cott. Gr. Rev.* **52**:295–339.

Ebbels, D. L. (1980) Cotton diseases. *Outlook on Agriculture* **10**:176–83.

Ebbels, D. L. and **Little, R.** (1975) Plant pathology. Cott. Res. Rep., Tanzania, 1973–74, pp 49–57.

Edlin, H. L. (1935) A critical revision of certain taxonomic groups of the Malvales. *New Phytol.* **1–20**:122–43.

Ehlig, C. F. and **Le Mert, R. D.** (1973) Effects of fruit load, temperature and relative humidity on boll retention in cotton. *Crop. Sci.* **13**:168–71.

Ellern, S. J. (1966) Ratoon (biennial) cotton in Israel. *Cott. Gr. Rev.* **43**:33–6.

Elliott, F. C., Hoover, M. and **Porter, W. K. Jr** (eds) (1968) *Advances in Production and Utilization of Quality Cotton: Principles and Practice.* Iowa State Univ. Press: Ames, Iowa.

Elliott, J. A. (1923) Cotton wilt; a seed borne disease. *J. Agric. Res.* **23**:397. (Quoted by Perry 1962a).

El Nur, E. (1969) Bacterial blight disease of cotton. In Siddig and Hughes (1969) pp 179–88.

El Nur, E. and **Abu Salih, H. S.** (1969) Cotton leafcurl. In Siddig and Hughes (1969) pp 196–206.

El-Zik, K. M. and **Yamada, H.** (1981) Effects of row spacing, irrigation scheduling and nitrogen rate on *Verticillium* wilt, yield and fiber quality of two Acala cotton cultivars. *Beltwide Conf.*: 31.

Engledow, F. L. (1947) *Agricultural development in the British colonial empire.* Agric. Economics Society: London.

Erickson, R. O. and **Michelini, F. J.** (1957) The plastochron index. *Amer. J. Bot.* **44**:297–305.

Evenson, J. P. (1969) Effects of floral and terminal bud removal on the yield and structure of the cotton plant in the Ord Valley, N.W. Australia. *Cott. Gr. Rev.* **46**:37–44.

Evenson, J. P. (1970) Ratooning of cotton: a review. *Cott. Gr. Rev.* **47**:1–7.

Evenson, J. P. and **Basinski, J. J.** (1973) Bibliography of cotton pests and diseases in Australia. *Cott. Gr. Rev.* **50**:79–86.

Falcon, L. A. (1971) Microbial control as a tool in integrated control programs. In Huffacker (1971) Ch. 15.

Falconer, D. S. (1960) *Introduction to quantitative genetics.* Oliver and Boyd: Edinburgh and London.

Farbrother, H. G. (1951) A note on the measurement of rainfall intensity. *E. Afr. Agric. For. J.* **17**:82–4.

Farbrother, H. G. (1961) Growth analysis and soil moisture. Cott. Res. Rep., Uganda, 1960–61, pp 26–31.

Farbrother, H. G. (1969) The pattern of soil moisture changes under irrigated cotton. In Siddig and Hughes (1969) pp 105–17.

Farbrother, H. G. (1972) Field behaviour of Gezira clay under irrigation. *Cott. Gr. Rev.* **49**:1–27. (Res. Mem. No 82).

Farbrother, H. G. (1973) Crop physiology. Cott. Res. Rep., Sudan, 1970–71, pp 36–101. (Res. Mem. No 89).

Farbrother, H. G. and **Harrison, L. E.** (1957) On an electrical resistance technique for the study of soil moisture problems in the field. *Cott. Gr. Rev.* **34**:71–92. (Res. Mem. No 27).

Farbrother, H. G. and **Munro, J. M.** (1970) Water. In Jameson (1970) Ch. 4 pp 30–42. (Res. Mem. No 81).

Faulkner, R. C. (1972) Cotton seed multiplication in the northern states of Nigeria. *Cott. Gr. Rev.* **49**:126–48.

Feaster, C. V. and **Turcotte, E. L.** (1962) Genetic basis for varietal improvement of Pima cotton. ARS 34–31, CR, ARS, USDA.

Feaster, C. V. and **Turcotte, E. L.** (1970) Breeding methods for improving Pima cotton and their implications on variety maintenance. *Crop. Sci.* **10**:707–9.

Fernald, H. T. and **Shepard, H. H.** (1942) *Applied Entomology.* McGraw-Hill: New York and London.

Fielding, W. L. (1948) The determination of staple length in single plant selections and variety trials. *J. Agric. Sci. Camb.* **38**:158–62. (Res. Mem. No 6).

Finney, D. J. (1963) Plant breeding, variety trials and statistical methods. *Cott. Gr. Rev.* **40**:161–9.

Fisher, W. D. (1973) Association of temperature and boll set. *Beltwide Conf.*: 72.

Foster, H. L. (1972) The identification of potentially potassium deficient soils in Uganda. *E. Afr. Agric. For. J.* **37**:224–33.

Freeman, W. E. (1946) 'Samaru 26C': a new strain of cotton bred in Northern Nigeria. *Trop. Agric.* **23**:109–13.

Fryxell, P. A. (1965) Stages in the evolution of *Gossypium* L. *Adv. Frontiers Pl. Sc.* **10**:31–56.

Fryxell, P. A. (1968) A redefinition of the tribe Gossypieae. *Bot. Gaz.* **129**:296–308.

Fryxell, P. A. (1979) *The natural history of the cotton tribe*. Texas A & M University Press: College Station and London.

Garber, R. H., De Vay, J. E. and **Wakeman, R. J.** (1981) Effect of plant spacing and symptom expression in *Verticillium* wilt of cotton. *Beltwide Conf.*: 30.

Garnett, R. P. (1980) A low volume herbicide applioator for tropical small holder farmers. Proc. 1980 British Crop Protection Conf. – Weeds, pp 629–36.

Gausi, R. K. (1967) Plant breeding. Cott. Res. Rep., Malawi, 1965–66, pp 4–5.

Geering, Q. A. and **Baillie, A. F. H.** (1954) The biology of red bollworm, *Diparopsis watersi* (Roths.), in Northern Nigeria. *Bull. Ento, Res.* **45**:661–81. (Res. Mem. No 20).

Geus, J. G. de (1973) *Fertilizer guide for the tropics and subtropics*. Centre d'Etude de l'Azote, Zurich.

Gillham, F. E. M. (1965) Evolutionary significance of glands and their importance in cultivated cotton. *Cott. Gr. Rev.* **42**:101–3.

Gipson, J. R. (1970) Temperature-variety interrelationships in cotton. 2. Seed development and composition. *Cott. Gr. Rev.* **47**:264–71.

Gipson, J. R. (1980) Integration of events in boll development: temperature effects. *Beltwide Conf.*: 344–5.

Gipson, J. R. and **Joham, H. E.** (1968) Influence of night temperature on growth and development of cotton: I. Fruiting and boll development. II. Fibre properties. *Agron. J.* **60**:292–8.

Gipson, J. R. and **Joham, H. E.** (1969) Influence of night temperature on growth and development of cotton: IV. Seed quality. *Agron. J.* **61**:365–7.

Gipson, J. R. and **Ray, L. L.** (1970) Temperature-variety interrelationships in cotton. 1. Boll and fibre development. *Cott. Gr. Rev.* **47**:257–63.

Glover, J. and **Robinson, P.** (1953) A simple method for assessing the reliability of rainfall. *J. Agric. Sci.* **43**:275–80.

Goebel, S. (1968) Travaux de selection effectués sur les triple hybrides d'origine interspecifique HAR et ATH en Cote d'Ivoire (station de Bouake). *Coton Fibr. Trop.* **23**:212–18.

Goldsworthy, P. R. (1967) Responses of cereals to fertilizers in Northern Nigeria. I. Sorghum. II. Maize. *Expl. Agric. Camb.* **3**:29–40; 263–73.

Goodman, A. (1955) The effect of cloudiness upon the shedding of fruiting points from cotton at Tokar Delta. *Cott. Gr. Rev.* **32**:24–30.

Goodman, A. (1956) The effect of leaf, bud and fruit pruning upon X 1730 A cotton at Tokar, Sudan. *Cott. Gr. Rev.* **33**:24–34.

Gower, J. and **Matthews, G. A.** (1971) Cotton development in the southern region of Malawi. *Cott. Gr. Rev.* **48**:2–18.

Green, J. M. and **Brinkerhoff, L. A.** (1956) Inheritance of three genes for bacterial blight resistance in Upland cotton. *Agron. J.* **48**:481–5.

Green, M. B., Hartley, G. S. and **West, T. F.** (1979) *Chemicals for Crop Protection and Pest Control* (reprinted). Pergamon Press: Oxford.

Gregory, F. G., Crowther, F. and **Lambert, A. R.** (1932) The interrelation of factors controlling the production of cotton under irrigation in the Sudan. *J. Agric. Sci.* **22**:617–38.

Griffith, G. ap (1938) A note on termite hills. *E. Afr. Agric. For. J.* **4**:70–1.

Grimes, D. W., Miller, R. J. and **Dickens, W. L.** (1970) Water stress during flowering of cotton. *Calif. Agric.* **24**:4–6.

Gulati, A. M. and **Turner, A. J.** (1928) A note on the early history of cotton. Tech. Lab. Bull. No 17. Indian Central Cott. Cttee. (Quoted by Hutchinson *et al.* 1947).

Halley, R. J. (ed.) (1982) *Primrose McConnell's The Agricultural Notebook* (17th edn). Butterworth: London.

Halliwell, R. S. (1966) Virus diseases of cotton. *Beltwide Conf.*: 121–8.

Halloin, J. M. (1976) Inhibition of cottonseed germination with abscisic acid and its reversal. *Plant Physiol.* **57**:454–5.

Halloin, J. M. (1981) Weathering: changes in planting seed quality between ripening and harvest. *Beltwide Conf.*: 286–9.

Hardingham, M. (1978) *Illustrated Dictionary of Fabrics.* Macmillan (Studio Vista): London.

Hardwick, D. F. (1965) The corn earworm complex. *Mem. Ento. Soc. Canada* **40**:1–248. (Cited by Siddig and Hughes 1969).

Hargreaves, H. (1948) *List of recorded cotton insects of the world.* Comm. Inst. of Entomology: London.

Harland, S. C. (1917) Manurial experiments with Sea Island cotton in St. Vincent. *West Indian Bulletin* **16**:169–202.

Harland, S. C. (1921) In discussion of cotton crop forecasting. World Cotton Conf., Liverpool and Manchester, pp 194–5.

Harland, S. C. (1932) The genetics of Gossypium. *Bibl. Genet.* **9**:107–82.

Harland, S. C. (1936) The genetical conception of the species. *Biol. Rev.* **11**:82.

Harland, S. C. (1939) *The Genetics of Cotton.* Jonathan Cape: London.

Harland, S. C. (1940) Taxonomic relationships in the genus Gossypium. *J. Wash. Acad. Sci.* **30**:426–32.

Harland, S. C. (1944) Institute of Cotton Genetics, Sociedad Nacional Agraria, Peru: Bull. No 1. (Quoted by Harland 1949).

Harland, S. C. (1949) Selection experiments with Peruvian Tanguis cotton. *Cott. Gr. Rev.* **26**:163–74; 247–55.

Harland, S. C. (1953) Some impressions of cotton in Chile. *Cott. Gr. Rev.* **30**:267–72.

Harmer, E. G. (1957) Grading and marketing of cotton in Northern Nigeria. West African Cott. Res. Conf. (App. XXVII). Min. of Agric.: Samaru, Nigeria.

Harrington, J. B. (1937) The mass-pedigree method in the hybridization improvement of cereals. *J. Amer. Soc. Agron.* **34**:270–4.

Hartley, C. W. S. (1977) *The oil palm* (2nd edn). Longman: London.

Hartstack, A. W. Jr *et al.* (1980) New trap designs and pheromone bait formulations for *Heliothis. Beltwide Conf.*: 132–6.

Harvey, P. N. (1952) *The practice of arable crop experimentation.* Norfolk Agricultural Station: Sprowston, Norwich.

Hassan, H. M. (1969) Progress in chemical control of pests of cotton in the Gezira. In Siddig and Hughes (1969) pp 232–46.

Hawtree, J. N. (1980) Weeds and cotton. *Outlook on Agriculture* **10**:184–90.

Hayward, A. C. (1964) Bacteriophage sensitivity and biochemical group in *Xanthomonas malvacearum. J. General Microbiol* **35**:287–98.

Hayward, J. A. (1975) General review. Cott. Res. Rep., Nigeria, 1973–74, pp 4 and 7.

Hearn, A. B. (1970) Crop physiology. Cott. Res. Rep., Uganda, 1969–70, pp 36–40.

Hearn, A. B. (1972a) Cotton spacing experiments in Uganda. *J. Agric. Sci. Camb.* **78**:13–25. (Res. Mem. No 83).

Hearn, A. B. (1972b) The growth and performance of raingrown cotton in a tropical upland environmental. I. and II. *J. Agric. Sci. Camb.* **79**:121–45. (Res. Mem. No 86).

Hearn, A. B. (1975) Ord Valley cotton crop: development of a technology. *Cott. Gr. Rev.* **52**:77–102.

Hearn, A. B. (1976) Crop physiology. In Arnold (1976) Ch. 4.

Hearn, A. B. (1980) Water relationships in cotton. *Outlook on Agric.* **10**:159–66.

Hearn, A. B. (1981) Cotton nutrition. *Field Crop Abstr.* **34**:11–34.

Hector, J. M. (1936) *Introduction to the Botany of Field Crops* vol. 2. Central News Agency: Johannesburg.

Hesketh, J. D., Baker, D. N. and **Duncan, W. C.** (1972) Simulation of growth and yield in cotton. II. Environmental control of morphogenesis. *Crop Sci.* **12**:436–9.

Hesketh, J. D. and **Low, A.** (1968) Effect of temperature on components of yield and fibre quality of cotton varieties of diverse origin. *Cott. Gr. Rev.* **45**:243–57.

Hesse, P. R. (1955) A chemical and physical study of the soils of termite mounds in East Africa. *J. Ecol.* **43**:449–61.

Higgins, G. M. (1961) Experimental farm coverage in Northern Nigeria. *Nigerian Geographical J.* **4**:11–25.

Hinkle, D. A. and **Brown, A. L.** (1968) Secondary nutrients and micronutrients. In Elliott *et al.* (1968) Ch. 10.

Hise, B. R. and **Ethridge, D. E.** (1980) Cottonseed processing costs and returns under alternative technologies. *Beltwide Conf.*: 242–5.

Holman, L. E. and **Snitzler, J. R.** (1961) Transporting, handling and storing seeds. *Seeds, the Yearbook of Agriculture 1961.* USDA: Washington DC. pp 338–47.

Holstun, J. R. Jr and **Wooten, O. B.** (1968) Weeds and their control. In Elliott *et al.* (1968) Ch. 6.

Huffacker, C. B. (ed.) (1971) *Biological Control.* Plenum Publ. Corp.: New York and London.

Hughes, L. C. (1966) Factors affecting number of ovules per loculus in cotton. *Cott. Gr. Rev.* **43**:273–85.

Hughes, L. C. (1968) Motes in varieties of *Gossypium barbadense. Cott. Gr. Rev.* **45**:266–80.

Hughes, L. C. (1969) Varieties in the Gezira environment. 1. Long staple varieties. In Siddig and Hughes (1969) pp 164–72.

Hunter, R. E., Brinkerhoff, L. A. and **Bird, L. S.** (1968) The development of a set of upland cotton lines for differentiating races of *Xanthomonas malvacearum. Phytopathology* **58**:830–2.

Husain, M. A. *et al.* (1931) Studies on *Platyedra gossypiella* Saunders in the Punjab. Pt. II. The sources of *P. gossypiella* infestation. *Ind. J. Agric. Sci.* **1**:204–85.

Hussain, M. (1963) Land reclamation operation. West Pakistan Engineering Congress, Lahore.

Hutchinson, J. (1959) *The Families of Flowering Plants. I. Dicotyledons* (2nd edn). Oxford Univ. Press: London.

Hutchinson, J. B. (1938) Some problems in genetics, whose solution would help the plant breeders. 3rd Conf. on Cotton Growing Problems. ECGC: London. pp 132–3.

Hutchinson, J. B. (1940) The application of genetics to plant breeding. I. The genetic interpretation of plant breeding problems. *J. Genet.* **40**:271–82.

Hutchinson, J. B. (1947) Notes on the classification and distribution of genera related to Gossypium. *New Phytol.* **46**:123–41. (Res. Mem. No 2).

Hutchinson, J. B. (1950) A note on some geographical races of asiatic cottons. *Cott. Gr. Rev.* **27**:123–7.

Hutchinson, J. B. (1959) *The Application of Genetics to Cotton Improvement.* Camb. Univ. Press: London.

Hutchinson, J. B. (1962) The history and relationships of the world's cottons. *Endeavour* **21**:5–15.

Hutchinson, J. B. and **Ghose, R. L. M.** (1937a) Studies in crop ecology I. The composition of the cotton crops of Central India and Rajputana. *Ind. J. Agric. Sci.* **7**:1–34.

Hutchinson, J. B. and **Ghose, R. L. M.** (1937b) The classification of the cottons of Asia and Africa. *Ind. J. Agric. Sci.* **7**:233–57.

Hutchinson, J. B., Knight, R. L. and **Pearson, E. O.** (1950) Response of cotton to leafcurl disease. *J. Genet.* **50**:100–11. (Res. Mem. No 8).

Hutchinson, J. B., Manning, H. L. and **Farbrother, H. G.** (1958a) On the characterisation of tropical rainstorms. *Quart. J. Royal Met. Soc.* **84**:250–8. (Res. Mem. No 30).

Hutchinson, J. B., Manning, H. L. and **Farbrother, H. G.** (1958b) Crop water requirements of cotton. *J. Agric. Sci. Camb.* **51**:177–88. (Res. Mem. No 33).

Hutchinson, J. B. and **Panse, V. G.** (1937) Studies in plant breeding techniques: II. The design of field tests of plant breeding materials. *Indian J. Agric. Sci.* **7**:531–64.

Hutchinson, J. B., Silow, R. A. and **Stephens, S. G.** (1947) *The Evolution of Gossypium*. Oxford Univ. Press: London.

Ibrahim, F. M. (1966) A new race of cotton-wilt *Fusarium* in the Sudan Gezira. *Cott. Gr. Rev.* **43**:296–9.

Ibrahim, F. M. and **Khalifa, O.** (1969) Fusarium wilt of cotton in the Gezira. In Siddig and Hughes (1969) pp 189–95.

ICAC (1959 etc.) *Cotton – World Statistics* (published quarterly). International Cotton Advisory Committee: Washington DC.

ICAC (1977) Bale standardization. ICAC: Washington DC.

ICAC (1979) Cotton production survey. ICAC: Washington DC.

ICAC (1983) *Cotton – World Statistics* (special base book issue, Oct 1983). ICAC: Washington DC.

ICCC (1960) *Cotton in India* (4 vols). Indian Central Cotton Committee: Bombay.

ICI (1979) *Cotton pest identification manual – American continent*. ICI Plant Protection Division: Fernhurst, England.

ICI (1981) *Cotton pest identification manual* (Old World). ICI Plant Protection Division: Fernhurst, England.

Idris, H. (1969) Chemical control of weeds in cotton. In Siddig and Hughes (1969) pp 213–31.

IIC (1974) (International Institute for Cotton) Open end spinning of cotton. *Cott. Gr. Rev.* **51**:167–76.

IIC (1982) Cotton. *Textiles, Manchester* **11**:58–64.

Indonesia (1979) Proposal for a nucleus estate and small-holders cotton development in South Sulawesi. Dept. of Agric., Indonesia (mimeographed).

INIAP (1980) Estación experimental Boliche, Programa de Algodón, Informe Anual 1980. Instituto Nacional de Investigaciones Agropecuarias, Ecuador (mimeographed).

Innes, N. L. (1961) Natural crossing in cotton in the Sudan. *Cott. Gr. Rev.* **38**:17–21.

Innes, N. L. (1971) Impressions of cotton poduction and research in India. *Cott. Gr. Rev.* **48**:163–74.

Innes, N. L., Gridley, H. E. and **Busulwa, L. N.** (1973) Cotton breeding. Cott. Res. Rep., Uganda, 1971–72, pp 13–28.

IRCT (1955) L'Institut de Recherches du Coton et des Textiles Exotiques. *Cott. Gr. Rev.* **32**:31–6.

IRCT (1978) *Variétés récentes de cotonniers*. Division de Génétique, IRCT: Paris.

Isaac, I. (1967) Speciation in *Verticillium*. *Ann. Rev. Phytopath.* **5**:201–22.

Ivy, E. E. (1944) Tests with D.D.T. on the more important cotton insects. *J. Econ. Ento* **37**:142.

Jameson, J. D. (ed.) (1970) *Agriculture in Uganda*. Oxford Univ. Press: London.

Jameson, J. D. and **Stephens, D.** (1970) Crop estimation. In Jameson (1970) Ch. 16.

Joham, H. E. (1951) The nutritional status of the cotton plant as indicated by tissue tests. *Plant Physiol.* **26**:76–89.

Johnson, S. H. III (1982) Large scale irrigation and drainage schemes in Pakistan – a study of rigidities in decision making. *Food Research Institute Studies* **18**:149–80.

Johnson, W. H. (1926) *Cotton and its Production*. Macmillan: London.

Johnstone, D. R. (1971) Droplet size for low and ultra low volume aerial spraying. *Cott. Gr. Rev.* **48**:218–33.

Jones, E. (1972) Principles for using fertilizers to improve red ferralitic soils in Uganda. *Expl. Agric.* **8**:315–32. (Res. Mem. No 87).

Jones, E. (1976) Soil productivity. In Arnold (1976) Ch. 3.

Jones, G. B. (1973) Improving the SATU variety in Uganda. *Cott. Gr. Rev.* **50**:218–44.

Jones, J. E. and **Andries, J. A.** (1969) Effect of frego bract on the incidence of cotton boll rot. *Crop Sci.* **9**:426–8.

Jones, J. K. and **Slater, G. A.** (1976) Dilute sulphuric acid cotton planting seed delinting process. *Agro-Industrial Report* **3**:1–26. Cotton Inc.: Memphis, Tennessee.

Jong, J. de (1965) *Work study on selected cotton operations.* Planning Branch, Dept. of Conservation and Extension, Rhodesia.

Jong, J. de (ed.) (1982) *Farm Management Handbook* (3rd edn) Part 1. Government Printer: Harare, Zimbabwe.

Joyce, R. J. V. and **Roberts, P.** (1959) The determination of the size of plot suitable for cotton spraying experiments in the Sudan Gezira. *Ann. Appl. Biol.* **47**:287–305.

Kabaara, A. M. (1965) Soil phosphate. Cott. Res. Rep., Uganda, 1964–65, pp 29–34.

Kammacher, P. (1965) *Etudes des relations génétique et, caryologiques entre genomes voisin du genre Gossypium.* IRCT: Paris.

Kasasian, L. (1969) Weed control in cotton. *Cott. Gr. Rev.* **46**:165–73.

Kearney, T. H. (1930) Cotton plants, tame and wild. *J. Hered.* **21**:195.

Keller, K. R. (1981) Trial evaluations and decision schedules. *Beltwide Conf.*: 141–3.

Kemp, D. C. and **Matthews, M. D. P.** (1982) The development of a stalk pulling machine. *J. Agric. Engng. Res.* **27**:201–13.

Kenya (1978, 1979) Annual Reports of Tebere Cotton Research Station. Dept. of Agric.: Nairobi.

Kerkhoven, G. J. (1963) Increase of yield potential of raingrown cotton in South Africa. *Cott. Gr. Rev.* **40**:83–97.

Kerkhoven, G. J. (1964) Cotton on tropical black clay, Kafue Flats, Northern Rhodesia. *Cott. Gr. Rev.* **41**:2–12.

Kibukamusoke, D. E. B. (1958) A note on a more precise method of estimation of the Uganda cotton crop. *Cott. Gr. Rev.* **35**:91–100.

King, H. E. (1953) Uniformity trial. Cott. Res. Rep., Northern Nigeria, 1951–52, pp 9–10.

King, H. E. (1957) Cotton seed multiplication in Northern Nigeria. West African Cott. Res. Conf. (App. XXII). Min. of Agric.: Samaru, Nigeria.

King, H. E. and **Lawes, D. A.** (1957) Breeding systems used for cotton at Samaru. West African Cott. Res. Conf. (App. X). Min. of Agric.: Samaru, Nigeria.

King, H. E. and **Lee, B. J. S.** (1957) Genetic differences in vitality of cotton plants cut back at the end of a season. West African Cott. Res. Conf. (App. XI). Min. of Agric.: Samaru, Nigeria.

Knight, R. L. (1944) The genetics of blackarm resistance IV. *G. punctatum* crosses. *J. Genet.* **46**:1–27.

Knight, R. L. (1945) The theory and application of the backcross technique in cotton breeding. *J. Genet.* **47**:76–86.

Knight, R. L. (1946) Breeding cotton resistant to blackarm disease. *Emp. J. Exp. Agric.* **14**:153–74. (Res. Mem. No 1).

Knight, R. L. (1954) The genetics of blackarm resistance. XI. *Gossypium anomalum. J. Genet.* **52**:466–72.

Knight, R. L. (1963) The genetics of blackarm resistance. XII. Transference of resistance from *Gossypium herbaceum* to *G, barbadense. J. Genet.* **58**:328–46.

Knight, R. L. and **Clouston, T. W.** (1939) The genetics of blackarm resistance. I. Factors B_1 and B_2. *J. Genet.* **38**:135–59.

Konzak, C. F. (1956) Induction of mutations for disease resistance in cereals. *Brookhaven Symposia in Biology* No 9, pp 157–76. (Quoted by Allard 1960).

La Croix, E. A. S. (1966) Stainer bugs (*Dysdercus* spp.) in Coast Province, Kenya. *Cott. Gr. Rev.* **43**:41–55.

Lagière, R. (1960) *La bacteriose du cotonnier dans le monde et en République Centrafricaine* (Oubanquichari). IRCT: Paris.

Last, F. T. and **Dransfield, M.** (1959) Measurement of the resistance of cotton to

Xanthomonas malvacearum (E.F.Sm.) Dowson using a leaf abrasion technique. *Cott. Gr. Rev.* **36**:196–9.

Lawes, D. A. (1955) A note on the effect of 2,4-D weedkiller on cotton. *Cott. Gr. Rev.* **32**:274–6. (Drawing by J. H. Saunders).

Lawes, D. A. (1961) Rainfall conservation and the yield of cotton in Northern Nigeria. *Emp. J. Exp. Agric.* **29**:307–18. (Res. Mem. No 44).

Lawes, D. A. (1964) Crop physiology. Cott. Res. Rep., Northern Nigeria, 1962–63, pp 8–12.

Leakey, C. L. A. and **Perry, D. A.** (1966) The relation between damage caused by insect pests and bollrot associated with *Glomerella cingulata* on upland cotton in Uganda. *Ann. Appl. Biol.* **37**:337–44.

Lee, B. J. S. (1962) Modal and positive bulks in plant breeding. *Cott. Gr. Rev.* **39**:181–7.

Lee, B. J. S. (1966) Cotton in Western Nigeria. 1. History and recent developments. *Cott. Gr. Rev.* **43**:85–97.

Lee, B. J. S., Walton, I. C. and **Jackson, A. C.** (1974) Breeding and Agronomy. Cott. Res. Rep., Tanzania, 1971–72 pp 8–24.

Lee, J. A. (1966) Some prospects for breeding more glandular cottons. *Beltwide Conf.*: 209–14.

Leffler, H. R. (1981) Developmental aspects of cotton seed planting quality. *Beltwide Conf.*: 283–6.

Le Mare, P. H. (1963) A comparison of the minus-one and 2^n designs for the exploratory investigation of fertilizer requirements. *E. Afr. Agric. For. J.* **28**:213–8.

Le Mare, P. H. (1968) Experiments on the effects of phosphate applied to a Buganda soil. *J. Agric. Sci. Camb.* **70**:265–70 (Pot expts); 271–9 (Field expts); 281–5 (Soil phosphate). (Res. Mem. No 70).

Le Mare, P. H. (1972) A long term experiment on soil fertility and cotton yield in Tanzania. *Expl. Agric.* **8**:299–310. (Res. Mem. No 88).

Le Mare, P. H. (1974) Responses of cotton to ammonium sulphate and superphosphate, and relationships between yield and some soil constituents in Tanzania. *J. Agric. Sci. Camb.* **83**:47–56. (Res. Mem. No 92).

Lewis, C. F. (1970) Concepts of varietal maintenance in cotton. *Cott. Gr. Rev.* **47**:272–84.

Lewton, F. L. (1912) *Kokia*: a new genus of Hawaiian trees. *Smithsonian Misc. Coll.* **60**(5):1–4. (Quoted by Hutchinson *et al.* 1947).

Lewton, F. L. (1915) The Australian Fugosias. *J. Wash. Acad. Sci.* **5**:303. (Quoted by Hutchinson *et al.* 1947).

Lineham, S. (1965) *Expected dates of planting rains in Zambia and N. Mashonaland.* Dept. of Met. Services, Zimbabwe.

Little, E. C. S. (1965) *PANS* (C) **11**:113. (Quoted by Hawtree 1980).

Logan, C. (1958) Bacterial boll rot of cotton I. A comparison of two inoculation techniques for the assessment of host resistance. *Ann. Appl. Biol.* **46**:230–42. (Res. Mem. No 31).

Lord, E. (1945) The production and characteristics of the world's cotton crops. I. The West Indies. *Shirley Inst. Mem.* **19**:315–40.

Lord, E. (1947) On the grade and staple of cotton. *Cott. Gr. Rev.* **24**:186–96.

Lord, E. (1961) Characteristics of raw cotton. In *Manual of Cotton Spinning* vol. 2 part 1. Textile Institute and Butterworths: London.

Lord, E. (1962) Difficulties in making cotton and spinning tests and pitfalls in interpreting the results. *Cott. Gr. Rev.* **39**:38–56.

Lord, E. (1981) *The origin and assessment of cotton fibre maturity.* International Institute for Cotton: Manchester.

Lord, E. and **Anthony, K. R. M.** (1960) Effect of exposure to the Aden climate on cotton lint strength. *Cott. Gr. Rev.* **37**:10–14.

Low, A. (1955) A selfing technique especially suitable for raingrown cotton. *Cott. Gr. Rev.* **32**:222–3.

Low, A. (1968) Developments toward tufted seed in varieties of *G. hirsutum. Cott. Gr. Rev.* **45**:101–14.

Low, A., Hesketh, J. and **Muramoto, H.** (1969) Some environmental effects on the varietal node number of the first fruiting branch. *Cott. Gr. Rev.* **46**:181–8.

Low, A. and **McMahon, J. P.** (1973) Development of narrow row high density cotton in Australia. *Cott. Gr. Rev.* **50**:130–49.

Luke, W. J. and **Pinckard, J. A.** (1970) The role of the bract in boll rots of cotton. *Cott. Gr. Rev.* **47**:20–8.

Lukefahr, M. J., Bottger, G. T. and **Maxwell, F.** (1966) Utilization of gossypol as a source of insect resistance. *Beltwide Conf.*: 215–22.

Lukefahr, M. J., Houghtaling, J. E. and **Cruhm, D. G.** (1975) Suppression of *Heliothis* spp. with cottons containing combinations of resistant characters. *J. Econ. Ent.* **68**:743–6.

Lukefahr, M. J., Jones, J. E. and **Houghtaling, J. E.** (1976) Fleahopper and leafhopper populations and agronomic evaluations of glabrous cottons from different genetic sources. *Beltwide Conf.*: 84–6.

Lyon, D. J. de B. (1970) Fleabeetle attack on glandless cotton in Nigeria, *Cott. Gr. Rev.* **47**:198–202.

Lyon, T. L. and **Buckman, H. O.** (1949) *The nature and properties of soils* (4th edn). Macmillan: New York.

McClelland, C. K. (1916) On the regularity of blooming in the cotton plant. *Science* **44**:578–81.

McClelland, C. K. and **Neely, J. W.** (1931) The order, rate and regularity of blooming in the cotton plant. *J. Agric. Res.* **42**:751–63.

MacDonald, D., Fielding, W. L. and **Ruston, D. F.** (1947) Effects of gapfilling on development and yield. *J. Agric. Sci.* **37**:297–300. (Res. Mem. No 6).

McKinlay, K. S. (1953) Use of repellants to simplify insecticide field trials. *Nature, London* **171**:658.

McKinlay, K. S., Geering, Q. A. and **Coaker, T. H.** (1957) Studies of crop loss following insect attack on cotton in East Africa. *Bull. Ent. Res.* **48**:833–66. (Res. Mem. No 28).

McKinley, D. J. (1965) Occurrence of *Paurocephala gossypii* on cotton in Malawi. *Cott. Gr. Rev.* **42**:209–10.

McKinley, D. J. (1968) Key to some larvae of Lepidoptera attacking cotton in Central Africa. *Cott. Gr. Rev.* **45**:184–97.

McKinstry, A. H. and **Prentice, A. N.** (1937) Progress Rep. Exp. Stas., Southern Rhodesia, 1935–36, p 60.

McMahon, J. and **Low, A.** (1972) Growing degree days as a measure of temperature effects on cotton. *Cott. Gr. Rev.* **49**:39–49.

McMichael, B. L. and **Powell, R. D.** (1971) Effect of temperature regimes on flowering and boll development in cotton. *Cott. Gr. Rev.* **48**:125–30.

McMichael, S. C. (1960) Combined effects of the glandless genes gl_2 and gl_3 on the pigment glands in the cotton plant. *Agron. J.* **52**:385–7.

Main, T. F. (1912) Cambodia cotton in Bombay. *Agric. J. India* **7**:373. (Quoted in Hutchinson *et al.* 1947).

Malawi (1976) *Cotton handbook of Malawi.* Min. of Agric. and Natural Resources: Lilongwe, Malawi.

Malm, N. R. and **Kerby, T. A.** (1981) Association between cotton yields and heat units in New Mexico over a 25-year period. *Beltwide Conf.*: 55–7.

Manning, H. L. (1948) Agronomy – Sowing date trials. Progress Rep. Exp. Stas., Uganda, 1946–47, pp 84–92.

Manning, H. L. (1951) Confidence limits of expected monthly rainfall. *J. Agric. Sci.* **40**:169–76. (Res. Mem. No 9).

Manning, H. L. (1952) Forecasting the Uganda cotton crop. *Cott. Gr. Rev.* **29**:241–57.

Manning, H. L. (1955) Response to selection for yield in cotton. Cold Spring Harbour Symposia on Quantitative Biology **20**:103–10.

Manning, H. L. (1956a) The statistical assessment of rainfall probability and its application in Uganda agriculture. *Proc. Royal Society* Ser. B **144**:460–80. (Res. Mem. No 23).

Manning, H. L. (1956b) Yield improvement from a selection index technique with cotton. *Heredity* **10**:303–22. (Res. Mem. No 26).

Manning, H. L. (1963) Realized yield improvement from twelve generations of progeny selection in a variety of upland cotton. In W. D. Hanson and H. F. Robinson (eds) *Statistical Genetics and Plant Breeding*. Publication No 982. NAS-NRC: Washington DC. pp 329–51. (Res. Mem. No 50).

Manning, H. L. and **Kibukamusoke, D. E. B.** (1958) The cotton crop. Cott. Res. Rep., Uganda, 1957–58, pp 3–9.

Marani, A. and **Dag, J.** (1962) Germination of seeds of cotton varieties at low temperature. *Crop Sci.* **2**:267.

Marani, A. and **Horwitz, M.** (1963) Growth and yield of cotton as affected by the time of a single irrigation. *Agron. J.* **55**:219–22.

Marr, J. C. and **Hemphill, R. G.** (1928) *Irrigation of cotton*. USDA Tech. Bull. No 72.

Martin, H. and **Worthing, C. R.** (eds) (1977) *Pesticide manual* (5th edn). British Crop Protection Council: London.

Martin, R. D., Ballard, W. W. and **Simpson, D. M.** (1923) Growth of fruiting parts in the cotton plant. *Agric. Res.* **25**:195–208.

Mason, T. G. (1938) *Cott. Gr. Rev.* **15**:113. (Quoted by Hutchinson 1959).

Matthews, G. A. (1972) Agronomy – Loss of stand. Cott. Res. Rep., Malawi, 1969–70, pp 11–12.

Matthews, G. A. (1981) Development in pesticide application for the small-scale farmer in the tropics. *Outlook on Agriculture* **10**:345–9.

Matthews, G. A. (1982) New developments in pesticide-application technology. *Crop Protection* **1**:131–45.

Matthews, G. A. and **Tunstall, J. P.** (1968) Scouting for pests and the timing of spray applications. *Cott. Gr. Rev.* **45**:115–27.

Matthews, G. A., Tunstall, J. P. and **McKinley, D. J.** (1965) Outbreaks of pink bollworm, *Pectinophora gossypiella* Saund., in Rhodesia and Malawi. *Cott. Gr. Rev.* **42**:197–208.

Matthews, M. D. P. and **Pullen, D. W. M.** (1976) Cultivation trials with ox-drawn implements using N'dama cattle in the Gambia. Overseas Dept. Report, National Inst. of Agric. Engineering, Silsoe.

Mauney, J. R. (1966) Some effects of environment on floral initiation in Upland cotton. *Beltwide Conf.*: 261–5.

Mauney, J. R. (1979) Production of fruiting points. *Beltwide Conf.*: 256–61.

Maxwell, F. G. (1980) Advances in breeding for resistance to cotton insects. *Beltwide Conf.*: 141–7.

Maxwell, F. G. and **Jennings, P.** (1980) *Breeding Plants Resistant to Insects*. Wiley: New York.

Maxwell Darling, R. C. (1958) The use of insecticides on cotton in Africa. Appendix to Pearson (1958).

Meadows, W. R. (1921) Universal standards for American cotton. World Cotton Conf., Liverpool and Manchester, pp 119–20.

Megie, C. (1963) Pluviometrie, date de semis et productivité du cotonnier (*G. hirsutum*) dans la région de Tikem (Tchad). *Coton Fibr. Trop.* **18**:251–62.

Mercer, W. B. and **Hall, A. D.** (1911) The experimental error in field trials. *J. Agric. Sci.* **4**:107–32.

Meredith, W. R. Jr (1980) Use of insect resistant germplasm in reducing the cost of production in the 1980s. *Beltwide Conf.* :307–10.

Mestanza, S. (1978) Suelos y fertilizantes. Informe Técnico, 1977. INIAP: Quito, Ecuador.

Meyer, J. R. and **Meyer, V. G.** (1961) Origin and inheritance of nectariless cotton. *Crop Sci.* **1**:167–9.

Meyer, V. G. (1974) Interspecific cotton breeding. *Econ. Bot.* **28**:56–60.
Miles, L. (1980) *Cotton*. World Resources Series. Wayland Publishers: Hove, England.
Miller, A. W. D. (1969) Spraying cotton – problems of the aerial applicator. *Rhod. Farmer* **40**:12–15.
Miller, E. (1973) *Textiles – Properties and Behaviour*. Batsford: London.
Miller, E. C. (1938) *Plant Physiology* (2nd edn). McGraw-Hill: New York.
Mirza, M. A. and **Shaikh, A. L.** (1980) The currently recognised species of *Gossypium*, their relationship and description. *Pakistan Cotton* **24**:283–98.
Monteith, J. L. (1973) *Principles of Environmental Physics*. Edward Arnold: London.
Moore, L. (1973) Lygus in cotton: management practices and control. *Cotton Gin and Oil Mill Press*, Sept 1973, p 14.
Morris, D. A. (1962) Elongation of lint hairs in Upland cotton in Uganda. *Cott. Gr. Rev.* **39**:270–6.
Morris, W. J. (1973) Ratoon cotton in Rhodesia. *Cott. Gr. Rev.* **50**:316–26.
Morton, N. (1973) The use of a hand-held ULV atomiser for cotton pest control. *PANS* **19**:548–56.
Moss, B. L. (1914) *The boll weevil problem*. The Progressive Farmer: Birmingham, Alabama.
Mowlam, M. D., Nyirenda, G. K. C. and **Tunstall, J. P.** (1975) Ultra low volume applications of water-based formulations of insecticides to cotton. *Cott. Gr. Rev.* **52**:360–70.
Munro, J. M. (1958) The summer cropping regime in the Lower River, Nyasaland. *Cott. Gr. Rev.* **35**:101–6.
Munro, J. M. (1966) Micronaire values. *Cott. Gr. Rev.* **43**:240–1.
Munro, J. M. (1968) Controls in a series of field experiments. *Cott. Gr. Rev.* **45**:42–53.
Munro, J. M. (1971) An analysis of earliness in cotton. *Cott. Gr. Rev.* **48**:28–41.
Munro, J. M. (1973) Cotton breeding. Cott. Res. Rep., Malawi, 1970–71, p 9.
Munro, J. M. (1974) Cotton breeding. Cott. Res. Rep., Malawi, 1971–72, pp 5–12.
Munro, J. M. and **Farbrother, H. G.** (1969) Composite plant diagrams in cotton. *Cott. Gr. Rev.* **46**:261–82.
Munro, J. M. and **Wood, R. A.** (1964) Water requirements of irrigated maize in Nyasaland. *Emp. J. Expt. Agric.* **32**:141–52. (Res. Mem. No 52).
National Cotton Council (1965) *Cotton counts its customers* (special edn 1956–64). Market Research Section, National Cotton Council of America: Memphis Tennessee.
National Cotton Council (1973) *Cotton counts its customers* (special edn 1965–71). Market Research Section, National Cotton Council of America: Memphis, Tennessee.
National Cotton Council (1983) *Cotton counts its customers* (1983 edn). Economic Section, National Cotton Council of America: Memphis, Tennessee.
Neal, D. C. (1953) Bacteria and fungi on seedlings. *Plant Diseases, the Yearbook of Agriculture* 1953. USDA: Washington DC. pp 311–14.
Nelson, R. R. (ed.) (1973) *Breeding plants for disease resistance*. Pennsylvania State University Press.
Niles, G. A. and **Richmond, T. R.** (1962) *Performance trials with selected strains of storm resistant cotton*. Texas Agric. Exp. Sta., Misc. Publ. 577. College Station, Texas.
Noggle, G. R. (1973) Cotton seed quality. *Cott. Gr. Rev.* **50**: 43–62.
Nonglak, S. and **Sompark, S.** (1980) Transmission of the casual agent of cotton leaf roll disease. 2nd S. E. Asian Symposium on Plant Diseases in the Tropics. Bangkok, Thailand.
Norman, D. W., Hayward, J. A. and **Hallam, H. R.** (1974) An assessment of cotton growing recommendations as applied by Nigerian farmers. *Cott. Gr. Rev.* **51**:266–80.
Nour, M. A. and **Nour, Jane J.** (1964) Identification, transmission and host range of leafcurl viruses infecting cotton in the Sudan. *Cott. Gr. Rev.* **41**:27–37.

Nowell, W. (1930) Conference on cotton growing problems. Empire Cotton Growing Corporation: London. p 141.

Nye, G. W. and **Hosking, H. R.** (1940) History and development of the cotton industry in Uganda. In Tothill (1940).

Ordish, G. (1969) Integrated control in Peruvian cotton. *PANS* **15**:37–40.

Pakistan (1979) *Cotton handbook of Pakistan* (revised edn). Pakistan Central Cotton Committee: Karachi.

Palomo, A. and **Davis D. D.** (1983) Response of an F_1 interspecific (*hirsutum & barbadense*) cotton hybrid to plant density in narrow rows. *Crop. Sci.* **23**:1053–6.

Pandey, J. (1984) Control of nutsedge in arable land. *Pesticides* **18**:38–46.

Parnell, F. R. (1933) Good picking. *Cott. Gr. Rev.* **10**:282–5.

Parnell, F. R. (1935) The origin and development of U4 cotton. *Cott. Gr. Rev.* **12**:177–82.

Parnell, F. R., **King, H. E.** and **Ruston, D. F.** (1949) Jassid resistance and hairiness of the cotton plant. *Bull. Ent. Res.* **39**:539–75. (Res. Mem. No 7).

Parry, G. (1982) *Le cotonnier et ses produits*. Maisonneuve et Larose: Paris.

Parsons, F. S. and **Hutchinson, H.** (1939) Investigations in Swaziland on the red bollworm. Progress Rep. Exp. Stas., South Africa, 1937–38, pp 35–9.

Parsons, F. S. and **Ulyett, G. C.** (1934) Investigations on the control of the American and red bollworms of cotton in South Africa. *Bull. Ent. Res.* **25**:349–81.

Parsons, F. S. and **Ulyett, G. C.** (1936) Investigations on *Trichogramma lutea*, Gir., as a parasite of the cotton bollworm, *Heliothis obsoleta*, Fabr. *Bull. Ent. Res.* **27**:219–35.

Paterson, D. D. (1939) *Statistical Technique in Agricultural Research*. McGraw Hill: New York and London.

Patti, J. H. and **Carner, G.R.** (1974) *Bacillus thuringiensis* investigations for the control of *Heliothis* spp. on cotton. *J. Econ. Ent.* **67**:415–18.

Peacock, H. A. and **Hawkins, B. S.** (1970) Effect of seed source on seedling vigour, yield and lint characteristics of Upland cotton. *Crop Sci.* **10**:667–70.

Pearson, E. O. (1954) The relationship between the African and the South American red bollworms of cotton, *Diparopsis* and *Sacadodes*. *Cott. Gr. Rev.* **31**:171–7.

Pearson, E. O. (1958) *The insect pests of cotton in tropical Africa*. ECGC and Comm. Inst. of Ent.: London.

Pearson, E. O., **Geering, Q. A.** and **McKinlay, K. S.** (1952) Entomology. Cott. Res. Rep., Uganda, 1951–52, pp 15–22.

Pearson, E. O. and **Mitchell, B. L.** (1945) *A report on the status and control of insect pests of cotton in the Lower River districts of Nyasaland*. Govt. Printer: Zomba, Malawi.

Pearson, N. L. (1949) *Mote types of cotton and their occurrence as related to variety, environment, position in lock, lock size and number of locks per boll*. USDA Tech. Bull. No 1000.

Peat, J. E. and **Brown, K. J.** (1961) A record of cotton breeding for the Lake Province of Tanganyika. *Emp. J. Exp. Agric.* **29**:119–35. (Res. Mem. No 42).

Peat, J. E. and **Brown, K. J.** (1962) The yield responses of raingrown cotton. Parts I and II. *Emp. J. Exp. Agric.* **30**:215–31; 305–14. (Res. Mem. No 48).

Peat, J. E. and **Prentice, A. N.** (1949) The maintenance of soil productivity in Sukumaland and adjacent areas, Tanganyika. *E. Afr. Agric. J.* **15**:48–56.

Penman, H. L. (1948) Natural evaporation from open water, bare soil and grass. *Proc. Royal Soc., London* Ser. A **193**:120–46.

Perry, D. A. (1962a) *Fusarium* wilt of cotton in the Lake Province of Tanganyika. *Cott. Gr. Rev.* **39**:14–21.

Perry, D. A. (1962b) Phytotoxic symptoms on cotton after spraying with D.D.T. *Cott. Gr. Rev.* **39**:203–5.

Phillips, L. L. (1966) The cytology and phylogenetics of the diploid species of *Gossypium*. *Amer. J. Bot.* **53**:328–55.

Phillips, L. L. (1974) Cotton (*Gossypium*). In King, R. C. (ed.) *Handbook of Genetics 2: Plants, plant viruses and protists*. Plenum Press: New York. pp 111–33.

Plimmer, J. R., Klun, J. A. and **Bierl-Leonhardt, B. A.** (1980) Pheromones for control of *Heliothis* on cotton. *Beltwide Conf.*: 148–51.

Pothecary, B. P. (1968) Destruction of old cotton for pest and disease control. *World Crops* Dec 1968:39–43.

Powell, R. D. (1969) Effect of temperature on boll set and development of *Gossypium hirsutum*. *Cott. Gr. Rev.* **46**:29–36.

Prentice, A. N. (1972) *Cotton*. Longman: London.

Presley, J. T. (1953) *Verticillium* wilt of cotton. *Plant Diseases, the Yearbook of Agriculture 1953*. USDA: Washington DC. pp 301–3.

Presley, J. T. (1958) Relation of protoplast permeability to cottonseed and predisposition to seedling disease. *Plant Dis. Reptr.* **42**:852.

Presley, J. T. and **Bird, L. S.** (1968) Diseases and their control. In Elliott *et al.* (1968) Ch. 12.

Proctor, J. H. (1962) The biology and control of the Sudan bollworm in the Abyan Delta, West Aden Protectorate. *Bull. Ent. Res.* **53**:311–35. (Res. Mem. No 47).

Pullman, G. S., de Vay, J. E., Garber, R. H. and **Weinhold, A. R.** (1979) Control of soil-borne fungal pathogens by plastic tarping of soil. In Schippers and Gams (1979) part 6, section 39.

Purseglove, J. W. (1960) Review of Hutchinson, J. B. (1959) *Trop. Agric., Trinidad* **37**:245–8.

Purseglove, J. W. (1963) Some problems of the origin and distribution of tropical crops. *Genetica Agraria* **17**:105–22.

Purseglove, J. W. (1968) *Tropical Crops: Dicotyledons*. Longman: London.

Quaintance, A. L. and **Brues, C. T.** (1905) *The cotton bollworm*. Bull. US Bur. Ent. No 50. (Quoted by Pearson 1958).

Rainey, R. C. (1948) Observations on the development of the cotton boll, with particular reference to changes in susceptibility to pests and diseases. *Ann. Appl. Biol.* **35**:64–83. (Res. Mem. No 5).

Ramachandran, N. (1983) Cottonseed oil milling. Annex 6 in Thailand Cotton Development Project. FAO: Rome.

Ramey, H. H. (1966) Historical review of cotton variety development. *Beltwide Conf.*: 310–26.

Ramey, H. H. Jr (1982) *The meaning and assessment of cotton fibre fineness*. International Institute for Cotton: Manchester.

Reed, W. (1965) *Heliothis armigera* in Western Tanganyika, Part 2. *Bull. Ent. Res.* **56**:127–40. (Res. Mem. No 61).

Reed, W. (1972) Uses and abuses of unsprayed controls in spraying trials. *Cott. Gr. Rev.* **49**:67–72.

Reed, W. (1974a) Populations and host-plant preferences of *Earias* spp. in East Africa. *Bull. Ent. Res.* **64**:33–44. (Res. Mem. No 91).

Reed, W. (1974b) The false codling moth as a pest of cotton in Uganda. *Cott. Gr. Rev.* **51**:213–25.

Reed, W. (1976) Entomology. In Arnold (1976) Ch. 5.

Reeves, J. D., Abernethy, J. R. and **Gipson, J. R.** (1982) Efficacy of selected herbicides on yellow and purple nutsedge. Abstract from Proc. 35th Ann. Meeting, Southern Weed Sci. Soc.

Revelle, R. (1964) *Report on land and water development in the Indus Plain*. US Govt. Printing Office: Washington DC.

Riggs, T. J. (1967a) Herbicide investigations. Cott. Res. Rep., Uganda, 1965–66, pp 56–60.

Riggs, T. J. (1967b) Response to modal selection in Upland cotton in northern and eastern Uganda. *Cott. Gr. Rev.* **44**:176–83.

Riggs, T. J. (1970) Trials of cotton seed mixtures in Uganda. *Cott. Gr. Rev.* **47**:100–11.

Rijks, D. A. (1965) The use of water by cotton crops in Abyan, South Arabia. *J. Appl. Ecol.* **2**:317–43. (Res. Mem. No 62).

Rijks, D. A. (1976) Agrometeorology. In Arnold (1976) Ch. 2.

Roberts, E. T. (1975) Cotton ginning. FAO Consultant Report (PAK/73/026). Central Cotton Research Institute: Multan.

Robinson, G. H., Magar, W. Y. and **Rai, K. D.** (1969) Soil properties in relation to cotton growth. In Siddig and Hughes (1969) pp 5–16.

Roe, H. B. (1950) *Moisture Requirements in Agriculture*. McGraw-Hill: London.

Rothwell, A., Bryden, J. W., Knight, H. and **Coxe, B. J.** (1967) Boron deficiency of cotton in Zambia. *Cott. Gr. Rev.* **44**:23–8.

Rounce, N. V. and **Thornton, D.** (1956) Ukara Island and the agricultural practices of the Wakara. *Cott. Gr. Rev,* **33**:255–63.

Rushtapakornchai, W. and **Vattanatungum, A.** (1981) *The 1981 Thailand pesticides handbook*. Dept. of Agric.: Bangkok.

Russell, E. W. (1950) *Soil conditions and plant growth* (8th edn). Longmans Green: London.

Russell, E. W. (1967) Climate and crop yields in the tropics. *Cott. Gr. Rev.* **44**:87–99.

Russell, M. B. (1939) Soil moisture sorption curves for four Iowa soils. *Proc. Soil Sci. Soc. Amer.* IV:51–4. (Quoted in Lyon and Buckman 1949).

Ruston, D. F. (1954) Main types of empire cottons. *Cott. Gr. Rev.* **31**:118–24.

Ruston, D. F. (1962) Effects of delay in sowing cotton. *Cott. Gr. Rev.* **39**:10–13.

Santhanam, V. and **Hutchinson, J. B.** (1974) Cotton. In Hutchinson, J. B. (ed.) *Evolutionary Studies in World Crops*. Camb. Univ. Press: London. pp 89–100.

Sappenfield, W. P. and **Dilday, R. H.** (1980) Breeding high terpenoid cottons – the 1978 regional tests. *Beltwide Conf.*: 92–5.

Satyanarayan, Y. and **Ranadive, J. P.** (1960) Ecological studies of the weeds of the cotton fields of Bombay State. *J. Univ. Bombay* **28**:42–7. (Quoted by Kasasian 1969).

Saunders, J. H. (1956) Bagging cotton flowers for breeding. II. Endrin treatment of paper bags. *Cott. Gr. Rev.* **33**:226–30.

Saunders, J. H. (1961) *The Wild Species of Gossypium*. Oxford Univ. Press: London.

Saunders, J. H. (1964) Genetics of hairiness transferred from *G. anomalum* to *G. barbadense*. *Cott. Gr. Rev.* **41**:16–22.

Saunders, J. H. (1972) The cytogenetics of Gossypium. In Prentice (1972) Ch. 5.

Schippers, B. and **Gams, W.** (eds) (1979) *Soil-borne Plant Pathogens*. Academic Press: New York.

Shenfield, A. A. (Chairman) (1962) *Report of the commission of inquiry into the cotton-ginning industry of Uganda*. Govt. Printer: Entebbe, Uganda.

Siddig, M. A. (1969) Breeding for leafcurl resistance in Sakel cotton. In Siddig and Hughes (1969) pp 153–8.

Siddig, M. A. and **Hughes, L. C.** (eds) (1969) *Cotton growth in the Gezira environment*. Agric. Res. Corp.: Wad Medani, Sudan.

Sikka, M. A. and **Dastur, R. H.** (1960) Climate and soils. In ICCC (1960) Vol. I. Ch. 2.

Silow, R. A. (1944) The genetics of species development in the Old World cottons. *J. Genet.* **46**:62–77.

Simmonds, N. W. (ed.) (1976) *Evolution of Crop Plants*. Longman: London.

Simpson, D. M. (1935) Dormancy and maturity of cottonseed. *J. Agric. Res.* **50**:429–34.

Simpson, I. G. (1969) An economic evaluation of cotton in the Gezira rotations. In Siddig and Hughes (1969) pp 58–66.

Singh, S., Singh, M., Singh, R. and **Barar, K. S.** (1970) Response to micronutrients of cotton in Northern India. *Cott. Gr. Rev.* **47**:191–7.

Singleton, W. R. (1955) The contribution of radiation genetics to agriculture. *Agron. J.* **47**:113–17.

Sión, F. and **Carvajal, T.** (1976) *Guia práctica para el cultivo de algodón*. Boletín divulgativo No 87. INIAP: Quito, Ecuador.

Skovsted, A. (1935) Chromosome numbers in the Malvaceae. *J. Genet.* **31**:263–96.

Skovsted, A. (1937) Cytological studies in cotton. IV. Chromosome conjunction in interspecific hybrids. *J. Genet.* **34**:97–134.

Smith, A. L. (1953) Anthracnose and some blights. *Plant Diseases, the Yearbook of Agriculture 1953*. USDA: Washington DC. pp 303–11.

Smith, C. E. Jr (1968) Plant remains. In Byers, D. S. (ed.) *Prehistory of the Tehuacan Valley*, I.: Austin, Texas. pp 220–55. (Quoted in Simmonds 1976).

Smith, C. W. and **Varvil, J. J.** (1984) Hybrid cotton yield trials in Arkansas. *Arkansas Farm Research* 33:7.

Smith E. F. (1920) *Bacterial Diseases of Plants*. W. B. Saunders: Philadelphia and London. (Quoted by Dowson 1957).

Smith, H. F. (1936) A discriminant function for plant selection. *Ann. Eugen.* 7:240–50.

Smith, K. M. (1972) *A Textbook of Plant Virus Diseases* (3rd edn). Longman: London.

Smith, K. M. (1976) *Virus-insect Relationships*. Longman: London.

Smith, R. and **Brown, R. A.** (1963) Improved methods of cotton growing in Eastern Region, Tanganyika. *Cott. Gr. Rev.* 40:268–77.

Smith, R. F. and **Falcon, L. A.** (1973) Insect control for cotton in California. *Cott. Gr. Rev.* 50:15–27.

Smith, R. L. and **Flint, H. M.** (1977) A bibliography of the cotton leaf perforator, *Bucculatrix thurberiella*, and a related species, *B. gossypiella*, that also feeds on cotton. *Bull. Ent. Soc. Amer.* 23:195–8.

Smithson, J. B. and **Heathcote, R. G.** (1977) Effect of rate and time of nitrogen application on yields of cotton in Northern Nigeria. *Expl. Agric.* 13:1–8.

Snedecor, G. W. (1946) *Statistical Methods* (4th edn). Iowa State College Press: Ames, Iowa.

Sorenson, J. W. and **Wilkes, L. H.** (1973) Seed quality and moisture relationships in harvesting and storing seed cotton. Seed Cotton Handling and Storing Seminar, Cotton Inc., Raleigh, N. C. (Quoted by Baskin 1981).

Soyer, D. (1947) Une nouvelle maladie du cotonnier, la psyllose provoquée par *Paurocephala gossypii* Russell. Publ. Inst. Nat. Etude Agron., Congo Belge, Ser. Sci. No 33. (Quoted by Pearson 1958).

Squire, F. A. (1940) On the nature and origin of the diapause in *Platyedra gossypiella* Saund. *Bull Ent. Res.* 31:1–6.

Stebbins, G. L. Jr. (1947) Evidence on rates of evolution from the distribution of existing and fossil plant species. *Ecol. Monogr.* 17:149–58. (Quoted by Fryxell 1979).

Stephens, D. (1969) Changes in yields and fertilizer responses with continuous cropping in Uganda. *Expt. Agric.* 5:263–9.

Stephens, D. (1970) Soil fertility. In Jameson (1970) Ch. 6.

Stephens, S. G. (1944) Canalization of gene action in the *Gossypium* leaf-shape system and its bearing on certain evolutionary systems. *J. Genet.* 46:345–57.

Stern, V. M. *et al.* (1969) Lygus bug control in cotton through alfalfa interplanting. *California Agric.* 23:8–10. (*Field Crop Abs.* 22:416).

Stewart, J. M. and **Hsu, C. I.** (1981) Interspecific hybrids of *Gossypium*. *Beltwide Conf.*: 83–4.

Stockton, J. R., Doneen, L. D. and **Walhood, V. T.** (1961) Boll shedding and growth of the cotton plant in relation to irrigation frequency. *Agron. J.* 53:272–5.

Stockton, J. R. *et al.* (1962) Precision tillage for cotton production. *California Agric.* 16:10–11.

Stride, G. O. (1964) Studies on the chemical basis of hostplant selection in the genus *Epilachna*. I. *J. Insect Physiol.* 11:21–.

Stride, G. O. (1968) On the biology and ecology of *Lygus vosseleri* with special reference to its hostplant relationships. *J. Ent. Soc., South Africa* 31:17–59. (Res. Mem. No 69).

Stride, G. O. (1969) Investigations into the use of a trap crop to protect cotton from attack by *Lygus vosseleri*. *J. Ent. Soc., South Africa* 32:469–77. (Res. Mem. No 79).

Sweeney, R. C. H. (1960) Cotton pest investigations in the Federation of Rhodesia and Nyasaland. II. Cotton stainer investigations. *Cott. Gr. Rev.* **37**:32–44.

Symonds, J. S. (1980) An evaluation of cotton versus synthetic fibres. *Outlook on Agriculture* **10**:210–14.

Taha, M. A., Malik, N., Choudrey, F. I. and **Makhdum, I.** (1981) Heat induced sterility in cotton sown during early April in West Punjab. *Expl. Agric.* **17**:189–94.

Tanzania (1953) *Tanganyika – the cotton industry, 1939–53.* Government Printer: Dar es Salaam.

Tarr, S. A. J. (1960) The effects of fungicide insecticide seed treatment on emergence, growth and yield of irrigated cotton in the Sudan Gezira. *Ann. Appl. Biol.* **48**:591–600.

Tarr, S. A. J. (1964) *Virus diseases of cotton.* Misc. Publ. Common. Mycol. Inst., No 18.

Taylor, T. H. C. (1936) Report on a year's investigation of *Platyedra gossypiella* (pink bollworm) in Uganda. Ann. Rep. Dept. Agric. Uganda, 1935–36, pt. 2, pp 19–39.

Taylor, T. H. C. (1945) *Lygus simonyi* Reut. as a cotton pest in Uganda. *Bull. Ent. Res.* **36**:121–48.

Templeton, J. (1925) *Ratoon cotton in Egypt.* Egypt. Min. Agric. Bull. No 55. (Quoted by Evenson 1970).

Textile Organon (1983a) World man-made fibre survey. *Textile Organon*, **54**(6).

Textile Organon (1983b) Textile fibre end-use survey. *Textile Organon*, **54**(9).

Tharp, W. H. (1953) Nonparasitic disorders of cotton. *Plant Diseases, the Yearbook of Agriculture 1953.* USDA: Washington DC. pp 318–20.

Tharp, W. H. (1960) *The cotton plant.* USDA Agric. Handbook No 178.

Thaxton, P. and **Bird, L. S.** (1981) Flowering intervals, boll set and earliness for TAM-MAR cotton cultivars. *Beltwide Conf.*: 96–8.

Thomas, D. G. (1970) Cotton: variety testing, seed multiplication and marketing. In Jameson (1970) Ch. 12.

Thomas, P. E. L. and **Schwerzel, P. J. A.** (1968) A cotton-weed competition experiment. Proc. 9th Br. Weed Control Conf., 1968, pp 737–43. (*Fld. Crop Abstr.* **22**:415).

Thomas, W. D. (1968) Weed control. Cott. Res. Rep., Sudan, 1966–67, pp 27–34.

Thomas, W. D. (1969) Some effects of weeds upon the growth and yield of cotton. In Siddig and Hughes (1969) pp 207–17.

Thomson, H. (1978) *Fibres and Fabrics of Today.* Heinemann: London.

Thomson, N. J. and **Basinski, J. J.** (1962) Cotton in the Ord Valley of Northern Australia. *Cott. Gr. Rev.* **39**:81–92.

Thornthwaite, C. W. and **Mather, J. R.** (1955) The water balance. *Publ. in Climatology* **8**(1), Centerton, NJ. (Quoted by Doorenbos and Pruitt 1977).

Thorp, T. K. (1973) Agricultural meteorology. Cott. Res. Rep., Uganda, 1971–72, pp 9–11.

Thorp, T. K. (1975) New computer programme for the calculation of confidence limits of expected rainfall. *Cott. Gr. Rev.* **52**:386–7.

Tiley, G. E. D. (1970) Weeds. In Jameson (1970) Ch. 18.

Tisdale, S. L. and **Dick, J. B.** (1942) Cotton wilt in Alabama as affected by potash supplements and as related to varietal behaviour and other important agronomic problems. *J. Amer. Soc. Agron.* **34**:405–26.

Tisdale, S. L. and **Nelson, W. L.** (1966) *Soil Fertility and Fertilizers.* Macmillan: London.

Todd, J. A. (1921) Statistical appendix. World Cotton Conf., Liverpool and Manchester, pp 405–26.

Tollervey, F. E. (1974) Weed control in cotton (CRC Summer Meeting). *Cott. Gr. Rev.* **51**:247–55.

Toms, A. M. (1957) Weed control in the Sudan Gezira. *Cott. Gr. Rev.* **34**:280–7.

Tothill, J. D. (ed.) (1940) *Agriculture in Uganda.* Oxford Univ. Press: London.

Tregear, T. R. (1965) *A Geography of China.* Univ. of London Press.

Trevelyan, G. M. (1942) *English Social History*. Longman: London.

Tunstall, J. P. (1958) The biology of the Sudan bollworm in the Gash Delta, Sudan. *Bull. Ent. Res.* **49**:1–23.

Tunstall, J. P. and **Matthews, G. A.** (1966) Large scale spraying trials for the control of cotton insect pests in Central Africa. *Cott. Gr. Rev.* **43**:121–39.

Tunstall, J. P., Matthews, G. A. and **Rhodes, A. A. K.** (1961) A modified knapsack sprayer for the application of insecticides to cotton. *Cott. Gr. Rev.* **38**:22–6.

Tunstall, J. P., Matthews, G. A. and **Rhodes, A. A. K.** (1962) Controlling cotton insects. *Rhod. Agric. J.* **59**:253–65.

Tunstall, J. P., Matthews, G. A. and **Rhodes, A. A. K.** (1965) Development of cotton spraying equipment in Central Africa. *Cott. Gr. Rev.* **42**:131–45.

Tunstall, J. P., Sweeney, R. C. H. and **Matthews, G. A.** (1959) Cotton insect pest investigations in the Federation of Rhodesia and Nyasaland. I. Cotton bollworm investigations. *Cott. Gr. Rev.* **36**:268–75.

USDA (1977) *Cotton ginners handbook*. USDA, Agricultural Research Service, Agric. Handbook No 503. USDA: Washington DC.

USDA (1980) *The classification of cotton*. USDA, Agricultural Marketing Service, Agric. Handbook No 566. USDA: Washington DC.

USDA (1982) *Cotton testing service*. USDA, Agricultural Marketing Service, Agric. Handbook No 594. USDA: Washington DC.

Valentine, E. W. (1954) Entomology. Cott. Res. Rep., Tanganyika Eastern Province, 1952–53, pp 6–8.

Valentine, E. W. (1955) Entomology. Ibid., 1953–54, pp 3–4.

Volkart (1984) *Volkart cotton*. Volkart Bros Ltd: Winterthur, Switzerland.

Walker, J. K. (1980) The development of short-season cotton production in Texas. *Beltwide Conf.:* 153–5.

Walker, J. T. (1960) The use of a selection index technique in the analysis of progeny row data. *Cott. Gr. Rev.* **37**:81–107. (Res. Mem. No 37).

Walker, J. T. (1963) Multiline concept and intra-varietal heterosis. *Cott. Gr. Rev.* **40**:190–215.

Walker, J. T. (1964) Modal selection in Upland cotton. *Heredity* **19**:559–83. (Res. Mem. No 55).

Walker, J. T. (1968) Bias in singling cotton seedlings. *Cott. Gr. Rev.* **45**:175–8.

Walker, J. T. and **Rijks, D. A.** (1967) A computer programme for the calculation of confidence limits of expected rainfall. *Expl. Agric.* **3**:337–41.

Wallach, D., Marani, A. and **Kletter, E.** (1978) The relation of crop growth and development to final yields. *Field Crop Res.* **1**:283–94.

Walter, H. *et al* (1980) Effect of mepiquat chloride on cotton plant leaf and canopy structure and dry weights of its components. *Beltwide Conf.:* 32–5.

Walton, P. D. (1962) The effect of ridging on the cotton crop in the Eastern Province of Uganda. *Emp. J. Exp. Agric.* **30**:63–76. (Res. Mem. No 46).

Ware, J. O. (1940) Relation of fuzz pattern to lint in an Upland cotton cross. *J. Hered.* **31**:489–96.

Watson, H. and **Helmer, J. D.** (1964) *Cottonseed quality as affected by the ginning process – a progress report*. USDA, ARS. 42–107.

Watson, J. S. (1975) Acid delinting cotton seed in Swaziland. *Cott. Gr. Rev.* **52**:146–64.

Watson, T. F. (1980) Short season cotton in the west. *Beltwide Conf.:* 151–3.

Watt, G. (1907) *The Wild and Cultivated Cotton Plants of the World*. Longmans Green: London.

Weatherley, P. E. (1950) Studies in water relations of the cotton plants. I. The field measurements of water details in the leaves. *New Phytol.* **41**:81–97.

Weaver, D. (1981) Effectiveness of glyphosate applications. *Beltwide Conf.:* 171–2.

Weaver, D. B. and **Weaver, J. B. Jr** (1977) Inheritance of pollen fertility restoration in cytoplasmic male sterile cotton. *Crop Sci.* **17**:497–9.

Weaver, J. B. Jr and **Weaver D. B.** (1979) Cracked root mutant in cotton: inheritance and linkage with fertility restoration. *Crop Sci.* **19**:307–9.

Webber, J. M. (1939) Relationships in the genus *Gossypium* as indicated by cytological data. *J. Agric. Res.* **58**:237–61.

Webster, C. C. and **Wilson, P. N.** (1966) *Agriculture in the Tropics.* Longman: London.

Weindling, R. (1944) A technique for testing resistance of cotton seedlings to the angular leaf spot bacterium. *Phytopath.* **34**:235–9. (Quoted by Wickens 1953).

Welford, T. (1966) *The Textile Students Manual.* Pitman: London.

Wells, W. G. (1942) Progress Rep. Exp. Stas., Australia, 1940–41, p 15.

White, M. H. (1957) Cotton growing in French territories in Africa. *Cott. Gr. Rev.* **34**:107–11.

Whitwell, T. and **Buchanan, G. A.** (1981) Comparison of preplant, pre-emergence and overlay treatments for weed control in cotton. *Beltwide Conf.*: 159–64.

Whitwell, T., Wells, L. W. and **Chandler, J. M.** (1981) Report of 1980 cotton weed loss committee. *Beltwide Conf.*: 175–84.

Wickens, G. M. (1950) *Verticillium* wilt disease. Progress Rep. Exp. Stas., South Africa, 1948–49, pp 17–21.

Wickens, G. M. (1953) Bacterial blight of cotton. *Cott. Gr. Rev.* **30**:1–23. (Res. Mem. No 15).

Wickens, G. M. (1957) Treatment of cotton seed against bacterial blight. *Cott. Gr. Rev.* **34**:170–6. (Res. Mem. No 29).

Wickens, G. M. (1958) Present practice in the treatment of cotton seed against bacterial blight. *Cott. Gr. Rev.* **35**:9–12.

Wilkes, L. H. and **Corley, T. E.** (1968) Planting and cultivation. In Elliot, Hoover and Porter (1968) Ch. 5.

Williford, J. R. (1981) Post emergence equipment available to growers. *Beltwide Conf.*: 170–1.

Wilson, F. D. and **George, B. W.** (1981) Breeding cotton for resistance to pink bollworm. *Beltwide Conf.*: 63–5.

Windel, D. (1946) The 1945–46 raw cotton situation in retrospect. *Cott. Gr. Rev.* **23**:182–8.

Wishart, J. (1940) *Field trials: their layout and statistical analysis.* Imperial Bureau of Plant Breeding and Genetics: Cambridge.

Wishart, J. and **Sanders, H. G.** (1955) *Principles and practice of field experimentation.* Commonwealth Bureau of Plant Breeding and Genetics: Cambridge.

Withrow, R. B. (1936) Light and its effects on plant growth. *Agr. Eng.* **17**:150–2. (Quoted by Miller 1938).

World Bank (1976) The Pakistan seed project (appraisal report). World Bank: Washington DC.

World Bank (1980) Cotton marketing and processing project, Kenya. World Bank: Washington DC.

Zaitzev, G. S. (1928a) A contribution to the classification of the genus *Gossypium* L. *Bull. Appl. Bot. Genet. Plant Breeding* **18**:39–65. (Russian, quoted by Fryxell 1979).

Zaitzev, G. S. (1928b) Effect of temperature on the development of the cotton plant. (Quoted by Hector 1936).

Zimbabwe (1980) *Cotton handbook.* Commercial Cotton Growers Association: Harare, Zimbabwe.